IEE ELECTROMAGNETIC WAVES SERIES 29

Series Editors: Professor P. J. B. Clarricoats,
Professor Y. Rahmat-Samii
Professor J. R. Wait,

Satellite-to-ground radiowave propagation

Theory, practice and system impact at frequencies above 1 GHz

Other volumes in this series

Satellite-to-ground radiowave propagation

Theory, practice and system impact at frequencies above 1GHz

J E Allnutt

Peter Peregrinus Ltd, on behalf of the Institution of Electrical Engineers

Published by: Peter Peregrinus Ltd., London, United Kingdom

© 1989: Peter Peregrinus Ltd.

British Library Cataloguing in Publication Data

Allnutt, J. A.
    Satellite-to-ground radiowave propagation
    1. Radio waves. Propagation
    I. Title   II. Institution of Electrical Engineers
    III. Series
    621.3841'1

ISBN 0 86341 157 6

Printed in Great Britain by BPCC Wheatons Ltd, Exeter

Contents

Preface

The original request to write this book was couched in the terms of 'Produce a definitive Earth-space propagation text'. Such a grandoise aspiration is probably not achievable, and certainly not by me. The questions then remained of the focus of the book and its intended audience.

I will admit to a number of biases in both respects. To those readers who delve past the opening general chapter it will soon be obvious that the essence of the book deals with radiowave communications to geostationary satellites. Low-earth-orbit satellites and deep-space probes will have different technological problems than fixed-service communications to geostationary satellites but they all have one thing in common: their signals must traverse the Earth's atmosphere. I have therefore set out to describe the propagation impairments introduced by the atmosphere of the Earth, with an emphasis on civilian/commercial geostationary telecommunications satellites.

The bulk of the theory and the experimental results deal with frequencies below 30 GHz, so again the emphasis is on this part of the frequency spectrum. If there is ever an updated version of this book, it will probably be appropriate to include optical communications. This aspect is presently receiving a lot of emphasis in the military sphere and it is possible that much of the technology will spill over into commercial endeavours in the near future.

The question of picking a target audience was far more difficult.

Radiowave propagation has the reputation for being a somewhat arcane science with a rather narrow specialisation. To anyone who has been involved in propagation experiments, especially with satellites, these are easy assertions to refute. A broad technical knowledge is required of earth-station technology, satellite transponder characteristics, antenna theory, meteorology, and co-ordination aspects, to name just a few topics, in addition to understanding the basics of propagation theory. I have therefore chosen to present a broad background in all the areas of relevance to radiowave propagation, as well as delving a little deeper into the specifics of propagation. To appreciate the various facets of propagation impairments, I felt that an almost historical overview would be useful and so I have included many details of the early experiments, together with the many pitfalls that can occur in conducting propagation experiments.

In setting out the text of this book, I have tried to separate out the various major impairment phenomena and present a discussion on each in as self-contained a manner as possible. After a general opening chapter, I have treated ionospheric scintillation, clear-air effects, attenuation and depolarisation in separate chapters with as little cross-referencing as possible. This has led to some duplication of text but, I hope, makes for an easier comprehension of the topics. Following these major topics, I have grouped together a number of aspects that did not seem to warrant a chapter in their own right, due either to the lack of a substantial data base in that area at present or to the topic being dealt with in more detail in other texts. In this chapter are sand and dust effects, multipath effects, and interference aspects, including site shielding. The final chapter contains a detailed review of methods of reducing propagation impairments. If anything, this area will be receiving increasing emphasis as the search for improved performance in the bands above 10 GHz intensifies.

In an attempt to make this book attractive to both an undergraduate in any of the applied sciences or engineering as well as to a fully fledged propagation expert, I have inserted a lot of descriptive text as an introduction to each of the topics and sub-topics together with many reference citations. The latter will enable the advanced reader to go to the original source material and take his or her inquiry beyond the level of this book. In this regard, it is worth pointing out that nearly all of the subject material for this book has been gleaned from the work of many experts in the field of propagation, in particular the combined pooling of talent that goes to make up the texts of the CCIR Study Group 5. It is with appreciation that I acknowledge permission received from the ITU for the reproduction of CCIR figures and diagrams utilised throughout this book. Other very useful starting points were the two Reference Publications of NASA. One problem in utilising such texts is picking the correct reference to cite. I have tried to include the correct citation wherever possible but sometimes this has been difficult to identify. I apologise for any errors made.

Most of the figures used in this book have been reproduced from other texts with the kind permission of the publishers and the authors. In addition to those of the ITU and NASA, I would like to acknowledge the permission granted to use text and figures originally published by AIAA (USA), AGARD, AT&T, American Geophysical Union, American Society of Photogrammetry and Remote Sensing, Bordas Dunod Gauthier-Villars, Bradford University Research Ltd., British Telecom International, British Telecom Research Labs, Butterworths (UK), CRC (Canada), CSRL (UK), CTR (USA), ESA, IECE (Japan), IEE, IEEE, International Journal of Satellite Communications, INTELSAT, KDD (Japan), Merrill Publishing Co. (USA), Nature, New Scientist, Ohio State University, Peter Peregrinus, and URSI. Among the many authors who have given me permission to reproduce figures I would especially like to thank Gert Brussaard, Dickson Fang, C. H. Liu, Jonathan Maas, Neil

McEwan, David Rogers, John Thirlwell, and Peter Watson. Private communications from these individuals, not referencible in the open literature, have been used in this book.

Finally, I would like to acknowledge the warmth and support that I received from my wife, Norma, throughout the preparation of this book. To her I would like to give my love and my thanks.

Washington D.C.
February 1989

Chapter 1

Radiowave earth–space communications

1.1 Introduction

The need to communicate complex instructions led to the development of language and, it is argued, to the development of *homo sapiens*. Once out of earshot, however, all long-distance communications were transmitted optically. The range was therefore limited to the visible horizon and so it remained for eons. To overcome this limit, relays of signallers could be set up. This reached its peak in the late eighteenth century when chains of fire towers were erected on hilltops to signal the sighting of an invasion fleet to the major cities inland. The system was digital: if the fire was out (level zero), all was safe; if the fire was lit (level one), the enemy was in sight.

Improvements in this digital optical communications system were made by the incorporation of codes [1] but the information rate was, to say the least, somewhat low. The introduction of electrical telegraphy and then wireless telegraphy in the late nineteenth and early twentieth centuries eclipsed all other forms of long-distance communications. Only recently, with the development of efficient, economical and reliable lasers together with the advances in low-loss optical fibre technology, have digital optical communications started to compete effectively with radiowave communications. Radio links, however, still carry the bulk of the world's telecommunications and, in the case of mobile or multi-point to multi-point services, will continue to do so for the foreseeable future.

With the introduction of electrical telegraphy, it became apparent that co-operative agreements would be needed between all countries in order to develop the new form of communications efficiently. This led to the formation of the International Telecommunication Union (ITU) on 17 May 1865. The ITU is the oldest of the intergovernmental organisations that now form the specialised agencies of the United Nations [1]. Detailed vocabularies and specifications have been agreed between the member countries of the ITU and are enshrined in the Articles [2] and Appendices [3] of the Radio Regulations. Article 1 defines the terminology and some of the relevant definitions are noted in Appendix 1.

The same volume of the Radio Regulations also allocates the frequencies between 9 kHz and 275 GHz within the three ITU regions (see Fig. 1.1). While a few experimental services receive some protection between 275 and 400 GHz, no allocations have been made above 275 GHz. The rapid development of high-power lasers may require an extension of the allocated frequencies into the terahertz region 10^{12} Hz but, for the present, most radiowave sensing or communications links using satellites are at frequencies well below 275 GHz.

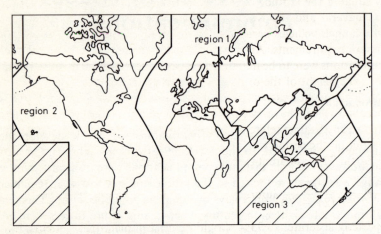

Fig. 1.1 *The three ITU Regions of the world*

1.2 Artificial earth satellites

The first artificial earth satellite used for communications between two earth stations was Echo 1 which was launched on 12 August 1960, less than three years after Sputnik 1. Echo 1 was an inflatable balloon approximately 30 m across with a thin metal skin covering to reflect the radiowaves transmitted to it. No amplification of the signals took place; the satellite simply acted as a passive repeater. Later in the same year, the satellite Courier 1B relayed back tape recorded transmissions that had been received in an earlier part of the orbit. This was an advance over the Project Score satellite that could not receive transmissions but merely transmitted a pre-recorded message for a few days over the Christmas period in 1958. The first active communications satellite was Telstar 1 launched on 10 July 1962. Telstar 1 provided intermittent voice and video communications across the Atlantic Ocean. Launched on 13 December 1962, Relay 1 demonstrated the same services over the Pacific Ocean.

At the same time as these rapid advances were being made in communications satellites, other new applications areas were being investigated that would revolutionise earth observation sciences. In the early 1960s, however, there were many critical aspects that still had to be resolved in the choice of technical parameters.

1.2.1 Choice of orbit

a) Equatorial orbits

Launched eastwards to take advantage of the angular velocity of their launch site with respect to the centre of the Earth, the first artificial earth-satellite launchers had little or no excess fuel to change the inclination of the orbit. The result was that the early satellites had orbital inclinations that were approximately equal to the latitude of their launch sites, i.e. 30° to 40°, the latitude of Cape Canaveral and the Kazakh region of the USSR. The advent of more powerful launch vehicles permitted the optimum inclination to be selected for the mission requirements. An equatorial orbit seemed to be a logical choice if communications were to be maintained on an equal basis north and south of the equator. The height of the orbit was a more difficult choice.

A satellite launched eastwards in an equatorial orbit (inclination zero degrees to the equator) will have two periods depending on the point of reference. Referred to an absolute reference, the period will be T hours but, to an observer on the equator, the period will be longer than T because of the eastward rotation of the Earth. This apparent period is P hours. P and T are related by

$$P = \frac{24T}{(24 - T)} \text{ hours} \tag{1.1}$$

To be strictly accurate, 23·9344 should be used in place of 24 in eqn. 1.1. One sidereal day is 23 hours 56 minutes 4 seconds, not 24 hours. Table 1.1 illustrates the difference between P and T for a number of orbital heights and also shows the time a satellite is visible to an observer on the equator, neglecting any atmospheric refraction and assuming that communications can be maintained down to an elevation angle of zero degrees. The increase in available observing time as the orbital height increases is evident in Table 1.1. The height of 35 786 km corresponds to that proposed by Clarke in his classic paper [4]. At this height, the angular rotation of the satellite about the centre of the Earth is the same as that of the Earth and the orbit is said to be geosynchronous. If the inclination of the geosynchronous orbit is close to zero, the orbit is said to be geostationary.

In 1962, no satellite had ever been put into a geosynchronous orbit. This led to the proposal [5] to establish a world-wide communications satellite system using 12 satellites, phased in their spacing, orbiting above the equator at an altitude of 13 800 km. There were many problems with this approach. Each operator required at least two earth stations so that communications could be established with the next satellite in the sequence before losing contact with the first; the transmitted power of the earth stations would have to be regulated closely as the elevation angle changed; very-long-distance communications would require complex, multiple hops; and frequency re-use would be very difficult. Another crucial argument against such a system of many low-earth-orbit satellites was the fact that at least 12 successful satellite launches were required before a 24 h/day communications system could be established between any two points. Coupled with this was the fact that only one successful launch into a geostationary orbit was required to establish 24 h/day communications over approximately one third of the Earth. The decision was therefore taken by COMSAT, and confirmed later by the Interim Communications Satellite Committee [to become INTELSAT (the International Telecommunications Satellite Organisation) in 1964], to proceed with the deployment of a geostationary communications satellite system. INTELSAT I F–1, also known as 'Early Bird', was launched on 6 April 1965 and positioned over the Atlantic Ocean. Within three years, all three ocean regions − Atlantic, Indian and Pacific − had been covered and world-wide communications established. By 1970, two-thirds of all international telephone traffic was being carried by satellite. Over 150 geosynchronous/geostationary satellite launches had taken place by early 1986 and the failure rate was down to about 10% [6]. Perhaps of more long-term significance, in 1983 domestic satellite communications traffic exceeded that carried over international satellites, although the bulk of the domestic satellite growth was in video services.

Table 1.1 *Orbital periods and observing time*

Orbital height (km)	Orbital period		Observing time (h)
	True (h)	Apparent (h)	
500	1·408	1·496	0·183
1 000	1·577	1·688	0·283
5 000	1·752	1·890	0·587
10 000	5·794	7·645	2·849
35 786	23·934	∞	∞

The orbits are all equatorial in an eastward direction with the observer located on the equator. The radius of the Earth is assumed to be 6378·15 km

b) Inclined orbits
There are two fundamental limitations of the geostationary orbit: latitude coverage and transmission delay. Figure 1.2 illustrates the time delay, ignoring the component introduced by the atmosphere, for elevation angles between 0° and 90°.

Typical two-way transmission delay times exceed 0·5 s which can be disruptive in some situations. To ameliorate this effect, some telephone companies try to split the two-way link, sending one path by satellite and the return path by cable. The latitude coverage problem can only be solved by using an inclined orbit. One solution is to have a geosynchronous satellite with an inclination of 30° or more. Examples of this are the ATS–3 satellite that has been used for communications links to the Antarctic and the two small LES–8 and LES–9 satellites that have been used for occasional links to the Arctic region. The problem with this solution is that the satellite is only visible for about half of each day; the remainder of the time it will be below the local horizon.

A second solution, which has been put into full-time operational use, was initiated with the launch of Molniya 1 on 23 April 1965 into what has now been called the 'Molniya orbit'. This particular orbit is unique amongst the inclined non-geosynchronus orbits, in that it repeats the same ground track on alternate orbits. The inclination is 65°, the perigee 500 km, and the apogee 39 152 km. By arranging for the apogee to occur over the region of interest, more than 60% of the 11 hour 38 minute orbit is usable for communications between latitudes of 30° and 90°. Four such satellites, phased in the same Molniya orbit, would provide continuous communications to regions well north of the 76° latitude limit imposed at present on most geostationary satellite systems.

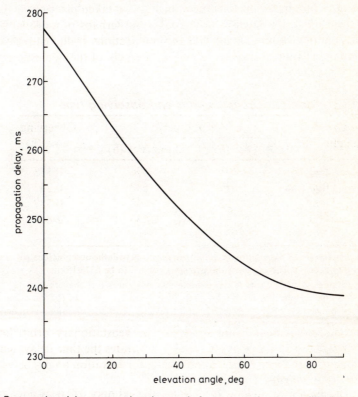

Fig. 1.2 *Propagation delay versus elevation angle for a geostationary satellite link. The delay time ignores any atmospheric effects and is for a one-way link. (Adapted from Fig. 1.3 of* [7])

More recently, a novel orbit that combines the advantage of a highly elliptical orbit (long dwell times at apogee) with that of a sun-synchronous orbit (available during the daytime for most of the orbit) has been proposed [68]. This orbit, called an Apogee at Constant time-of-day Equatorial (ACE) orbit by its inventors, the Ford Aerospace Comany, has a fixed orientation with respect to the sun and the following characteristics:

Period	4·8 h
Perigee	1 030 km
Apogee	15 100 km
Inclination	0°

The ACE orbit derives its sun-synchronous capability by virtue of a precessional force that is applied to orbits which lie in the plane of the equator. By suitably arranging the orbit period, the precessional motion matches the apparent movement of the Sun around the Earth. Since the ACE orbit is in the plane of the equator, it cannot provide coverage to high latitudes like the Molniya orbit but, by timing the apogee to coincide with the business day of the country beneath it, the satellite in an ACE orbit can provide relatively cheap domestic communications capabilities in some services when compared with a geostationary satellite.

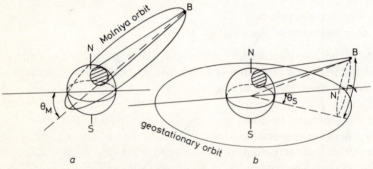

Fig. 1.3 *Illustration of high-latitude communications links using (a) a Molniya orbit and (b)*
a highly inclined geosynchronous orbit
Point B in both cases is the extreme northerly point of the orbit. In (*a*), the orbit
swings close to the Earth with an inclination of θ_M = 65°. In (*b*) the nominal centre
of movement of the satellite is point N. The satellites describes a quasi-elliptical
movement about this point with an inclination given by θ_S.

1.2.2 Choice of antenna

At frequencies above 1 GHz, the electrical resistance of transmission wires becomes appreciable. Even well designed coaxial cable has significant losses at frequencies above 10 GHz. The use of waveguides, particularly for high-power applications, is therefore almost universal at frequencies of 4 GHz and above and this has an impact on the choice of antenna.

An antenna has two basic requirements: it must match the characteristic impedance of the coaxial cable or waveguide to that of the transmission medium as exactly as possible and it must radiate the radiowave signals with the correct characteristics and in the desired direction. Meeting the first requirement will eliminate reflections within the antenna system that could lead to serious deficiencies in the transmitted signal. Meeting the second requirement will ensure that the signals are successfully received at the desired location.

Wire-fed dipole and helical antennas give way to waveguide-fed reflector antennas at frequencies of around 2 or 3 GHz when the need is for percentage bandwidths in excess of about 5% of the carrier frequency, a low loss feed, or high gain, or any combination of the three.

The gain of an antenna is a measure of the directivity of that antenna in the desired direction with respect to an antenna that radiates equally in every direction, a so-called isotropic antenna. The directivity of an antenna is usually defined by the width of the beam between points that are at half the power of the peak gain. This is the half-power, or 3 dB, beamwidth. For a circularly symmetric parabolic reflector antenna with a uniform aperture distribution, the 3 dB beamwidth, θ_{Bu}, is given by:

$$\theta_{Bu} = 1 \cdot 02 \left(\frac{\lambda}{D} \right) \text{ radians} \tag{1.2}$$

where λ = wavelength, m
         D = diameter of the antenna, m

A uniform aperture distribution means that the power density of the energy radiated from the feed measured at the rim of the reflector antenna is the same as that measured on the axis of the reflector. The uniform distribution makes it very difficult to prevent some of the energy from the feed from 'spilling over' the edge of the reflector. In addition, the high energy density at the rim of the reflector will cause appreciable edge diffraction, thereby causing interference to the main beam and increasing the amount of energy that is not radiated in the desired direction. The energy that falls outside the main beam generates what are known as sidelobes.

The characteristics of the sidelobes will be a function, amongst other things, of the aperture distribution. With a uniform aperture distribution, the first sidelobe will have an amplitude that is $-17 \cdot 6$ dB relative to the peak of the main lobe. The first sidelobe can be suppressed, and the spill-over greatly reduced, by introducing a non-uniform aperture distribution. A typical non-uniform distribution is one in which the power density falls off as $(\cos)^2$ with respect to the peak power on the axis of the reflector. If the edge illumination of the reflector is 10 dB less than the axial illumination, the first sidelobe amplitude reduces to -24 dB relative to the peak of the main lobe. The beamwidth of the antenna with this illumination has also changed to

$$\theta_B = 1 \cdot 2 \left(\frac{\lambda}{D} \right) \text{ radians} \tag{1.3}$$

Fig. 1.4 illustrates the above effects. In general, what is 'gained' in sidelobe suppression and reduced spill-over is 'lost' in antenna directivity (beamwidth) and gain. Examples of the effect of various aperture distributions can be found in Reference 8.

An antenna that is 100% efficient, i.e. all the feed power incident upon the reflector is contained in the main lobe, will have a transmit gain G_T given by

$$G_T = (4\pi A)/\lambda^2 \tag{1.4}$$

where A is the area of the reflector normal to the direction of transmission. The gain is the increase in the signal power in the desired direction over that of an isotropic antenna. For a circularly symmetric antenna, eqn. 1.4 reduces to

$$G_T = (\pi D/\lambda)^2 \tag{1.5}$$

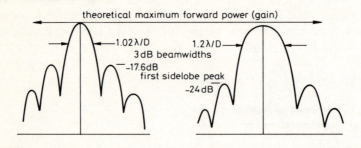

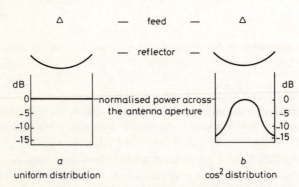

Fig. 1.4 *Differences between antennas having (a) uniform aperture illumination and (b) cos^2 aperture illumination*
The antennas shown are front-fed symmetrical paraboloids, but similar differences occur with other types of reflector antennas.

where D is the diameter of the antenna. Since an antenna will not be 100% efficient, eqn. 1.5 needs to be multiplied by an efficiency factor, η. The modified equation is as follows:

$$G_T = \eta(\pi D/\lambda)^2 \tag{1.6}$$

A good antenna will have an efficiency of between 60 and 75%, i.e. η lies between 0·6 and 0·75.

Given that a parabolic reflector antenna will be used, there are several choices that can be made in the configuration of the antenna. The antenna can be front-fed, cassegrainian, or gregorian, and it can be symmetrical or offset-fed. A gregorian antenna has a sub-reflector that has an elliptical surface shape while a cassegranian antenna has a sub-reflector with a hyperbolic surface shape. In all other mechanical respects, they are the same. Fig. 1.5 illustrates the difference between front-fed and offset-fed antennas.

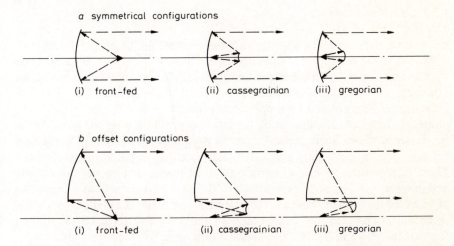

Fig. 1.5 *General Schematic of the three major classes of antennas with (a) symmetrical configurations and (b) off-set configurations*
In each case, the surface of the main reflector is part of a paraboloid. The sub-reflectors have different shapes, however, with the cassegrainian antenna's sub-reflector having a hyperbolic surface and that of the gregorian antenna's sub-reflector an elliptical surface. Note the blockage of the main antenna beam with the symmetrical configurations. This blockage is removed in the offset configurations, leading to improved polarisation isolation and sidelobe suppression performance.

The advantage of a cassegranian antenna is that the transmitter and receiver can be located right next to the feed behind the main reflector. Feed losses are therefore reduced and maintenance is eased. The sub-reflector, however, has to be larger than approximately 10 wavelengths in diameter to be efficient. A rule-of-thumb limit in deciding between cassegranian or front-fed designs is about 100 wavelengths in main reflector diameter. If the main reflector diameter exceeds 100 wavelengths, it is normal to choose a cassegranian design; below 100 wavelengths, a front-fed design is usually better. Of more importance nowadays is the selection between symmetrical and offset-fed configurations.

The major advantage of an offset-fed design is that there is no aperture blockage. The elimination of blockage greatly reduces the chance of interference between rays emanating from the main reflector and rays that originate from undesired reflections within the antenna system, e.g. off a strut holding the sub-reflector. These constructive and destructive interferences give rise to radiation outside of the desired direction. Eliminating the blockage, and hence the major source of mutual interference within the antenna system, greatly reduces the design problems inherent in antennas to be used in regions where interference into other systems is the major obstacle. Recent advances have also led to the universal adoption of beam waveguides for large earth stations where feed losses are of great concern [7].

1.2.3 Choice of frequency

The larger the effective aperture of an antenna, the more directed the signals become; i.e. the gain of the antenna increases. Increasing the gain of the antenna will improve the resolution of a sensing device attached to the antenna, increase the transmission rate of a communications link using that antenna, or reduce the power requirements of the transmission system. The effective aperture of a given antenna can be increased by simply increasing the carrier frequency since, by inference from eqn. 1.5, increasing the frequency will increase the gain. As the frequency increases, so, in general, does the complexity of equipment to produce the same level of reliablility and EIRP (Equivalent Isotropic Radiated Power). There is, therefore, at least one trade-off to be made, that between directivity and complexity. The early experiments with Echo 1 were made at a frequency of 1 and 2·5 GHz while those with Courier 1B were made at 2 GHz. Operational communications satellites, beginning with Early Bird of INTELSAT, utilised frequencies of 4 and 6 GHz. If increased gain or EIRP were the only criteria, with each advance in the availability of reliable higher-frequency equipment, a corresponding movement to utilise those higher frequencies on commercial satellite systems would have resulted. As will be seen in later chapters, moving to higher frequencies, expecially those above 10 GHz, introduces many additional loss mechanisms that have to be overcome in the design of a satellite system. Sometimes these loss mechanisms are beneficial, for example in the design of earth observation satellites. A loss mechanism that is unique to a certain characteristic of the Earth or its environment can be used as a sensing tool. The choice of transmission or sensing frequency is therefore a more involved trade-off than simply balancing the EIRP requirements against the complexity of equipment. Sensing devices tend to use frequencies that are close to molecular or atomic absorption lines or are sensitive to a particular radiation characteristic. Communications systems try to use frequencies that are well away from absorption bands. In addition, because of the international nature of communications, most frequenceis below 275 GHz have already been allocated to certain services, as will be seen in Section 1.4.

1.2.4 Choice of polarisation

Electromagnetic waves that are sufficiently far from their source are usually represented by plane waves. That is the electric, E, and magnetic, H, fields are orthogonal to each other and are contained by a plane that is normal to the propagation direction (see Fig 1.6).

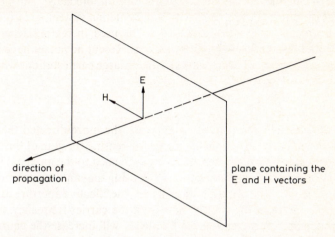

direction of
propagation

plane containing the
E and H vectors

Fig. 1.6 *Plane waves*
The E and H vectors are orthogonal to each other and both lie in the same plane. The plane containing the E and H vectors is orthogonal to the direction of propagation.

In Fig. 1.6, the electric field is represented by a single vector E. A randomly polarised radiowave will have no such preferred orientation of E and an antenna that can detect the polarisation orientation will detect no averaged maxima or minima about the direction of propagation. In general, however, radiowaves used in communications or sensing systems will be transmitted and received with a preferred polarisation orientation. The two major classes of polarisation are linear and circular. In linear polarisation, the electric vector has a fixed orientation while propagating in free space, as depicted in Fig. 1.6. In circular polarisation, the electric vector rotates about the axis of propagation. The rotation is set up by first splitting a linearly polarised vector into two equal vectors at angles of plus and minus 45° with respect to the original vector, and then either delaying or advancing the phase of one of the 45° vectors with respect to the other. Linear and circular polarisation are two special cases of elliptical polarisation.

The general case of elliptical polarisation is illustrated in Fig. 1.7 [7]. In Fig. 1.7, the elliptical polarisation has been resolved into two orthogonal circular polarisations: left-hand circular polarisation E_L and right-hand circular polarisation E_R. The ellipticity r, or as it is sometimes called in the case of antennas, the axial ratio, is defined as

$$r = a/b \tag{1.7}$$

where $2a$ and $2b$ are the major and minor axes of the polarisation ellipse, respectively. By geometry

$$r = \frac{E_L + E_R}{E_L - E_R} \qquad (1.8)$$

In decibels, this is expressed as

$$R = 20 \log |r| \text{ decibels} \qquad (1.9)$$

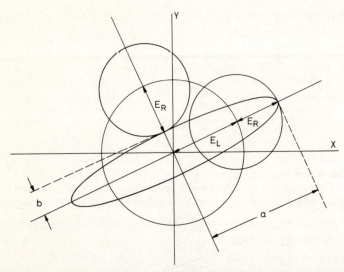

Fig. 1.7 *Resolution of the general case of elliptical polarisation into two circular polarisation elements with opposite polarisation senses (adapted from Fig. 5.14 of Reference 7) The suffices L and R refer to left and right hand circular polarisation.*

The ratio between the voltage amplitudes in the two polarisations ϱ, when referenced to the amplitude in the left hand polarisation, is given by

$$\varrho = E_L/E_R \qquad (1.10)$$

In a practical system, no polarisation is ever absolutely pure; there will always be a residual amplitude in the opposite polarisation sense. A measure of the purity of the polarisation is the cross-polarisation discrimination, or XPD. The XPD can be found from

$$\text{XPD} = 20 \log |\varrho| \text{ dB} \qquad (1.11)$$

In eqn. 1.11, if the XPD is 40 dB, this means that the voltage amplitude of the unwanted polarisation is 100 times lower than the voltage amplitude of the wanted polarisation. Two useful expressions relating ellipticity r and circular polarisation ratio ϱ are [7]

$$r = \frac{\varrho + 1}{\varrho - 1} \qquad (1.12)$$

and

$$\varrho = \frac{r + 1}{r - 1} \tag{1.13}$$

The angle between the major axis of the polarisation ellipse to the X-axis is called the tilt angle. The significance of the tilt angle will be seen in Chapter 5.

The choice between linear and circular polarisation depends on a number of factors. Propagation impairments on linear polarisation are generally less severe than on circular polarisation. Linearly polarised antenna feeds are also simpler than circularly polarised feeds and hence less expensive. However, if the propagation medium causes a significant rotation of the E vector on transmission through it, a circularly polarised system is to be preferred. In a like manner, the axial symmetry of a circularly polarised antenna reduces the alignment problems that are inherent in linearly polarised systems. The eventual decision as to which polarisation to use will therefore be based upon the severity of the more important propagation impairments applicable to the carrier frequency selected and the importance of antenna cost and axial alignment in the overall system design. The diameter of the antenna will also determine the required degree of tracking accuracy.

1.2.5 Choice of tracking

In order to maintain good communications with an earth satellite or a planetary probe, it is necessary to track the spacecraft accurately. In a like manner, the spacecraft must direct its antenna towards the desired receiving point. Eqn. 1.3 gave the angle of the 3 dB beamwidth. It is normally necessary to track to a much finer tolerance than this, ± 1 dB being the usual requirement. Tracking to within ± 1 dB normally requires active antenna tracking; i.e. the desired signal is sensed and its amplitude used as an error-correcting signal. Multiple-horn or higher-mode tracking systems [7] are used for continuous tracking. If the spacecraft has only a small angular velocity with respect to the earth station, the pointing of the antenna need only be adjusted periodically when the received signal level has fallen a pre-set amount. This type of tracking is called 'hill climbing' or 'step tracking', for obvious reasons. It is a fairly low cost method but it is subject to large errors in the presence of scintillation or any other type of signal-attenuation phenomenon.

In propagation measurements, where the changing level of the received signal must not be due to any imperfections in the tracking system, two methods are normally employed: program tracking or fixed pointing. Program tracking utilises a computer to aim the antenna towards the predicted position of the satellite. The latter method of 'tracking' requires the use of a geostationary satellite to transmit the beacon or carrier signal, otherwise the satellite will quickly move out of the antenna beam of the earth station which has been fixed in one direction.

A geostationary satellite is usually not precisely stationary for any length of time with respect to a point on the surface of the Earth. To be stationary, the period must be one sidereal day and both the inclination and the eccentricity of the orbit must be zero. If the eccentricity is zero, but there is a small component of inclination, the azimuth and elevation angles from the earth station to the satellite — the look angles — will describe a figure-of-eight [10] with the elevation-angle movement close to the inclination excursion. Usually the eccentricity is non-zero and, as a result, the figure-of-eight degrades into an open ellipse. Some examples of this are shown in Fig. 1.8 taken from Reference 11.

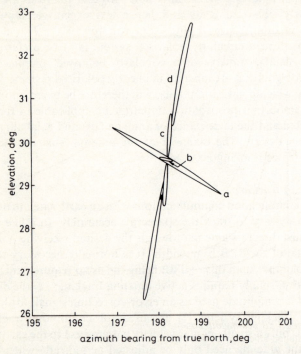

Fig. 1.8 *Look-angle predictions for geostationary satellite (Fig. 1 of Reference 11)*
In all cases the subsatellite point is 15° W longitude and the argument of perigee west is 0°. The earth station is assumed to be in Slough (UK). The parameters are as follows:

	Eccentricity	Inclination
a	0·01	0·5°
b	0·001	0·005°
c	0·001	1·0°
d	0·001	3·0°

(Copyright © 1977, IEE reproduced with permission)

Communications satellites are normally launched with eccentricities below 0·001 and with residual inclinations of less than 0·1°. The maximum perceived excursion of the satellite within its station keeping 'box' is therefore less than

0·3° when viewed from an earth station. If the 1 dB beamwidth of the earth station antenna is not less than 0·3° (18′ arc) then no tracking is required and the antenna can be fixed in one position. Fig. 1.9 illustrates the variation of 1 dB beamwidths with antenna diameter and frequency.

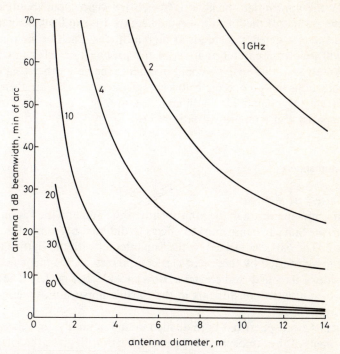

Fig. 1.9 *Antenna 1 dB beamwidth against diameter*
An antenna efficiency of 50% is assumed with a −10 dB edge taper, $(\cos)^2$ aperture distribution.

If the 1 dB beamwidth is smaller than the satellite excursions then some method of tracking must be used. For propagation measurements, this can only take the form of passive tracking; i.e. no use can be made of the signal being sensed to update the tracking. Passive tracking can take the form of a long-time-constant type of tracking [12] that essentially remembers where the satellite was exactly one sidereal day ago and predicts the change required for the present look angles, or a continuous computer prediction can be generated from the satellite ephemeris data. The former system is best for situations requiring active tracking at all times except in adverse propagation conditions. The latter is essential for propagation experiments where the potential errors of quasi-active tracking could contaminate the data.

1.2.6 Choice of service
The service to be provided by a particular satellite naturally has the major influence in the design of that satellite [9]. Once the primary mission is selected, most of the other important decisions regarding the payload are immediately bounded by certain limits. If the spacecraft is to be an earth resources satellite, for example, the sensing frequencies will be selected to detect the required samples at the appropriate heights: infra-red for vegetation; low microwave to reach the surface of the Earth or even detect the ice-cap thickness; and high-millimetre wave to detect the clouds at high altitudes. Each of the frequencies selected will involve a careful analysis of the propagation effects at that frequency. An error in this analysis could invalidate the data processing or cause a communications link to operate below specification. To assist in this analysis, it is important to understand the different features of the Earth's atmosphere and their spatial and temporal variations.

1.3 The atmosphere

1.3.1 Atmospheric divisions
The Earth is moving through the atmosphere of the Sun and the varying energy levels from the Sun can cause marked effects to different portions of the Earth's atmosphere. The particles ejected by the Sun interact with the Earth's magnetic field. The force exerted by these particles compresses the magnetic field on the sunward side of the Earth and creates a 'Bow shock' (see Fig. 2.1). The particles also tend to ionise the constituents in the Earth's upper atmosphere, creating ionised layers.

The Earth's atmosphere has been divided up into a number of spheres, which are really shells of the atmosphere around the Earth with given thicknesses, and regions. The two primary categories are the neutral atmosphere and the ionised atmosphere. These categories are themselves divided up as shown in Fig. 1.10.

Below a height of about 80 km, the gases are well mixed and keep approximately the same proportions [12]. This is known as the homosphere. Table 1.2 gives the major atmospheric constituents in the homosphere.

Above 80 km the gases tend to stratify according to their weights, and this region is known as the heterosphere. To give an indication of the tenuousness of the atmosphere at a height of about 80 km, only 0·0002% of the total atmosphere remains above this height [12].

The thermal energy arriving from the Sun amounts to an average of about 1·4 kW/m^2 [5] and it is this energy that is the driving force of the neutral atmosphere. The average temperature through the neutral atmosphere is neither constant nor monotonic with height. The general distribution with height is shown in Fig 1.11.

The heating of the atmosphere at the stratopause is the result of direct absorption of the Sun's ultraviolet rays by ozone [12]. This region is important since it largely prevents any vertical motion between the layers on either side. It is also quite dry [12]. As will be seen in later Chapters, it is the ionised layers high up in the ionised atmosphere and the moist, turbulent layers way down in the lower reaches of the neutral atmosphere that are the principal factors in radiowave propagation. The variations in the moist, turbulent layers of the lower neutral atmosphere, the troposphere and the stratosphere, are part of the science of meteorology. Understanding the fundamentals of meteorology is important since the weather patterns in the lower atmosphere play a major part in determining the likely propagation impairments to be encountered by radiowave signals on Earth–space paths.

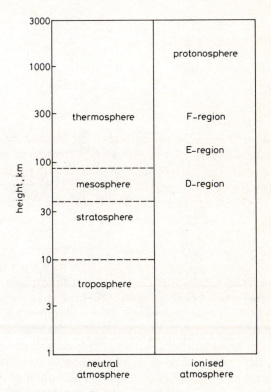

Fig. 1.10 *Primary categories of the Earth's atmosphere*

1.3.2 Weather patterns
If the potential energy contained within the atmosphere were uniformly distributed over the globe there would be no movement within the atmosphere. The energy stored in the atmosphere due to the energy reaching the Earth from the Sun is, however, not uniformly distributed. The bulk of the energy is received in the Earth's equatorial regions, leading to much higher average temperatures

within the tropics compared to the polar regions. Fig. 1.12 gives the isotherms of the mean temperatures in July and January. The temperature imbalance leads to two principal displacement mechanisms: horizontal flow and vertical flow.

Table 1.2 *Normal composition of clean, dry atmospheric air near sea level. (from US standard atmosphere, 1962) (from Table 1.1 of Reference 48)*

Constituent gas	Gas symbol	Content (% by volume)	Molecular weight*
Nitrogen	N_2	78·084	28·0134
Oxygen	O_2	20·9476	31·9988
Argon	Ar	0·934	39·948
†Carbon dioxide	CO_2	0·0314	44·00995
Neon	Ne	0·001818	20·183
Helium	He	0·000524	4·0026
Krypton	Kr	0·000114	83·80
Xenon	Xe	0·0000087	131·30
Hydrogen	H_2	0·00005	2·01594
†Methane	CH_4	0·0002	16·04303
Nitrous oxide	N_2O	0·00005	44·0128
†Ozone	O_3	Summer: 0 to 0·000007	47·9982
		Winter: 0 to 0·000002	47·9982
†Sulfur dioxide	SO_2	0 to 0·0001	64·0628
†Nitrogen dioxide	NO_2	0 to 0·000002	46·0055
†Ammonia	NH_3	0 to trace	17·03061
†Carbon monoxide	CO	0 to trace	28·01055
†Iodine	I_2	0 to 0·000001	253·8088

* On basis of C^{12} isotope scale for which C^{12} equals 12·0000.
† The content of these gases may undergo significant variations from time to time or from place to place relative to the normal.
(Copyright © 1979 Ohio State University, reproduced with permission)

a) Horizontal flow

Hot air masses are at higher average pressures when compared with cooler air masses. There is a consequent flow of air from the high-pressure regions to the low-pressure regions. In the absence of any external forces, the direction of flow will be as illustrated in Fig. 1.13, i.e. perpendicular to the lines of constant pressure. In practice, this does not happen because the spin of the Earth applies an external rotational force. This force is known as the Coriolis effect and it causes the wind to flow almost parallel to the lines of equal pressure. It also causes opposite rotational forces to be applied in the northern hemisphere to those experienced in the southern hemisphere. The wind flow around high-pressure areas (anticyclones) and low-pressure areas (cyclones or depressions) for the two hemispheres is illustrated in Fig. 1.14.

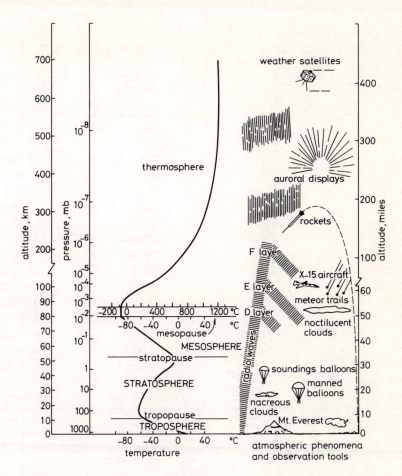

Fig. 1.11 *Pressure and temperature distribution of the Earth's atmosphere* (*Fig. 1.2 of Reference 12*)
(Copyright © 1981 Merrill Publishing Co., reproduced with permission)

The horizontal flow generated by the high-pressure region can affect a very small region, due to some local irregularity in the heating of the atmosphere, or encompass a million square kilometres. The smaller the scale size, the shorter is the life of the effect. Conversely, the larger the scale size, the more enduring is the result. Fig. 1.15 encapsulates the horizontal and temporal scale sizes. Note the microclimate which we shall return to in later Sections and Chapters.

The relative movement of anticyclonic air masses sets up unstable boundary

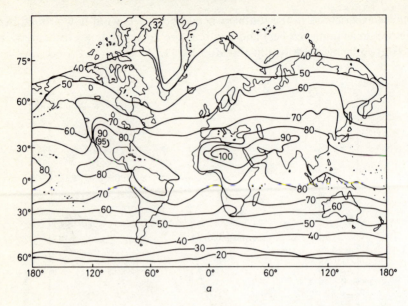

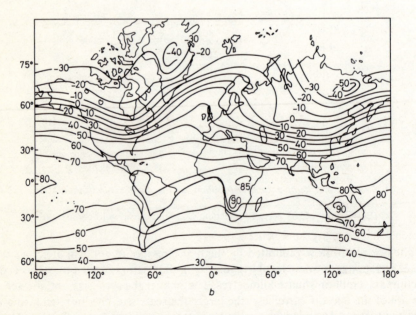

Fig. 1.12 *Isotherms giving the mean temperatures on the surface of the Earth for (a) July and (b) December (Fig. 3.10 of Reference 12)*
(Copyright © 1981 Merrill Publishing Co., reproduced with permission)

conditions between the air masses. The demarcation between recognisably different air masses is known as a front. In general, if the warmer air mass is overtaking a colder air mass, the front is referred to as a warm front; if the colder air mass is overtaking the warmer air mass, the front is referred to as a cold front. When the air masses meet, the cooler air from one air mass will tend to

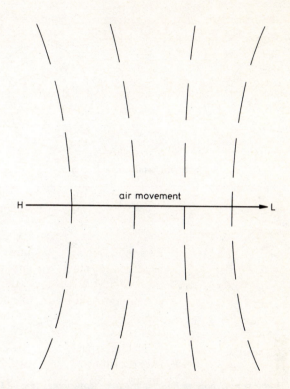

Fig. 1.13 *Direction of air movement in the absence of external forces*
H represents a high pressure area and *L* a low pressure area.
The broken lines are the isopleths of equal pressure.
The arrow is the direction of air flow at right angles to the isopleths in the absence of any other external forces.

slide under the warmer air from the other air mass. This will create a wedge which will tend to lift the warmer air mass. The resultant cooling of this air mass will usually lead to precipitation. Fig. 1.16 illustrates the effect schematically.

On some occasions, two cold air masses can trap a warm air mass between them. This happens when a cold front overtakes a warm front. The result is that the warm air mass is completely lifted from the surface of the Earth. This is referred to as an occluded front. Occluded fronts tend to dissipate energy more violently than warm or cold fronts. The width of the various fronts is usually less than 100 km and the band of precipitation is generally confined to this region. For an earth station, the relative orientation of the front to the azimuth look angle can make a significant difference to the amount of precipitation that falls along the transmission path. The speed of movement of the frontal region over the earth station will also determine the length of time that precipitation stays in the path.

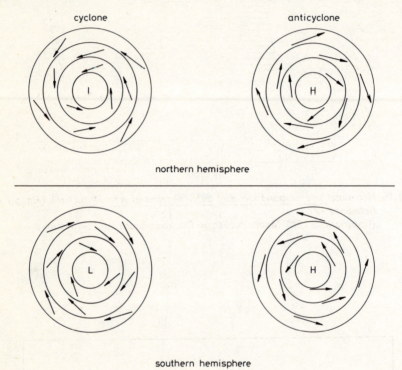

Fig. 1.14 *Cyclonic and anticyclonic air flows in the two hemispheres (Fig. 4.14 of Reference 12)* (Copyright © 1981 Merrill Publishing Co., reproduced with permission)

The weather patterns that contain the highest horizontal wind velocities are tropical cyclones, the hurricanes of North America and the typhoons of Asia. The major paths of the tropical cyclones are shown in Fig. 1.17.

b) Vertical flow
The vertical movement of an air mass is accompanied by a pressure and temperature change within the air mass to compensate for the change in external forces. When the vertical movement is slow, the process will usually be a dry adiabatic process, that is the water content of the air mass will not change its chemical phase. For this to come about, the temperature change of the air mass must be exactly compensated for by a volume change. The temperature change

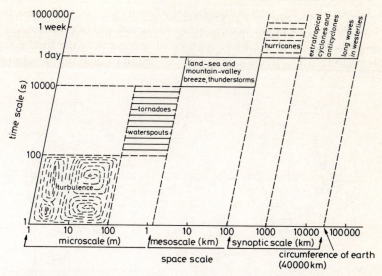

Fig. 1.15 *Horizontal and temporal scales of different types of weather patterns (Fig. 5.1 of Reference 12)*
(Copyright © 1981 Merrill Publishing Co., reproduced with permission)

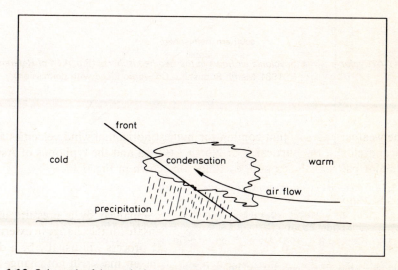

Fig. 1.16 *Schematic of the vertical cross-section of a front between a warm and a cold air mass*
The arrow denotes the relative movement of the warm air mass that is being lifted up by the incursion of the cold air mass.

with height, the lapse rate, of a dry adiabatic process is a constant 9·8°/km [12]. If the vertical movement results in a change in the chemical phase of the water content, the process is said to be moist adiabatic. Since the condensation of the

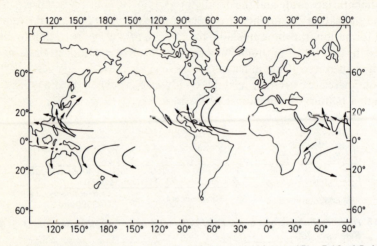

Fig. 1.17 *Major paths followed by tropical cyclones, or hurricanes (Fig. 5.16 of References 12)* (Copyright © 1981 Merrill Publishing Co., reproduced with permission)

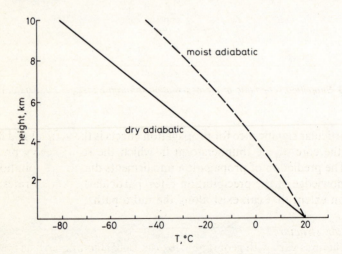

Fig. 1.18 *Variation of temperature with height for adiabatic processes (Fig. 4.15 of Reference 12)*
The dry adiabatic lapse rate is a constant 9·8°C/km while the moist adiabatic rate varies from about 6°C/km at sea level to 9·8°C/km at 10 km
(Copyright © 1981 Merrill Publishing Co., reproduced with permission)

water vapour will release latent heat, thereby keeping the air warmer than the surroundings, the lapse rate will be less than that of a dry adiabatic process. It will also not be a constant value [12]. Fig. 1.18 shows the two lapse rates.

The lapse rates are significant in determining whether or not a particular air mass and its vertical movement are stable or unstable. Gently ascending air, leading to stratiform clouds, is a stable condition. The lapse rate is less than the dry adiabatic lapse rate and the depth of the clouds is not very extensive, usually a good indication of stability. Conversely, the greater the vertical extent of a cloud formation, the more unstable the process is likely to be. Cumulo–nimbus clouds, the generators of thunderstorms and the most energetic of the convective systems, can exceed 10 km in height and have vertical wind velocities of 100 km/h [12, 46]. These are extremely unstable conditions. A schematic of a cross-section through a mature thunderstorm is shown in Fig. 1.19.

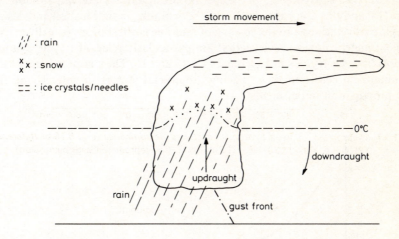

Fig. 1.19 *Simplified schematic of a cross-section through a mature thunderstorm*

Of particular significance for propagation effects is the vertical and horizontal size of the core of the thunderstorm in which the really heavy precipitation exists. The prediction of propagation impairments due to precipitation needs a good knowledge of the precipitation rates, particularly rainfall rates, and precipitation extent that can exist along the radio path.

1.3.3 Rain structures

Rain structures vary with geographic region. Local factors, such as precipitation accumulation, types and relative occurrence of different rain structures (i.e. stratiform, convective and mixed), vertical and horizontal extents of rain cells, and perhaps even the existence of preferred axes of cell anisotropy, all influence the prediction of propagation impairments along a given satellite–ground link. Local terrain features may also play a role, especially as regards the generation of mountain-region rain (orographic rain) or the blockage of rain structures [12]. Such microclimate effects can cause radical changes in the measured rainfall accumulation over small distances. An appreciation of what can cause such effects will enable the judicious siting of earth stations in those locations where below average propagation impairments should be experienced.

Nevertheless, the physical processes responsible for the generation of rain presumably must operate similarly from region to region (excepting extraordinary events such as typhoons, hurricanes and tornadoes). For example, thunderstorms are expected to form, yield rain, then dissipate in a somewhat similar fashion worldwide. The synthesis of various meteorological parameters to form the basis of sufficiently accurate predictive models for propagation impairments and potential restoration techniques should therefore be possible.

Many variations in rainfall over regions that are generally less than $10\,000\,km^2$ (i.e. subsynoptic, or mesoscale precipitation) are not random in nature [14] and this non-randomness can be exploited in the location and design of single earth stations and the spacing of multiple earth stations (site diversity). Excellent meteorological reviews that are particularly relevant to propagation concerns have been given in References 13, 15 and 16. The characteristics of the individual rain cells and their spacing are crucial factors in determining the likely propagation impairments.

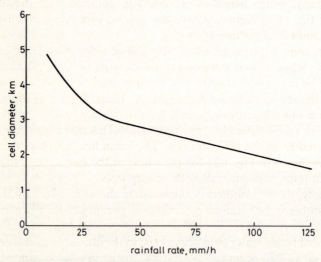

Fig. 1.20 *Representative model of average rain cell diameter versus rain rate (from Reference 22)* (Copyright © 1978 ITU, reproduced with permission)

a) Individual rain-cell characteristics
Rainfall rate measurements from extensive rain-gauge networks [17, 18, 19, 20] and radar investigations [21, 22] have shown that, statistically, the diameters of intense rain areas decrease as the average rain rate within the area increases. Average sizes of intense rain cells appear to be typically 2–5 km; an example is shown in Fig. 1.20.

There is evidence that the equivalent cell diameters follow a log-normal distribution [23] or perhaps an exponential distribution [24]. Some variations in average cell size with locality have been observed, but available data indicate that the extent of heavy rain along a given path is usually less than 10 km [15]. Occasional exceptions might be due to 'super-cell' thunderstorms [16], hurricanes or typhoons, particularly for paths at low elevation angles (below 10°).

The lifetimes of individual rain cells are usually rather brief, often only 10–20 min [13, 25], although, on occasion, intense cells can persist for periods as long as an hour [25]. The lifetimes appear to be distributed approximately exponentially [26]. There is a tendency for new cells to form adjacent to existing cells [13] so that rainstorms may have lifetimes much longer than the average lifetime of single cells within a storm.

The height of rain structures is an important parameter for slant-path propagation applications since it bounds the vertical extent of significant attenuation. Radar data show that, on average, the reflectivity profiles of cells are almost constant from the Earth's surface to the height of the peak reflectivity; they then decrease with additional increases in height by several dB/km [27]. Rain cells can therefore be modelled as uniform cores up to the height of peak reflectivity, which height is presumed to coincide with the top of the liquid-water (i.e. rain) regime. Above this height, precipitation is assumed to exist as ice, snow and melting snow.

Stratiform rain displays a characteristic, high-reflectivity layer (called the 'bright band') when a radar is used to probe through rain [47]. The bright band is known to be composed of snow and melting snow that can be presumed to exist above the region of liquid precipitation. Therefore, both convective cells and stratiform rain structures appear to have maximum rain heights that can be determined by radar-reflectivity measurements. This maximum rain height is now considered to be the 0°C isotherm. The mean height of the 0°C isotherm in summer is shown in Fig. 1.21 for the whole of the Earth [28]. Note that the summer data refer to summer in both hemispheres.

Occasionally, the 0°C isotherm is higher than shown in Fig. 1.21. The height of the 0°C isotherm, with annual percentage as parameter, is shown in Fig. 1.22 [27].

Super-cooled liquid water (which can exist in the atmosphere at temperatures as low as $-40°C$ [29]) occurs above the 0°C isotherm in some convective cells, especially in updraft regions of developing or mature thunderstorm cells. Therefore, the correspondence between rain height and the 0°C isotherm is uncertain for the intense convective cells of most concern in radiowave propagation predictions. Similarly, the tops of clouds, and hence rainfall, are sometimes well below the 0°C isotherm. Observations in some regions [30] have shown that, for almost 15% of the time, clouds are completely below the 0°C isotherm. To predict propagation impairments through precipitation, it is important not only to know the average height and extent of the cells, but their average separations.

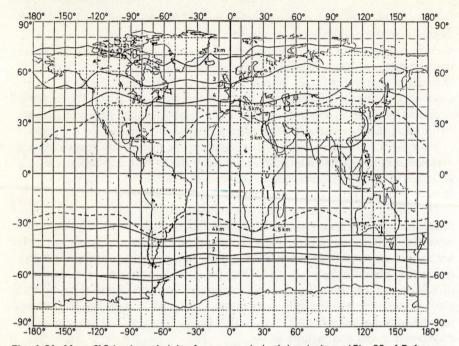

Fig. 1.21 *Mean 0°C isotherm heights for summer in both hemispheres (Fig. 22 of Reference 65)* (Copyright © 1988 ITU, reproduced with permission)
The numbers associated with the isotherms are the isotherm heights in kilometres.

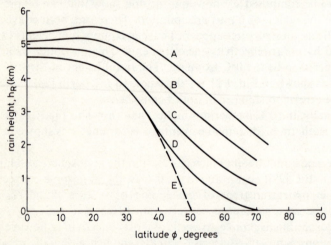

Fig. 1.22 *Rain height h_R (km) versus latitude φ (degrees) below which attenuation due to hydrometeors is expected (Fig. 17 of Reference 27)*
A: probability of occurrence of associated rain intensity: 0·001%
B: probability of occurrence of associated rain intensity: 0·01%
C: probability of occurrence of associated rain intensity: 0·1%
D: probability of occurrence of associated rain intensity: 1·0%
E: includes rain and snow occurrences
(Copyright © 1982 ITU, reproduced with permission)

b) *General areal rainfall characteristics*

Convective and stratiform rain components generally accompany each other, and the convective elements are typically multicellular in character [13, 16]. Common summertime 'air mass' thunderstorms are composed of a mixture of developing, mature and dissipating cells that individually move in an essentially random manner, while severe 'multi-cell' thunderstorms contain cells that develop and progress through the structure in a systemmatic way [16]. Statistics of the joint effects of multiple cells is very useful for propagation considerations.

Investigations of rain-producing convective clouds [23, 26, 30] have shown that rain cells tend to align along preferred azimuths for a given locale and that preferred spacings between adjacent rain cells also exist. New cells tend to develop adjacent to existing cells [13] forming cell clusters. One experiment [24] showed that, on average, a cluster was made up of three cells. Furthermore, cells often occur in rainbands (frontal precipitation) or squall lines. Typical separations of bands of rain cells are about 25 km [26] with average rain cell (or cluster) separations along the band of 15–30 km [30]. These values appear to be fairly representative [31], at least in sub-tropical and temperature climates, but slightly smaller separations occur in occlusions [32]. In a radar study of approximately 750 rain cells for the Ottawa, Canada, area, Strickland [23] found that the separations between individual cells (not just the adjacent cells) were Rayleigh distributed with a median separation of 28·8 km. A similar study for Wallops Island, Virginia [33], showed an average intercell separation of 33·3 km, but no definite correlation with a probability distribution could be established.

The forming of intermediate-scale rain structures into bands may be caused by gravity waves [26, 34]. This thesis is supported by observations of land/sea interface effects which appear to create wavelike cloud structures [35] and by radar precipitation studies [26]. Tropical precipitation systems (hurricanes and typhoons) also appear to form bands [13, 45], arranged in a spiral fashion about the storm centre. The position of the storm centre relative to the location of the earth station therefore determines the orientation of the rain band as it passes over the earth station (see Fig. 1.23). In this case, the prevailing wind direction cannot be used to predict the orientation of the band with respect to the propagation path.

The generation and dissipation of cells within storm systems inevitably leads to changes in the areal character of the precipitation structures. Zawadski [36] analysed the spatial and temporal autocorrelation functions of radar returns from a widespread convective storm system in the Montreal vicinity and concluded that spatial structure patterns were maintained for periods of about 40 min as the storm travelled, although the lifetime of the storm itself was several hours. In a companion study [37], the rate of spatial decorrelation of the radar patterns was found to be nominally independent of radial direction for distances of about 10 km, implying that the joint probabilities of rain rate for two locations will be independent of baseline orientation for similar separation distances. Pronounced topographical features (e.g. a big hill) would naturally have altered this conclusion owing to the concomitant orographic effects induced by such features.

Convective cells tend to cluster, with intercellular separations of the order of 5–6 km and with separations between cluster centres of roughly 11–12 km [25]. The spatial arrangement of clusters is typically related to the mechanism that initiated the rainfall process (e.g. clusters tend to align along the weather fronts, or adjacent to pronounced terrain features in orographically-induced rain). The most intense cells in the rain structure sometimes elongate in the direction of motion, i.e. perpendicular to a weather front, thereby creating an axis of cell anisotropy. Indeed, most rain patterns tend to become asymmetrical once they exceed about 20 km in size [31]. Radar data indicate that the number of convective cells within storms decreases exponentially with cell width and cell height [38], and that shower sizes [39] and cell lifetimes [26] are distributed exponentially.

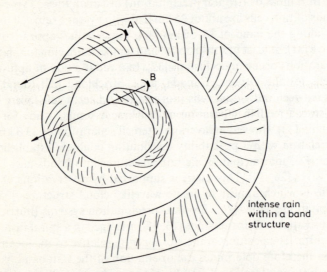

Fig. 1.23 *Effect of path orientation with respect to the weather system*
The signals from earth station A experience a longer path through intense rain than those from earth station B. The propagation impairments on path A are worse than on path B.

Measurements have been made of the joint probabilities of rainfall at two separated points [e.g. 18, 40]. The probability that a given point rainfall rate is exceeded jointly at two locations decreases as the site separation increases. An example of the behaviour of the normalised rainfall rate probability (i.e. the joint probability divided by the single-point probability) based on three years of data from Japan [40] is shown in Fig. 1.24.

Figure 1.25 presents some joint probability data from North America [18]. Note the slight increase in the joint probability beyond a certain site separation. This could seem to indicate that the likelihood of two independent rain cells

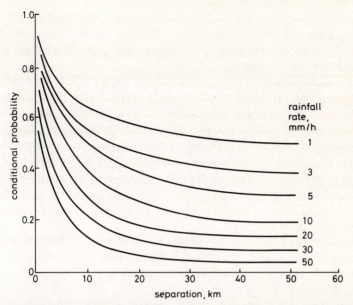

Fig. 1.24 *Normalised conditional probability for jointly exceeding rain rate at two points versus separation between the points (from Fig. 3 of Reference 40)*
(Copyright © 1975 IECE, reproduced with permission)

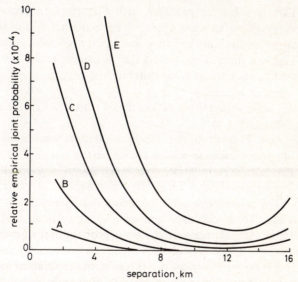

Fig. 1.25 *Relative empirical joint probability that the rain rate at both sites exceeds the value noted on the curve versus the separation of the sites (from Fig. 23(b) of Reference 18)*

Minimum rainfall rate (mm/h)

A	B	C	D	E
170	140	110	90	70

(Copyright © 1969 AT&T, reproduced with permission)

raining simultaneously at two widely separated sites can exceed the probability of a single, large rain cell raining simultaneously at the two sites with the same intensity. This supposition would support the idea of preferred separations of rain cells and rain-cell clusters.

Terrain factors can play a role in modifying the areal behaviour of rainfall accumulations and precipitation rates [13, 31, 41]. There is evidence that large urban areas can enhance rainfall accumulation [42], often 20–50 km down wind from the main urban centre [43, 44]. The areal distribution of rainfall due to microclimate [12], and its spatial and temporal variability, are expected to be complex and difficult to characterise completely. Within these rainfall structures, it is important to recognise the different types of precipitation and their effect on radiowave propagation.

1.3.4 Precipitation types

The vast majority of precipitation generated in weather systems never reaches the ground. Some of the precipitation is formed into such small particles that aerodynamic forces keep them aloft. In other cases, the particles evaporate on their way down to Earth.

Precipitation can take the form of hail, sleet, graupel, freezing rain, snow, and rain. A seventh crucial precipitation type that does not usually extend to sea level is ice crystals. The general characteristics of each are summarised below:

Hail: This is solid ice particles with diameters in excess of 5 mm. Sometimes hail particles can exceed 10 cm in diameter. If the particles are less than 5 mm in diameter, they are strictly referred to as ice pellets.

Sleet: This is a form of ice pellets that is usually translucent and hard. Sleet usually exists originally as rain but freezes on its way down through a cold layer.

Graupel: This is a form of ice pellets that is white, rather like a compacted snow pellet, and is soft.

Freezing rain: This is rain that freezes on impact when it encounters the ground or other solid objects.

Snow: This is an accumulation of ice crystals into larger clusters, or snow flakes. The temperature at which the crystals form determines their type. Many of the crystals contain a hexagonal symmetry.

Ice crystals: These can exist individually at extreme altitudes in the stratosphere. Clouds above the zero-degree isotherm generally consist almost entirely of ice crystals.

Rain: This is liquid precipitation. If the drops are smaller than 0·5 mm in diameter, the precipitation is referred to as drizzle.

1.3.5 Rainfall characteristics

For most frequencies below 100 GHz, and in some cases at frequencies well into the optical spectrum, rain is the greatest propagation impairment. The average drop diameter in a non-precipitating cloud is about 0·02 mm in diameter [12].

In order to fall, and not evaporate before reaching the ground, the drops must increase their average diameter almost a hundredfold. The two principal mechanisms for doing this are through the formation of ice crystals and through the collision and coallescence of rain drops.

The growth mechanism due to ice crystals was established by Bergeron [49]. Ice crystals will grow at the expense of super-cooled rain drops owing to the significantly higher vapour pressure of the rain drops compared with the ice crystals. The simultaneous presence of ice crystals and super-cooled rain drops is not uncommon and rain generated from the so-called Bergeron process leads to much of the rainfall accumulation on the Earth.

The collision of rain drops becomes more likely the more turbulent, or mixed, the atmosphere is. Rain drops will tend to coalescence on collision and so the average rain-drop size will increase quite rapidly in turbulent conditions such as convective updrafts. The wind sheer effects from such updrafts have been shown to increase the collection efficiency of small raindrops [50], i.e. those less than 25 μm in diameter, compared to still air, although this effect was not apparent for larger rain drops. The collection efficiency refers to the apparent success the individual rain drop has in meeting, and 'capturing' or coalescing with, adjacent rain drops. The larger drops, however, as well as tending to coalesce on collision, will, if large enough, separate again and give rise to numerous 'satellite' rain drops [50]. These additional satellite drops could account for the rapid increase in rainfall rates in certain, highly turbulent conditions. In stable air conditions, the process is more predictable, depending, to a great degree, on the size of the rain drop.

Given that the rain drop is large enough to start falling, as the size of the rain drop increases, so the aerodynamic forces will start to overcome the surface-tension forces. In small drops, the latter force will keep the rain drops spherical. The larger the rain drop becomes, the more elliptical it tends to be, the faster it falls, and the more distorted the shape of the drop becomes. Terminal velocity and drop shape are an integral part of the rain drop-size distribution, usually abbreviated to drop-size distribution. The drop-size distribution is a measure of the number of drops of a given size there are in a unit volume and, from this, a rainfall rate can be inferred (or the converse). Segmenting the Earth into regions of similar rainfall rate distributions yields the climatic zones of the Earth. These factors are discussed below.

a) Terminal velocity
The pioneering work in this area was conducted by Laws in 1941 [51]. Later refinements in technique and range were made [29, 52, 53] for measurements on the ground, and the standard that is now used is due to Gunn and Kinzer [52]. They measured the fall velocities for drops ranging in size from 0·2 to 100 000 μg, the upper limit being set by the drop stability. Later measurements by Beard [54] were conducted in varying atmospheric pressures to assess the variation in fall velocity with height. The results are partially reproduced in Fig. 1.26 (after

Fig. 4.2 in Ref. 55) with the addition of a profile due to Davies [29, 53]. Meteorologists tend to use the Gunn and Kinzer profile while microwave engineers have traditionally used that due to Davies. For all practical purposes, the two are the same and they essentially form the ground-level curve for the set due to Beard.

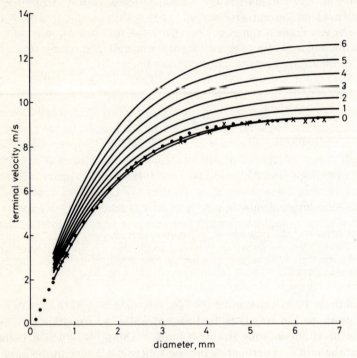

Fig. 1.26 *Terminal velocity profiles of rain drops*
● Data of Gunn and Kinzer [52]
x Data of Davies [53]
Parameter on each curve is the height above mean sea level.
(Reproduced with permission from the Communications Satellite Corporation)

As can be seen in Fig. 1.26, the limiting terminal velocity for most large rain drops is in the region of 10 m/s. There is a large difference in terminal velocity with height. For a more exact calculation of the terminal velocity with respect to drop diameter, temperature, pressure and relative humidity, reference can be made to Wobus *et al.* [56] in which a relatively simple set of equations is given that approximate, with good accuracy, to the exact, but complex, formulations of Beard [54].

b) Drop shapes
The exact shape of a rain drop at any instant of time is a complex mix of surface tension and aerodynamic forces. For very small drops (less than or equal to

$170\,\mu m$ in diameter), surface-tension forces will dominate under almost any wind conditions and the drops will be almost exactly spherical. Between 170 and $500\,\mu m$, the drops become elliptical in cross-section while, above $500\,\mu m$, the drops become progressively flattened at the base. Ultimately, the bases of the drops are hollowed out to form the so-called Prupacher and Pitter rain-drop shape [57]. Fig. 1.27 summarises these findings schematically. Note that when reference is made to the diameter of a non-spherical rain drop, it is the diameter of the equivalent-volume spherical rain drop that is being referred to.

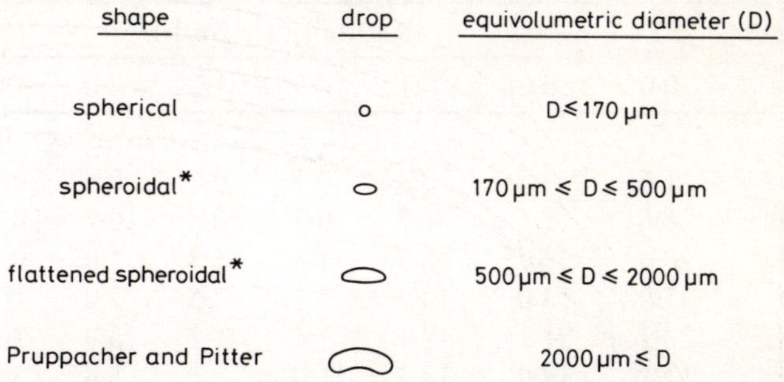

shape	drop	equivolumetric diameter (D)
spherical	o	$D \leqslant 170\,\mu m$
spheroidal*	⬭	$170\,\mu m \leqslant D \leqslant 500\,\mu m$
flattened spheroidal*	⬬	$500\,\mu m \leqslant D \leqslant 2000\,\mu m$
Pruppacher and Pitter	⬭	$2000\,\mu m \leqslant D$

Fig. 1.27 *Schematic of rain drop shapes with approximate size ranges*
*These shapes can be prolate as well as oblate, the case illustrated here. Typically, the rain drop will oscillate between the two shapes in periods of a few tens of milliseconds.

The rain drop, once it distorts out of the spherical form, is not stationary in terms of its shape but oscillates between an oblate and prolate spheroid. A spheroid is formed by rotating an ellipse about its shortest axis which, in the case of an oblate spheroid, is vertical. A prolate spheroid is one with the minor axis horizontal. The natural oscillations of rain drops have been measured [58]. The frequency of the natural oscillations f is related to the diameter d (in centimetres) by Reference 58

$$f = (11 \cdot 7) \times d^{-1 \cdot 47} \text{ hertz} \qquad (1.14)$$

c) Drop-size distributions
The relationship between terminal velocity and drop-size distribution is given by

$$N_{(D)} = N'_{(D)} / V_{(D)} \qquad (1.15)$$

where $N'_{(D)}$ is the number of drops per unit time per unit increment in diameter D crossing a unit area of a measuring device; $V_{(D)}$ is the terminal velocity; and $N_{(D)}$ is the drop-size distribution, i.e. the number of drops that exist between diameters D and $D + d\text{D}$ per unit volume. Usually the drop-size distribution is measured on the ground. A suffix g is normally employed for these measurements. To estimate the drop-size distribution at a height h above the ground, the following relationship can be used:

$$N_{h(D)} = N_{g(D)} \times [V_{g(D)}/V_{h(D)}]$$ (1.16)

The first systemmatic measurements of drop-size distribution that covered a range of drop sizes were conducted by Laws and Parsons [59] in the early 1940s. Their original results are given in Table 1.3.

Table 1.3 *Drop-size distribution of rain drops as a percentage of the total volume (from the original Table 3 of Reference 59)*

Drop-diameter limits	Rainfall-rate (in/h)							
	0·01	0·05	0·1	0·5	1·0	2·0	4·0	6·0
mm mm								
0·00–0·25	1·0	0·5	0·3	0·1	0	0	0	0
0·25–0·50	6·6	2·5	1·7	0·7	0·4	0·2	0·1	0·1
0·50–0·75	20·4	7·9	5·3	1·8	1·3	1·0	0·9	0·9
0·75–1·00	27·0	16·0	10·7	3·9	2·5	2·0	1·7	1·6
1·00–1·25	23·1	21·1	17·1	7·6	5·1	3·4	2·9	2·5
1·25–1·50	12·7	18·9	18·3	11·0	7·5	5·4	3·9	3·4
1·50–1·75	5·5	12·4	14·5	13·5	10·9	7·1	4·9	4·2
1·75–2·00	2·0	8·1	11·6	14·1	11·8	9·2	6·2	5·1
2·00–2·25	1·0	5·4	7·4	11·3	12·1	10·7	7·7	6·6
2·25–2·50	0·5	3·2	4·7	9·6	11·2	10·6	8·4	6·9
2·50–2·75	0·2	1·7	3·2	7·7	8·7	10·3	8·7	7·0
2·75–3·00	0	0·9	2·0	5·9	6·9	8·4	9·4	8·2
3·00–3·25	0	0·6	1·3	4·2	5·9	7·2	9·0	9·5
3·25–3·50	0	0·4	0·7	2·6	5·0	6·2	8·3	8·8
3·50–3·75	0	0·2	0·4	1·7	3·2	4·7	6·7	7·3
3·75–4·00	0	0·2	0·4	1·3	2·1	3·8	4·9	6·7
4·00–4·25	0	0	0·2	1·0	1·4	2·9	4·1	5·2
4·25–4·50	0	0	0·2	0·8	1·2	1·9	3·4	4·4
4·50–4·75	0	0	0	0·4	0·9	1·4	2·4	3·3
4·75–5·00	0	0	0	0·4	0·7	1·0	1·7	2·0
5·00–5·25	0	0	0	0·2	0·4	0·8	1·3	1·6
5·25–5·50	0	0	0	0·2	0·3	0·6	1·0	1·3
5·50–5·75	0	0	0	0	0·2	0·5	0·7	0·9
5·75–6·00	0	0	0	0	0·2	0·3	0·5	0·7
6·00–6·25	0	0	0	0	0·1	0·2	0·5	0·5
6·25–6·50	0	0	0	0	0	0·2	0·5	0·5
6·50–6·75	0	0	0	0	0	0	0·2	0·5
6·75–7·00	0	0	0	0	0	0	0	0·3

The tabular values for rainfall rates of 0·01, 4·0, and 6·0 in/h, and for raindrops smaller than 0·5 mm were derived by extrapolation·

(Copyright © 1943 American Geophysical Union, reproduced with permission)

An inspection of Table 1.3 will show that, as the rainfall rate increases, so the median drop diameter increases. This effect is shown schematically in Fig. 1.28.

What is not immediately apparent in Table 1.3 is the mathematical description of the results. Later measurements by Marshall and Palmer proposed an exponential mathematical model to fit the results; namely

$$N_{g(D)} = N_0 e^{-\lambda D} \tag{1.17}$$

where N_0 and λ are coefficients chosen to fit the measured drop-size distribution.

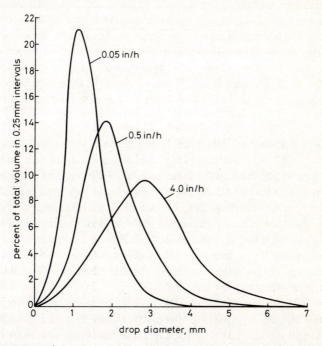

Fig. 1.28 *Percentage of the total rainfall volume contributed by rain drops of the given sizes with rainfall rate as a parameter (from Fig. 3 of Reference 59)*
(Copyright © 1943, American Geophysical Union, reproduced with permission)

The terms N_g and N_0 have the units $mm^{-1} m^{-3}$ while λ has the units mm^{-1}. The term N_0 was assigned the value 8000 and λ the value $4\cdot 1 R^{-0\cdot 21}$, where R was the rainfall rate in mm/h. Later results showed that, while the exponential distribution was a reasonable description of the shape of the data, particularly for rain drops above $1\cdot 5$ mm in diameter, the coefficients chosen did not give acceptable agreement.

A modified form of the exponential distribution, assuming spherical drop shapes, is

$$N_{g(D)} = N_0 e^{-3\cdot 67 D/D_0} \tag{1.18}$$

In this distribution, an upper (D_{max}) and lower (D_{min}) limit on the rain-drop sizes is postulated. The term D_0 is set to be at least four times the size of D_{min} and no greater than half the size of D_{max} to ensure that half the water content is from drops with diameters greater than D_0. That is, D_0 is the diameter of the median volume drop. For the Marshall and Palmer distribution

$$D_0 = 0.89 \, R^{0.21} \tag{1.19}$$

In general, larger rain drops will not have a spherical shape and so an equivalent diameter D_e is used. This diameter is that of a perfectly spherical rain drop that has the equivalent volume of the distorted rain drop. The equivalent diameter D_e has been measured [61] for a range of drops. An upper limit of about 10 mm was found for D_e; above this diameter, aerodynamic forces tended to break up the drop. With $D_e = 10$ mm, the ratio of the minor axis to the major axis of the actual rain drop was 0.41. This ratio approached unity in an approximately linear fashion as D_e dropped from 10 mm to 1 mm. If D_e is used in eq. 1.18, the general equation for a Marshall and Palmer drop-size distribution becomes

$$N_{g(D)} = N_0 \, e^{-3.67 D_e / D_0} \tag{1.20}$$

This drop-size distribution has been found to be a very good description of long-term data and, when applied to propagation-impairment calculations, to give theoretical results close to the experimental measurements in the microwave region of the spectrum. The difficulty comes when short-term predictions need to be made or the statistical results extended to frequencies above about 30 GHz. In the former case, the rapid fluctuations in the rainfall rate and drop-size distribution are not amenable to fitting to a long-term (i.e. average) distribution; in the latter case, the effect of smaller rain drops becomes significant and, for drops below 1.5 mm, the Marshall and Palmer drop-size distribution over-predicts.

Joss *et al.* [62] departed from the single, long-term distribution approach of Marshall and Palmer by splitting the distributions into one due to thunderstorms (J–T) and one due to drizzle (J–D). Later measurements [63] showed that there was yet a third distinct class of distributions: that due to showers. In showers, there are extremely few rain drops above 5 mm in diameter, the bulk of the rain drops being between 1.0 and 3.5 mm. For accurate prediction of drop-size distributions, therefore, three classifications of rain types should be used: drizzle (sometimes referred to as widespread rain), showers and thunderstorms. Since most rainfall data are not categorised into these three types, propagation scientists still tend to use the single, long-term distribution modelled after Marshall and Palmer to provide statistical results while recognising its shortcomings for frequencies above 30 GHz and for short-term predictions. The division of rainfall into convective and non-convective rain is, however, an established tool for climate divisions.

d) Rainfall rate distributions

A classic paper by Rice and Holmberg [64] divided rainfall into two types to permit the prediction of rainfall-rate statistics from the total rainfall accumulation measured in an average year. The two types were mode 1 rain (M_1) and mode 2 rain (M_2). Mode 1 contained the high rainfall rates associated with strong convective activity and thunderstorms. Mode 2 was simply everything else. The total annual average rainfall accumulation M was therefore

$$M = M_1 + M_2 \text{ millimetres} \tag{1.21}$$

A coefficient β was postulated that was equal to the ratio of convective rainfall accumulation to total rainfall accumulation, namely

$$\beta = \frac{M_1}{M} \tag{1.22}$$

With simply the values M and β, Rice and Holmberg proposed a formula that gives the number of hours in an average year that the rainfall rate in mm/h exceeds a given value R, namely [64]

$$\begin{aligned} T_1 = M \times \{0{\cdot}03\beta \times \exp{(-0{\cdot}03\ R)} \\ + 0{\cdot}2 \times (1 - \beta)\,[\exp{(-0{\cdot}258R)} \\ + 1{\cdot}86 \times \exp{(-1{\cdot}63R)]}\} \text{ hours} \end{aligned} \tag{1.23}$$

The subscript 1 for T indicates a time constant of 1 min is applied to the rainfall rate measurements. Fig. 1.29 gives the contours of β for the world.

The CCIR have generalised the situation even further than that used in the Rice–Holmberg formulation and set up a total of 15 climate zones, each with a particular rainfall rate distribution. The current distributions are given in Table 1.4 [65] and the global boundaries associated with this Table in Fig. 1.30 [28, 65].

The advantages of this approach are the simplicity of use and the relatively small amount of descriptive data required to give many general distributions. Clearly, the rainfall rates do not change abruptly at the zone boundaries and a potentially more accurate description would be to give the rainfall rate contours. The CCIR does indeed do this for some rainfall rates and an example for 0·01% rainfall rates is given in Fig. 1.31 [28]. While overcoming the 'step function' between zones apparent in Fig. 1.30, the data presentation in Fig. 1.31 lacks some vital oceanic data.

Before leaving the topic of rainfall rate distributions, it is interesting to note a relatively recent discovery of a trend in the rainfall rate distributions with solar sunspot cycle [66]. The wide variations possible between annual rainfall rate distributions has been observed in almost all long-term measurements. Typically, a minimum of seven years had been thought to be the required measurement period in order to achieve stable results; the test for stability being the effect on the long-term distribution of removing any one annual data file from the data bank. If no significant effect was observed, the results were considered to be stable. It now appears that at least an 11-year period is required to

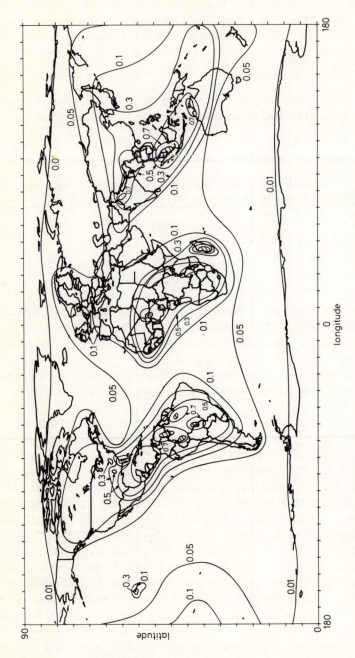

Fig. 1.29 *Contours of the coefficient β for use in the Rice–Holmberg model (after Fig. 3 of Reference 64)*
(Copyright © 1973 IEEE, reproduced with permission)

Table 1.4 *Rainfall rates in the 15 CCIR rainfall climatic zones (Table 1 of Reference 65)*

Percentage of time	A	B	C	D	E	F	G	H	J	K	L	M	N	P	Q
1·0	<0·1	0·5	0·7	2·1	0·6	1·7	3	2	8	1·5	2	4	5	12	24
0·3	0·8	2·0	2·8	4·5	2·4	4·5	7	4	13	4·2	7	11	15	34	49
0·1	2	3	5	8	6	8	12	10	20	12	15	22	35	65	72
0·03	5	6	9	13	12	15	20	18	28	23	33	40	65	105	96
0·01	8	12	15	19	22	28	30	32	35	42	60	63	95	145	115
0·003	14	21	26	29	41	54	45	55	45	70	105	95	140	200	132
0·001	22	32	42	42	70	78	65	83	55	100	150	120	180	250	180

Rainfall rates exceeded for the given percentage times in the various climatic zones
(Copyright © 1988 ITU, reproduced with permission)

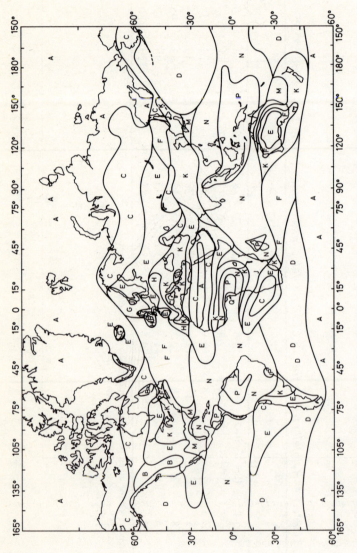

Fig. 1.30 *Rainfall elimatic zones of the Earth as given by the CCIR (combined from References 28 and 65)*

(Copyright © 1986 and 1988 ITU, reproduced with permission)

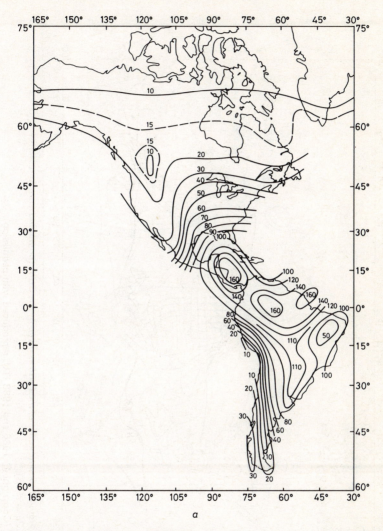

a

Fig. 1.31 *Rainfall rate contours for 0·01% of the time (from Figs. 15, 16 and 17 of Reference 28)*
a For the Americas
b For Europe, Africa and the Middle East
c For Asia, India and Australasia
(Copyright © 1986 ITU, reproduced with permission)

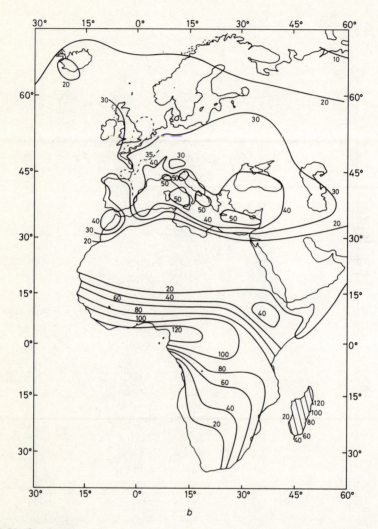

b

Fig. 1.31 *Continued*

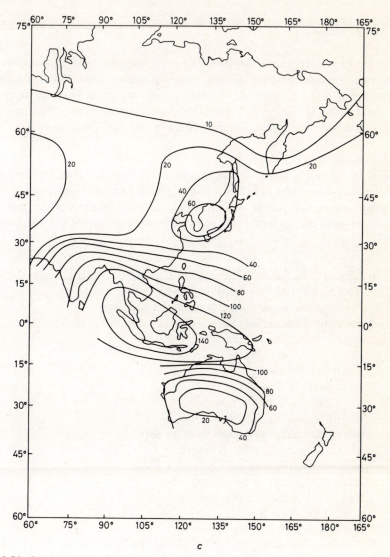

Fig. 1.31 *Continued*

average out completely the effects of the sunspot-induced variations in the given rain climate.

1.4 System planning

Any system that requires the communication or acquisition of information via radiowaves is subject to detailed, mandatory planning steps. The further the information is to be transmitted to or acquired from, the more rigorous the planning becomes, involving, as it does, the necessary agreement of both national and international authorities. The procedure is called co-ordination.

1.4.1 Earth-station co-ordination

To assist in the initial steps of planning and co-ordination, the International Telecommunications Union (ITU) has designated service types together with the agreed frequency bands within which they are permitted to operate. Table 1.5 sets out the applicable list of satellite services currently designated by the ITU.

Table 1.5 *ITU Designated services using space stations**

Aeronautical Mobile Satellite Service
Aeronautical Radionavigation Satellite Service
Amateur Satellite Service
Broadcasting Satellite Service
Earth Exploration Satellite Service
Fixed Satellite Service
Inter-Satellite Service
Land Mobile Satellite Service
Maritime Mobile Satellite Service
Maritime Radionavigation Satellite Service
Meteorological Satellite Service
Mobile Satellite Service
Radiodetermination Satellite Service
Radionavigation Satellite Service
Space Operation Service
Space Research Service
Standard Frequency and Time Signal Satellite Service

* Note that the term 'space station' denotes any spacecraft, manned or unmanned.

The appendices of the ITU Radio Regulations [3] set out the agreed frequencies within which each service may operate and the restrictions that have been agreed to in the ITU fora, e.g. maximum power flux density levels and permissible interference levels.

Interference is eventually the fundamental limiting factor in all telecommunications systems. The gain of the antenna can be increased, for example, as can the power of the transmitter and the sensitivity of the receiver, but no portion of the radio spectrum is free from other users, natural or man-made, and so unwanted signals will always enter the receiving system. The difference in power between the wanted and unwanted signals will ultimately determine the performance of the communications system. The strength of the unwanted signal P_i from the interfering transmitter will be determined by the transmitting power P_u and antenna gain G_u of the interfering station in the direction of the receiving antenna, the gain of the receiving antenna G_w along the same path, and the transmission loss A along the same path. Here, the suffix u refers to the system producing the unwanted signal, the suffix i refers to the interference, and the suffix w refers to the system trying to receive only the wanted signal. Expressed mathematically in decibels, the interference power level P_i is

$$P_i = P_u + G_u + G_w - A \text{ dBW} \tag{1.24}$$

Fig. 1.32 sets out the geometry of this simple interference path.

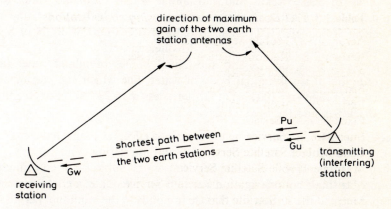

Fig. 1.32 *Schematic of the geometry of a simple interference path*
    G_w is the component of gain of the receiving station along the shortest path between the two stations
    G_u and P_u are the components of gain of the interfering station along the same path and the maximum power of the unwanted transmission, respectively.

Most systems are designed to operate satisfactorily with a certain amount of interference entering the receiver. The level of permissible interference varies both with the type of system and with the percentage of the operating time that the permissible level of interference can be tolerated. The worse (i.e. the higher) the level of permissible interference, the smaller is the percentage time that that level of interference will be allowed. Using the suffix p to denote the percentage time that a certain interference power is permitted and re-arranging eqn 1.24

gives

$$A = P_u + G_u + G_w - P_i(p) \text{ decibels} \tag{1.25}$$

From a knowledge of P_u, G_u, G_w and $P_i(p)$, the required value of A can be found. If the interference path is along the great-circle path and it is in clear-sky conditions, this particular value of A is called the minimum permissible basic transmission loss. The reason it is called the minimum permissible basic transmission loss is that, if the loss is greater than this value, no unacceptable interference will occur for the given percentage time. In addition, the word *basic* is used to describe the mode of propagation; since no other mechanism of loss is involved (e.g. rain attenuation), it is essentially the free-space propagation loss. In calculating the co-ordination distance, two modes of propagation are used in which the previous mode of propagation is referred to as Mode (1) propagation. Mode (1) propagation can have very different values depending on the terrain. For this reason, three types of terrain, separated into zones, are defined, namely [3]:

> "*Zone A*: Entirely land.
> *Zone B*: Seas, oceans and substantial bodies of inland water (as a criterion of a substantial body of water, one which can encompass a circle of diameter 100 km) at latitudes greater than 23° 30′ N or S, but excepting the Black Sea and the Mediterranean.
> *Zone C*: Seas, oceans and substantial bodies of inland water (as a criterion of a substantial body of water, one which can encompass a circle of diameter 100 km) at latitudes less than 23° 30′ N or S, and the Black Sea and the Mediterranean."

Procedures exist [3] for calculating the combined effects of paths that cross more than one zone between the interfered-with and interfering systems.

When a transmitting or receiving station is subject to co-ordination procedures, a single co-ordination distance is not adequate; a contour must be developed around the station site that corresponds to the minimum permissible basic transmission loss in every direction. The area within the contour is referred to as the co-ordination area. An earth station that will be operating at all elevation and azimuth angles will have a co-ordination contour of Mode (1) propagation that is determined by the maximum gain of the antenna. Further, if that earth station is completely within one zone, the Mode (1) co-ordination contour will be a circle. On the other hand, if the earth station is operating to a geostationary satellite, the elevation and azimuth angles will be virtually constant. In this case, the maximum gain of the earth-station antenna is not used to derive the co-ordination contour; only that fraction of the gain that is directed towards the horizon is employed. Figs. 1.33a and 1.33b illustrate these two cases for Mode (1) propagation.

The second mode of propagation, Mode (2), is due to the scattering of radiowave energy by rain drops. There are both numerical and graphical methods for calculating the rain-scatter distance [3]. The distance is given the character d_r and is independent of the type of terrain, or zone. The rain-scatter co-ordination contour, of radius d_r, will therefore be a circle. The energy from the scatter, however, is larger in the forward direction than in other directions.

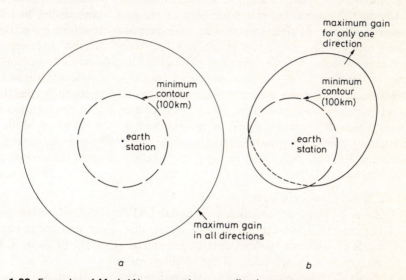

Fig. 1.33 *Examples of Mode (1) propagation co-ordination contours*
In (*a*), the contour is a circle due to the fact that the antenna will be pointed at any elevation or azimuth angle. The maximum gain is therefore used in all directions. In (*b*), since the antenna is fixed, the co-ordination distance will vary with azimuth as the gain of the antenna changes with azimuth angle.
In both cases, a minimum value of 100 km is used for Mode (1) propagation co-ordination distances. In (*a*), the 100 km minimum distance is exceeded in all directions but, in (*b*), the co-ordination contour is modified by the minimum distance contour of 100 km.

The centre of the rain-scatter co-ordination contour is therefore offset by an amount Δ_d in the direction of the maximum gain of the earth station, as shown in Fig. 1.34. Δd is related to the rain-scatter distance d_r by [3]

$$\Delta d \ = \ 5 \cdot 88 \times 10^{-5} \times \{d_r - 40\}^2 \cot \theta \text{ kilometres} \qquad (1.26)$$

where θ is the earth station elevation angle.

The complete co-ordination contour is a composite of the Mode (1) and Mode (2) co-ordination contours, where the larger distance given by either of the propagation-mode calculations is used. Fig. 1.35 illustrates the combined coordination contour for an earth station operating to a geostationary satellite.

In some cases, severe interference between terrestrial (e.g. line-of-sight) systems and earth–space systems can be reduced if a form of site shielding can be used.

1.4.2 Site shielding

Site shielding can take the form of natural shielding, such as mountains, or artificial shielding. Increasingly, special fences erected around earth-station complexes, particularly those sited near the centres of major cities, are being used to reduce the interference. The site shielding factor F_s, can be calculated (see Chapter 7) and A, the minimum permissible basic transmission loss for propagation Mode (1) given in eqn. 1.25, is decreased by this amount, namely

$$A = P_u + G_u + G_w - F_s - P_i(p) \text{ decibels} \tag{1.27}$$

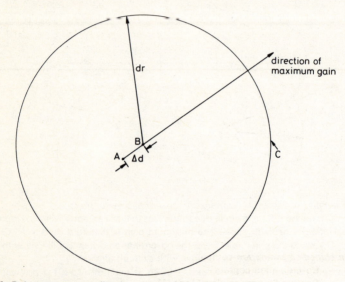

Fig. 1.34 *Rain-scatter co-ordination contour*
  A: earth station
  B: centre of rain-scatter co-ordination contour
  C: rain-scatter co-ordination contour
  The circular contour due to the rain-scatter mechanism (propagation Mode (2)) is offset by an amount Δd in the direction of maximum gain of the earth station.
  If the calculated value of dr is less than 100 km, a value of 100 km is used for dr in the co-ordination contour.

The larger F_s becomes, the smaller A needs to be and the closer two, potentially interfering, systems can be placed. Counteracting this is the possibility of scintillations causing the signal to go above the median level. Scintillation amplitudes can be predicted for a given percentage time (see Chapters 2 and 3), and a value G_s that corresponds to the positive scintillation amplitude at the given percentage time can be used to account for an apparent increase in system 'gain' due to scintillations [67]. If G_s is the effective gain due to scintillation at the appropriate percentage time, the value of A in eqn. 1.27 becomes

$$A = P_u + G_u + G_w - F_s + G_s - P_i(p) \text{ decibels} \tag{1.28}$$

The gain of the antennas and the powers to be transmitted in any radiotelecommunications system will be a function of the required signal level in the receiver to meet, or exceed, the performance and availablility levels set for that system. To obtain the required antenna gains and transmission powers, it is usual to calculate the end-to-end performance of the system for various power and gain

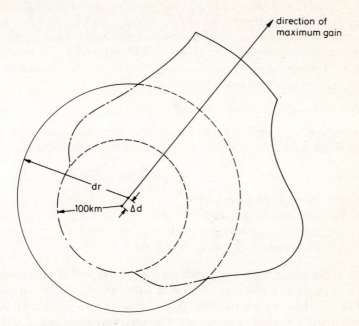

direction of
maximum gain

dr

100km

Δd

Fig. 1.35 *Combined co-ordination contour*
———— Co-ordination contour
–·–·– Contour for propagation Mode (1)
– – – – Contour for propagation Mode (2)
The co-ordination contour is a composite of the two propagation mode contours,
the larger distance being used for the combined coordination contour.

settings. Various allocations, or budgets, are made for each element of the link and what results is referred to as a link budget.

1.4.3 Link budget

The performance of a system is related to the ratio of the wanted power to the unwanted power in the receiver. The unwanted power is usually dominated by the thermal 'noise' generated by random atomic motions within the elements of the receiver. Other unwanted power contributions due to interference, intermodulation etc., can be reduced to an equivalent noise power level [7].

Considering for the moment only the thermal noise in the receiver, the noise power generated, N, is given by

$$N = k \times T_r \times B \tag{1.29}$$

where k = Boltzmann's constant = 1.38×10^{-23}, J/K
   T_r = noise temperature of the receiver, K
   B = bandwidth of the receiver, Hz.

To the internal noise temperature of the receiver must be added a contribution due to the area illuminated by the antenna. An earth station antenna, for example, will 'see' a portion of the sky in its main beam, with other lobes of the antenna intercepting the ground. The two effects are usually treated as a combined sky-noise temperature T_s which will vary with the weather (see Chapter 4). The combined receiver noise power is therefore made up of the additive noise temperature contributions giving

$$N = k \times (T_r + T_s) \times B \text{ watts} \qquad (1.30)$$

The received carier power C is then compared to the noise power N to give the carrier-to-noise ratio C/N. This can be expressed logarithmically as

$$C/N = P_t + G_t + G_r - L - 10 \log (N) \text{ decibels} \qquad (1.31)$$

where P_t = transmitted power, dBW
   G_t = transmitting antenna gain, dB
   G_r = receiving antenna gain, dB
   L = link losses = $L_w + L_g + L_{fs}$, dB

Eqn. 1.31 is the power budget equation for a given link and the parameters can be varied to establish a link loss that can be tolerated to give the required C/N ratio. The link loss, or attenuation, is made up of two essentially constant components and a variable component. The variable component L_w, is due to the weather (see Chapter 4) while the constant components are due to gaseous absorption L_g (see Chapter 3) and free-space loss L_{fs}.

The free-space loss is the decrease in power experienced as the signal travels over a distance d, and it will be proportional to the inverse of the square of the distance $(1/d^2)$. For an isotropic antenna transmitting with a power P_t, the power flux density p_{fd} at this distance will be given by

$$p_{fd} = P_t/(4 \times \pi \times d^2) \text{ watts/m}^2 \qquad (1.32)$$

If the transmitting antenna has a gain, G_t, then the power flux density is increased by this factor giving

$$p_{fd} = P_t \times G_t/(4 \times \pi \times d^2) \text{ watts/m}^2 \qquad (1.33)$$

The term $(P_t \times G_t)$ is called the Equivalent Isotropically Radiated Power, or EIRP. The amount of signal received P_r will be a function of the receiving antenna area A, namely

$$P_r = P_t \times G_t \times A/(4 \times \pi \times d^2) \text{ watts} \qquad (1.34)$$

But, from eqn. 1.4, the gain of an antenna is related to its area and so eqn. 1.34

can be re-written as

$$P_r = P_t \times G_t \times G_r \times (\lambda/(4 \times \pi \times d))^2 \text{ watts} \qquad (1.35)$$

The term $(\lambda/(4 \times \pi \times d))^2$ is referred to as the free-space loss. Although, strictly speaking, the decrease in power with distance is only due to $(1/d^2)$ losses, the free-space loss as defined above is frequency dependent. A sample link

Table 1.6 *Link budget for a satellite-to-ground propagation experiment using a beacon as the measurement source*

Satellite Parameters			
Frequency	11·7 GHz		
Transmit power		23 dBm	
Transmit antenna gain		20 dB	
Feed losses		2 dB	
EIRP	(a)	41 dBm	41 dBm
Transmission-path parameters			
Elevation angle	29°		
Free space loss		205·6 dB	
Gaseous loss		0·3 dB	
Net losses	(b)	205·9 dB	205·9 dB
Earth-station receiver parameters			
Antenna gain		53 dB	
Feed losses		0·8 dB	
Net receiver gain	(c)	52·2 dB	52·2 dB
Received signal power	$(a - b + c) = (d)$		−112·7 dBm
Threshold and signal margin calculations			
System noise temperature		450 K	
System noise power	(e)	−172·1 dB/Hz	−172·1 dB/Hz
Noise bandwidth (100 Hz)	(f)	20 dBHz	20 dBHz
C/N	$(d - (e + f)) = (g)$		39·4 dB
C/N threshold required	(h)	20 dB	20 dB
Net margin available	$(g - h) = (i)$		19·4 dB

budget for a satellite-to-ground link using the terms developed above is given in Table 1.6. (Note that no allowance is made for downlink degradation − see Chapter 4.)

The margin of 19·4 dB given in Table 1.6 is the amount the carrier/noise (C/N) can decrease while still giving acceptable performance in the receiver. This margin can be eroded by a large number of propagation impairments. Fig. 1.36 is a schematic representation of the various propagation impairments that can occur on an Earth–space path. Some of the impairments occur in more than one weather situation. For example, ice depolarisation, although an essentially lossless mechanism and therefore classified as a 'clear sky' impairment, will also occur during thunderstorms, a severe 'degraded-sky condition. In addition, some impairments can occur simultaneously, e.g. ionospheric scintillation and rain depolarisation. There is, therefore, a need to be able to predict accurately the propagation impairments likely to be expected on any path in order to achieve the required link performance at the requisite availability levels.

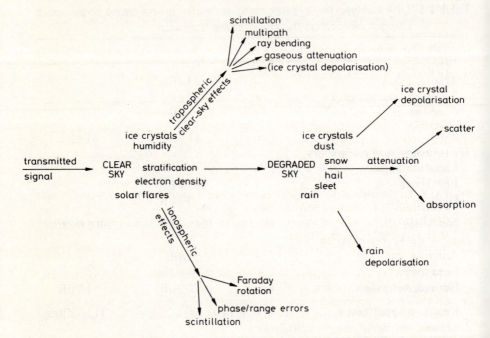

Fig. 1.36 *Schematic presentation of propagation impairment mechanisms*

1.5 References

1 'From semaphore to satellite', ITU, 2 Rue Varembé 1211, Geneva 20, Switzerland, 1965
2 Radio Regulations, Vol. 1, ITU, 2 Rue Varembé 1211, Geneva 20, Switzerland
3 *Ibid.*, Vol. 2, ITU, 2 Rue Varembé 1211, Geneva 20, Switzerland
4 CLARKE, A. C.:'Extra-terrestrial relays', *Wireless World*, 1945 pp. 305–308
5 BRAY, W. R.: 'Satellite communication systems', *Post Office Elec. Eng. J.*, 1962, **55**, pp. 97–104.
6 'Space statistics review', Aviation Information Services Limited, Cardinal Point, Newall Road, Heathrow Airport (London), Hounslow TW6 2AS, UK. This document is updated on at least a quarterly basis
7 MIYA, K., (Editor): 'Satellite communications technology' (KDD Engineering and Consulting, Inc., Tokyo, 1985, 2nd edn.) (English language edition)
8 PRATT, T., and BOSTIAN, C. W.: 'Satellite communications' (John Wiley, 1986)
9 AGRAWAL, B. N.: 'Design of geosynchronous spacecraft' (Prentice Hall, 1986)

10 FLURY, W.: 'Station-keeping of a geostationary satellite', *ELDO–Cecles/ESRO–CERS Scient. and Tech. Rev.*, 1973, **5**, pp. 131–156

11 ALLNUTT, J. E., and GOODYER, J. E.: 'Design of receiving stations for satellite-to-ground propagation research at frequencies above 10 GHz', IEE Jnl. on Micro., Optics, and Acoustics, 1977, 1, pp. 157–164

12 MILLER, A., and ANTHES, R. A.: 'Meteorology' (Charles E. Merrill Publishing Co., 4th. edn., 1981)

13 HARROLD, T. W., and AUSTIN, P. M.: 'The structure of precipitation systems – a review', *J. Rech. Atmos.*, 1974, **8**, pp. 41–57

14 AUSTIN, P. M.: 'Spatial characteristics of precipitation on the sub-synoptic scale', IUCRM Colloquium on the Fine Scale Structure of Precipitation and E.M. Propagation, Nice, France, October, 1973

15 ROGERS, R. R.: 'Statistical rainstorm models: Their theoretical and physical foundation', *IEEE Trans.*, 1976, **AP–24**, pp. 547–566

16 HOUZE, R. A. Jr.: 'Structure of precipitation systems: A global survey', *Radio Science*, 1981, pp. 671–689.

17 HOGG, D. C.: 'Path diversity in propagation of millimeter waves through rain', *IEEE Trans.*, 1967, **AP–15**, pp. 410–415

18 FREENY, E. E., and GABBE, J. D.: 'A statistical description of intense rainfall', *Bell Syst. Tech. J.*, 1969, **48**, pp. 1789–1851

19 HARDEN, B. N., NORBURY, J. R., and WHITE, W. J. K.: 'Measurements of rainfall for studies of millimetric radio attenuation', *IEE J. Microwaves, Optics & Acoustics*, 1977, **1**, pp. 197–202

20 FALK, L.: 'Statistics of the fine scale structure of rain in Stokholm 1977–1979'. Proc. URSI Commission F. Symposium, ESA SP-194, pp. 197–199

21 GOLDHIRSH, J.: 'Rain cell statistics as a function of rain rate for attenuation modeling', *IEEE Trans.*, 1983, **AP–31**, 799–801

22 Recommendations and Reports of the CCIR, XIVth. Plenary Assembly, Kyoto, 1978, Volume V (Propagation in non-ionized media); Report 563–1: 'Radiometeorological data'.

23 STRICKLAND, J. I.: 'Radar measurements of site diversity improvement during precipitation', *J. Rech. Atmos.*, 1974, **8**, pp. 451–464

24 CRANE, R. K.: 'An analysis of radar data to obtain storm cell sizes, shapes, and spacings', Report RWP-1, 1983, Thayer School of Engineering, Dartmouth College, Hanover, New Hampshire, USA

25 CRANE, R. K., and HARDY, K. R.: 'The HIPLEX program in Colby–Goodlands, Kansas: 1976–1980', ERT Document P–1552–F, 1981, prepared for the Water and Power Resources Service, US Dept. of the Interior, Denver, Colorado 80225, USA

26 CRANE, R. K.: 'Automatic cell detection and tracking', *IEEE Trans.*, 1979, **GE–17**, pp. 250–262

27 Recommendations and Reports of the CCIR, XVth. Plenary Assembly, Geneva, 1982, Volume V (Propagation in non-ionized media); Report 563–2: 'Radiometeorological Data'.

28 Ibidem, XVIth. Plenary Assembly, Dubrovnik, 1986, Volume V (Propagation in non-ionized media); Report 563-3: 'Radiometeorological Data'.

29 MASON, B. J.: 'The physics of clouds' (Clarendon Press, Oxford, 1977)

30 BRADLEY, J. H. S.: 'Rainfall extreme value statistics applied to microwave attenuation climatology', Scientific Report MW–66, 1970, Stormy Weather Goup of McGill University, Montreal, Canada

31 ROGERS, R. R. *et al.*: 'The mesoscale structure of precipitation and space diversity', *J. Res. Atmos.*, 1974, **8**, pp. 485–490 (Working Group IV Report: R. R. Rogers, Chairman)

32 WANG, P.-Y., and HOBBS, P. V.: 'The mesoscale and microscale structure and organization of clouds and precipitation in mid-latitude cyclones. X: Wavelike rainbands in an occlusion', *J. Atmos. Sciences*, 1983, **40**, pp. 1950–1964

33 KONRAD, T. G: 'Statistical models of summer rainshowers derived from fine-scale radar observations', *J. Appl. Meteor.*, 1978, **17**, pp. 171–188

34   MATSUMOTO, S.: 'Mesometeorological aspects of precipitation', *J. Rech. Atmos.*, 1974, **8**, pp. 205–212

35   BLACK, P. G.: 'Mesoscale cloud patterns revealed by Apollo–Soyuz photographs', *Bull. Amer. Meteor. Soc.*, 1978, **59**, pp. 1409–1419

36   ZAWADSKI, I. I.: 'Statistical properties of precipitation patterns', *J. Appl. Meteor.*, 1973, **12**, pp. 459–472

37   ZAWADSKI, I. I.: 'Statistics of radar patterns and EM propagation', *J. Rech. Atmos.*, 1974, **8**, pp. 391–397

38   NEWELL, R. E.: 'Some radar observations of tropospheric cellular convection', Proc. 8th. Weather Radar Conf., 1960, pp. 315–322

39   DENNIS, A. S., and FERNALD, F. G.: 'Frequency distribution of shower sizes', *J. Appl. Meteor.*, 1963, **2**, pp. 767–769

40   YAMADA, M., OGAWA, A., and YOKOI, H.: 'Precipitation attenuation in the low elevated earth-satellite path', IECE Tech. Group Antennas and Propagation, Japan, 1975, paper AP 75–66

41   SEGAL, B.: 'High-intensity rainfall statistics for Canada', CRC Report 1329-E, 1979, Communications Research Centre, Ottawa, Canada

42   BRAHAM, R. R., and WILSON, D.: 'Effects of St. Louis on convective cloud heights', *J. Appl. Meteor.*, 1978, **17**, pp. 587–592

43   CHAGNON, S. A. Jr.: 'The Laporte weather anomaly-fact or fiction', *Bull. Amer. Meteor. Soc.*, 1968, **49**, pp. 4–11

44   SEMONIN, R. G., and CHAGNON, S. A. Jr.: 'METROMEX: summary of 1971–1972 results', *Bull. Amer. Meteor. Soc.*, 1974, **55**, pp. 95–100

45   FURUKAWA, T.: 'A study of typhoon rainbands with quantized radar data', *J. Met. Soc. Japan*, 1980, **58**, pp. 246–260

46   RUSTLAND, W. D., and DOVIAK, R. J.: 'Radar research on thunderstorms and lightning', *Nature*, 1982, **297**, pp. 461–468

47   HALL, M. P. M.: 'Effects of the troposphere on radio communications' (Peter Perigrinus, 1979)

48   OMOURA, A. I., and HODGE, D. B.: 'Microwave dispersion and absorption due to atmospheric gases', Tech. Note 10, August 1979, Ohio State University, ElectroScience Lab., Dept. of Electrical Engg., Columbus, Ohio 43212, USA

49   BERGERON, T.: 'On the physics of cloud and precipitation', Proc. 5th. Assembly UGGI, Paris, 1935, Vol. 2, pp. 156 *et seq*

50   JONES, P. R., and GOLDSMITH, P.: 'The collection efficiencies of small droplets falling through a sheared air flow', *J. Fluid Mech.*, 1972, **52**, pp. 593–608

51   LAWS, J. O.: 'Measurements of the fall velocities of water drops and raindrops', *Trans. Amer. Geophys. Union*, 1941, **22**, pp. 709–712

52   GUNN, R., and KINZER, G. D.: 'The terminal velocities of fall for water droplets in stagnant air', *J. of Meteorology*, 1949, **6**, pp. 243–248

53   BEST, A. C.: 'Empirical formulae for the terminal velocity of water drops falling through the atmosphere', *Q. J. Roy. Met. Soc.*, 1950, **76**, 302–311

54   BEARD, K. V.: 'Terminal velocity and shape of cloud and precipitation drops aloft, *J. Atmos. Sciences*, 1976, **33**, pp. 851–864

55   FLOCK, W. L.: 'Propagation effects on satellite systems at frequencies below 10 GHz', NASA Reference Publication 1108, 1983

56   WOBUS, H. B., MURRAY, F. W., and KOENIG, L. R.: 'Calculation of the terminal velocities of water drops', *J. Appl. Meteorology*, 1971, **10**, pp. 751–754

57   PRUPPACHER, H. R., and PITTER, R. L.: 'A semi-empirical determination of the shape of cloud and rain drops', *J. Atmos. Science*, 1971, **28**, pp. 86–94

58   NELSON, A. R., and GOKHALE, N. R.: 'Oscillation frequencies of freely suspended water drops', *J. of Geophys. Research*, 1972, **77**, pp. 2724–2727

59   LAWS, J. O., and PARSONS, D. A.: 'The relation of rain drop-size to intensity, *Trans. Amer. Geophys. Union*, 1943, **24**, pp. 432–460

60   MARSHALL, J. S., and PALMER, W. M.: 'The distribution of rain drops with size', *J. Meteorology*, 1948, **5**, pp. 165–166

61   PRUPPACHER, H. R., and BEARD, K. V.: 'A wind tunnel investigation of the internal circulation and shape of water drops falling at terminal velocity in air', *Q. J. Roy. Meteorological Soc.*, 1970, **96**, pp. 247–256

62   JOSS, J., THAMS, J. C., and WALDVOGEL, A.: 'The variation of rain drop size distribution at Locarno'. Proc. International Conference on Cloud Physics, Toronto, Canada, 1968, pp. 369–373

63   BARCLAY, P. A.: 'Rain drop-size distributions in the Melbourne area', Institute of Engineers, Hydrology Symposium, Armidale, N.S.W, 1975, pp. 112–116.

64   RICE, P. L., and HOLMBERG, N. R.: 'Cumulative time statistics of surface-point rainfall rates', *IEEE Trans.*, 1973, **COM-21**, pp. 1131–1136

65   Conclusions of the Interim Meeting of Study Group 5 (Propagation in non-ionized media), Geneva, 11–26 April 1988, Document 5/204; Report 563-3 (MOD-I): 'Radiometeorological Data'.

66   POIARES BAPTISTA, J. P. V., ZHANG, Z. W., and McEWAN, N. J.: 'Stability of rain-rate cumulative distributions', *Electron. Lett.*, 1986, **22**, pp. 350–352.

67   Recommendations and Reports of the CCIR, XVIth. Plenary Assembly, Dubrovnik, 1986, Volume V (Propagation in non-ionized media); Report 885-2: 'Propagation data required for evaluating interference between stations in space and those on the surface of the Earth'.

68   TURNER, A. E., and PRICE, K. M.: 'The potential in non-synchronous orbit', *Satellite Commun.*, June 1988, pp. 27–31

Chapter 2

Ionospheric effects

2.1 Introduction

On 12 December 1901, Guglielmo Marconi succeeded in transmitting the letter 'S' in Morse code from St. Johns, Newfoundland to Poldhu, Cornwall, in England. Marconi's earlier wireless telegraphy work was based, in part, on the theory that the higher the transmitter, the further the communication distance could be. Clearly, some other mechanism was in force to permit radio signals to be transmitted across the Atlantic from Canada to England. The existence of a reflecting layer in the Earth's atmosphere was postulated and, when Marconi found in experiments the following year that he could transmit further at night than during the day, the influence of the Sun on radio communications became a subject of intense study. These studies have continued virtually unabated to the present day and have led to a new branch of science: ionospheric physics.

The ionosphere is a region of ionised plasma that extends from roughly 50 km to 2000 km above the surface of the Earth. Only a portion of the molecules are ionised; large quantities of neutral molecules remain. Above the ionosphere is the plasmasphere or protonsphere which still has a significant content of free electrons. Trapping these electrons is the Earth's magnetic field which is, itself, within the solar wind (see Fig. 2.1).

The boundary of the Earth's magnetic field within the solar wind is known as the magnetopause; between the magnetopause and the ionosphere is the magnetosphere. The magnetosphere extends about 10 earth radii in the direction of the Sun, acting as a 'bow wave' against the solar wind. On the other side of the Earth, the magnetosphere streams backwards for about 60 earth radii. Within the magnetosphere, the magnetic field dominates the motion of the charged particles, while, in the ionosphere, the collision recombination process plays the major role. The Van Allen radiation belts lie within the magnetosphere. The plasmasphere is below the Van Allen belts and forms the lower boundary of the magnetosphere with the ionosphere [40].

Electromagnetic radiation from the Sun ionises the atmospheric particles and the degree of ionisation is a function of the path length through the atmosphere and hence the zenith angle. The closer to zenith, the lower down into the atmosphere the ionisation process will take place (see Fig. 2.2).

There will therefore be a concentration of ionised particles around the Earth's equator, but, owing to the shift of about 12° in the Earth's magnetic axis away from the Earth's spin axis, the geomagnetic equator will not correspond exactly to the geographic equator. Fig. 2.3 illustrates this fact.

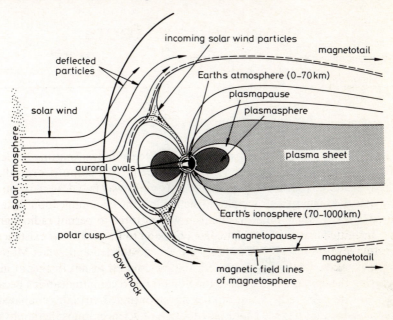

Fig. 2.1 *Interaction between the solar wind and the Earth's magnetic field* (*This first appeared in New Scientist, London, the weekly review of science and technology, as Fig. 2.1 of Reference 61*) The solar wind results chiefly from the expansion of gases in the corona of the sun, the outermost atmosphere of the sun. The high temperature of the corona, in the region of 1 700 000° C, heats the gases. In the expansion, many of the atoms collide and lose their electrons. The resultant ions make up much of the solar wind which has a velocity in the region of the Earth of about 500 km/s and an average density of 5 ions per cc [61]. The Earth's magnetic field deflects the solar wind which then stretches the magnetosphere, the region inside the magnetopause, behind the Earth in the form of a tail. The Van Allen radiation belts lie within the magnetosphere but above the ionosphere.

(Copyright © 1985 New Scientist, reproduced with permission)

Similarly, the Sun rotates about an axis which is tilted 7·3° from the normal to the plane of the Earth's orbit around the Sun. Energy emitted normal to the Sun's surface which strikes the Earth's atmosphere must therefore originate within 7·3° of the Sun's equator. It is this narrow band of 14·6° around the Sun's equator that has the greatest immediate influence on the Earth's ionosphere and, indeed, on all life on Earth (see Fig. 2.4).

The magnetic field of the Earth will interact with the ionised particles in the ionosphere and seek to align them with the field strength pattern of the Earth. Ionised particles which flow down these lines of force towards the poles of the Earth will recombine in the denser parts of the atmosphere and give rise to the aurora. The collision recombination of electrons and positive ions, and

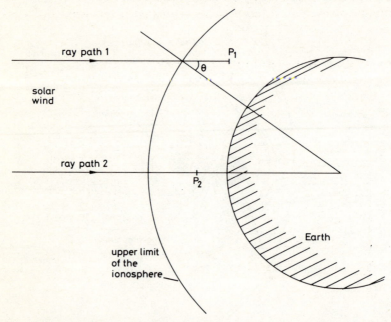

Fig. 2.2 *Penetration of rays into the atmosphere*
Two rays of equal intensity enter along two paths, one at a zenith angle of zero (directly overhead) and one at a zenith angle of θ. Both rays penetrate the same distance, reaching point P_2 and P_1, respectively. Point P_2, on the path with the smaller zenith angle, reaches closer to the surface of the Earth.

the attachment of electrons to neutral gas atoms and molecules, forms the de-ionisation process. The sunlit portion of the Earth's atmosphere, since it is receiving a strong flux of energy from the Sun in the ultra-violet and X-ray frequency bands, will have a much more strongly ionised region than the atmosphere on the night side of the earth. Fig. 2.5 shows how the ionosphere expands during the day and collapses back during the night. In periods of really intense solar activity, however, the de-ionisation process cannot neutralise the ionisation activity sufficiently during the night and remnants of the ionosphere can exist other than in the E and F layers for all of the night period.

The labelling of the various ionospheric layers, first proposed by E.V. Appleton, began at 'D' because it was felt at the time that other layers might exist both below and above this layer. As it turned out, none has been uniquely identified below the D region which is now considered to be the lowest region of the ionosphere. Among the first systematic studies conducted on the iono-sphere were those by T. L. Eckersley [1, 2] and a summary of research carried on up to 1950 can be found in Reference 3. Those early studies postulated two basic mechanisms for the apparent reflection of radio waves: critical reflection and partial reflection. The essential idea behind the critical reflection theory was

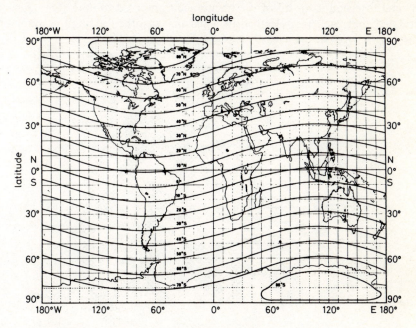

Fig. 2.3 *Geomagnetic latitudes (Fig. 2 of Reference 59)*
(Copyright © 1986 ITU, reproduced with permission)

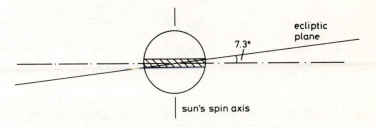

Fig. 2.4 *Principal band on the sun affecting the ionosphere*
As the Earth moves around the Sun, only those parts of the Sun within the shaded band will emit normal to the Sun and in the plane of the ecliptic.

that the index of refraction at a given frequency f depends directly on the plasma density N. Since N does not increase indefinitely with height but reaches a maximum value of about 10^{12} electrons/m^3 at approximately 300 km (see Fig. 2.6), an upper bound should also exist for the frequency above which no signal is returned by the ionosphere.

According to this theory, no reflected signals should be detected above the UHF (300–3000 MHz) frequencies. Since some return signals were detected, the second theory, that of partial reflection, was postulated. Unfortunately, this theory also could not sufficiently explain the magnitude of the signals returned above this range. A third mechanism was proposed later [4] which seemed at that time to account for all of the phenomena observed to date. This theory postulated that each free electron in a plasma absorbed a small amount of

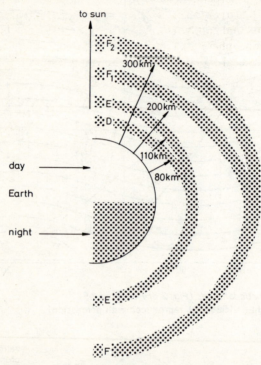

Fig. 2.5 *Ionospheric regions as a function of height above the Earth's surface (from Fig. 1 of Reference 59)*
(Copyright © 1986 ITU, reproduced with permission)

incident radiowave energy, irrespective of the frequency of the signal, and then re-radiated the energy coherently with the other free electrons. The strength of the return signal would be proportional to the root mean-square value of the fluctuating component of the refractive index with respect to the mean which, in turn, would be proportional to f^{-2}. The fact that the magnitude of the propagation effects due to the ionosphere were predicted to decrease as f^{-2} is an important point which will be returned to later. An understanding of some of the basic principles involved in electromagnetic wave propagation in the ionosphere is important before moving on to the current status of modelling the phenomena and their system impairments.

2.2 Some basic formulations

2.2.1 Critical frequency

Radiowave propagation through the ionosphere is a complex mix of inter-actions between the ionised constituents, the magnetic field and the parameters of the signal itself (frequency, polarisation, amplitude, bandwidth, direction, modulation etc.). The direction of propagation can be resolved into two direc-tions, parallel and perpendicular to the magnetic field, and the characteristic waves defined [6]. Derivation of the theory from first principles [7] is beyond the scope of this book and good texts on ionospheric radiowaves [8], the ionosphere [41], and the ionosphere and magnetosphere [40], will give more detailed mathematical physical explanations. For now, it is sufficient to consider just the characteristic waves.

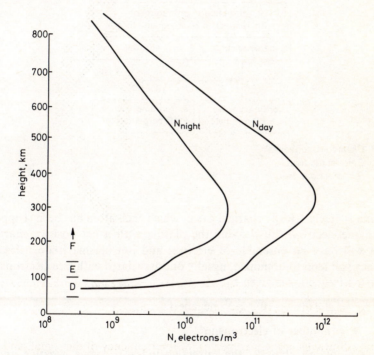

Fig. 2.6 *Number of electrons/m³ versus the height above the Earth during an average day and night (After Fig. 9.1 of the French edition of Reference 35)*
The approximate range of the three regions, D, E and F, is indicated
(Copyright © 1983 Bordas, Dunod, Gauthier-Villars; translated from the French edition by the author; reproduced with permission)

A characteristic wave is one which propagates without any change in the polarisation state. Those characteristic waves which propagate perpendicularly to the magnetic field can be further subdivided into ordinary waves and extra-ordinary waves. Figure 2.7 summarises these subdivisions.

The ordinary wave has its electric vector aligned along the magnetic field. The electrons are therefore moved in the same direction as the constant-force lines of the magnetic field and no interactions take place. The index of refraction n_0 in this case is [5]:

$$n_0^2 = 1 - f_p^2/f^2 \qquad (2.1)$$

where f_p = plasma frequency
        f = frequency of the radiowave

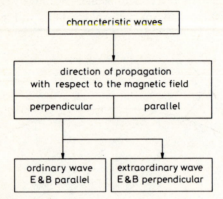

Fig. 2.7 *Characteristic waves*
        E = electric field
        B = magnetic field

A plasma is an electrically charged gas in which each atom has been stripped of at least one electron, thus leaving the plasma with a net positive charge. A plasma will have an ensemble of electron and ion plasma frequencies. One frequency will tend to dominate, usually one associated with free electrons, and the plasma frequency is given by

$$f_p = 8{\cdot}9788 \times 10^{-6} \sqrt{N} \text{ megahertz} \qquad (2.2)$$

where N = number of electrons/m^3.

If f is a lot less than f_p, the refractive index becomes negative and the radiowave is reflected [5]. Reflection and refraction are sometimes difficult to separate. In Fig. 2.8, a radiowave received at point B could equally well have been refracted by the ionosphere as it travelled from point A or it could have been reflected by an apparent layer at point C.

A critical frequency, f_c, will be reached at which only partial reflection will occur. Conversely, only frequencies above this critical frequency can traverse the ionosphere. The critical frequency is the plasma frequency and so

$$f_c = 8{\cdot}9788 \times 10^{-6} \sqrt{N} \text{ megahertz} \qquad (2.3)$$

Fig. 2.9 shows how the critical frequency varies both with the seasons (here referred to the northern hemisphere), the ionised region and the time of day. The highest critical frequency is generally about 12 MHz. As the radiowave frequency increases, so the absorption and refraction within the ionosphere decrease, until they are negligible at 1 GHz. Table 2.1 shows some typical values for frequencies between 1 and 30 GHz.

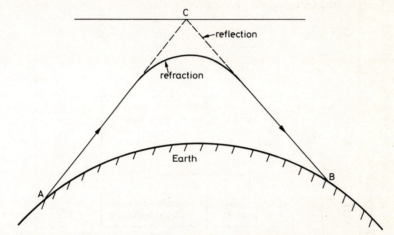

Fig. 2.8 *Reflection and refraction of a radiowave transmission from A to B*

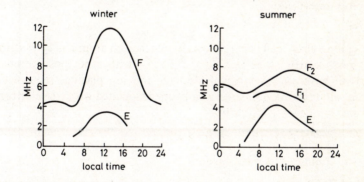

Fig. 2.9 *Typical critical frequency variations (from Fig. 9.2 of the French edition of Reference 35)*
(Copyright © 1983 Bordas, Dunod, Gauthier-Villars; translated from the French edition by the author; reproduced with permission)

2.2.2 Total electron content

A linearly polarised wave propagating in a plasma will generally suffer a rotation of the plane of polarisation due to the anisotropy of the medium. This effect is referred to as the Faraday rotation. The magnitude of the Faraday

Table 2.1 *Estimates of maximum ionospheric effects in the United States on paths with elevation angles of about 30°* (From Table 2.2 of Reference 5 and Table VII of Reference 60)

Effect	Frequency dependence	100 MHz	300 MHz	1 GHz	3 GHz	10 GHz
Faraday Rotation	$1/f^2$	30 rot.	3·3 rot.	108°	12°	1·1°
Propagation delay, μs	$1/f^2$	25	2·8	0·25	0·028	0·0025
Refraction	$1/f^2$	$\leqslant 1°$	$<7'$	$\leqslant 0·6'$	$<4·2''$	$\leqslant 0·36''$
Variation in the direction of arrival	$1/f^2$	20′ of arc	2·2′ of arc	12″ of arc	1·32″ of arc	0·12″ of arc
Absorption (auroral and polar cap), dB	$1/f^x$ $1 < x \leqslant 2$	5	1·1	0·2	0·04	0·008
Absorption (mid-latitude), dB	$1/f^2$	<1	0·1	$<0·01$	0·001	$<10^{-4}$
Dispersion, ps/Hz	$1/f^3$	0·4	0·015	0·0004	1.5×10^{-5}	4×10^{-7}

rotation will depend on the frequency of the radiowave, the magnetic field strength, and the electron density of the plasma. Since the first two parameters are essentially constant for any particular radio link, the rotation of the linearly polarised vector from its nominal orientation will give a good indication of the total number of electrons encountered by the radiowave on its passage through the plasma. The integrated number of electrons on the path is known as the Total Electron Content (TEC). Usually, the TEC value is quoted for a zenith path having a cross-section of $1 \, m^2$. The TEC of this vertical column can vary between 10^{16} and 10^{18} electrons/m² with the peak occurring during the sunlit portion of the day. Excellent reviews of TEC variability, the relationship between mean and extreme values of TEC, and the resultant impact of TEC on 6/4 GHz communications satellites are given in Reference 9 and 36. From a knowledge of the TEC, a lot of other transmission parameters can be calculated.

2.2.3 Faraday rotation

The rotation ϕ of a linearly polarised vector about its direction of propagation when passing through the ionosphere is given by [5]:

$$\phi = \frac{2 \cdot 36 \times 10^4}{f^2} \int NB \cos \theta_B \, dl \text{ radians} \tag{2.4}$$

where f = frequency, Hz
   N = number of electrons/m³
   B = earth's magnetic field, Wb/m²,
   θ_B = angle between the magnetic field and the direction of propagation,
   dl = incremental distance through the plasma

$B \cos \theta_B$ is sometimes replaced by an average value, B_{av}, and so eqn. 2.4 can reduce to

$$\phi = \frac{2 \cdot 36 \times 10^4}{f^2} B_{av} \int N \, dl \text{ radians} \tag{2.5}$$

But,

$$\text{TEC} = \int N \, dl \text{ electrons/m}^2 \tag{2.6}$$

and so the Faraday rotation ϕ becomes

$$\phi = (C/f^2) \times \text{TEC radians} \tag{2.7}$$

where C is a constant replacing $2 \cdot 36 \times 10^4 B_{av}$. Assuming average values for the constant C, Faraday rotation is plotted against frequency in Fig. 2.10 with TEC as parameter. Note from eqn. 2.6 and 2.7 that, if TEC is unknown, a measurement of ϕ will allow the TEC to be derived. Long-term measurements to derive TEC from ϕ have been reported [10]. Below 1 GHz, Faraday rotation can exceed many complete rotations of 360°. What is of more relevance to communications satellite systems, however, is the change in Faraday rotation, diurnally,

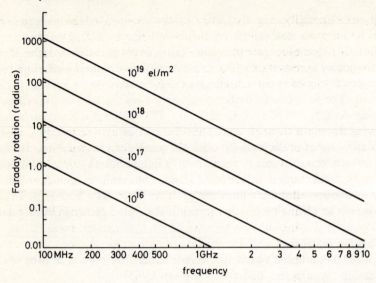

Fig. 2.10 *Faraday rotation as a function of TEC and frequency (extended from Fig. 2.9 of Reference 5)*

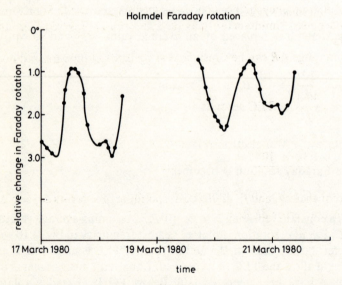

Fig. 2.11 *Diurnal variation in the Faraday rotation at 4 GHz as measured in March 1980 at Holmdel, NJ, USA (After Fig. 2 of Reference 10)*
(Copyright © 1985 John Wiley and Sons, Ltd., reproduced with permission)

seasonally and annually. Fig. 2.11 shows the change in Faraday rotation as perceived by an earth station in Holmdel, New Jersey, USA, operating to the COMSTAR D3 domestic satellite [10]. The frequency was 4 GHz and the elevation angle in excess of 30°. The impact of such rotations and their compensation are discussed in later Sections of this book.

2.2.4 Group delay

The phase path length through a medium does not equal the true path length if the refractive index of the medium does not equal unity. Since the phase path length is directly proportional to the refractive index and the refractive index is a function of frequency, a transmitted signal consisting of a bandwidth or spectrum of signals will suffer a group delay. The delay, when acting on a single frequency such as a ranging pulse, will cause the signal to arrive later than it would have done if transmitted *in vacuo*. That is, the inferred range R, will be greater than the true range since the velocity of light *in vacuo* c would have been assumed in calculating the range. The error ΔR will be:

$$\Delta R = (40 \cdot 3/f^2) \times \text{TEC metres} \tag{2.8}$$

If an average TEC of 10^{17} electrons/m^2 is assumed, the ranging error utilising a 4 GHz signal is approximately 2·5 m. The corresponding values for a TEC of 10^{16} and 10^{18} are 0·25 and 25 m. For geostationary communications satellites which need to maintain station keeping to within 0·1° of a given position, the maximum ranging error that can be tolerated is considered to be about 3 m [5]. To obtain this accuracy, either a knowledge of the TEC at a given time is required or a frequency higher than 4 GHz must be used. Sometimes two frequencies are used simultaneously as the TEC can be inferred from the two range errors.

An error in range ΔR can be considered as an error in time Δt, and inverting eqn. 2.8:

$$\Delta t = \frac{40 \cdot 3}{cf^2} \times \text{TEC seconds} \tag{2.9}$$

or

$$= \frac{1 \cdot 34 \times 10^{-7}}{f^2} \times \text{TEC seconds} \tag{2.10}$$

With two frequencies f_1 and f_2, giving corresponding time errors Δt_1 and Δt_2, the difference in time error δt, is:

$$\delta t = \frac{40 \cdot 3}{c} \times (1/f_2^2 - 1/f_1^2) \times \text{TEC seconds} \tag{2.11}$$

whence

$$\Delta t_1 = \frac{f_2^2}{(f_1^2 - f_2^2)} \times \delta t \text{ seconds} \tag{2.12}$$

By inserting a common timing reference onto the modulation applied to the signals 1 and 2, the difference in time error, δt, is easily extracted from the received signals. By inference, a knowledge of δt will give TEC. Solving eqn. 2.11 for TEC gives:

$$\text{TEC} = \frac{\delta t \times c}{40 \cdot 3} \frac{f_1^2 f_2^2}{f_1^2 - f_2^2} \text{ electrons/m}^2 \tag{2.13}$$

The difference in the time error is therefore yet another way to infer the TEC.

2.2.5 Phase advance

Delaying the arrival of a signal is similar to advancing the phase of the received signal; i.e. it appears to have travelled further than it actually has. Phase advance $\Delta\phi$ is given by [5]

$$\Delta\phi = \frac{(1 \cdot 34 \times 10^{-7})}{f} \times \text{TEC cycles} \tag{2.14}$$

or

$$= \frac{(8 \cdot 44 \times 10^{-7})}{f} \times \text{TEC radians} \tag{2.15}$$

For a frequency of 4 GHz and a TEC of 10^{16} electrons/m^2, $\Delta\phi = 21 \cdot 1$ rad or $3 \cdot 35$ cycles. One cycle corresponds to one complete wavelength and so $\Delta\phi = 3 \cdot 35$ wavelengths or $0 \cdot 25$ m; the wavelength at 4 GHz being $7 \cdot 5$ cm. This is the same result obtained using eqn. 2.8 to arrive at a range error.

The phase advance (or time delay) will vary across a band or spectrum of frequencies used in a typical communications channel, e.g. a 36 MHz transponder. The phase dispersion is given by the derivative of $\Delta\phi$ with respect to frequency; thus the phase dispersion $d\phi/dt$ is

$$\frac{d\phi}{dt} = \frac{(-8 \cdot 44 \times 10^{-7})}{f^2} \times \text{TEC radians/s} \tag{2.16}$$

2.2.6 Doppler frequency

The rate of change of phase is frequency.
If

$$f = \frac{1}{2\pi} \frac{d\phi}{dt} \text{ hertz} \tag{2.17}$$

and eqn. 2.14 is substituted into eqn. 2.17, the result is the Doppler frequency f_D, given by [5]

$$f_D = \frac{1 \cdot 34 \times 10^{-7}}{f} \frac{d\,(\text{TEC})}{dt} \text{ hertz} \tag{2.18}$$

where $d\,(\text{TEC})/dt$ is the change in the TEC over a given interval dt. In practice,

f_D is negligible compared to the Doppler shift due to satellite motion for frequencies above 1 GHz. The effect of varying Doppler shift, however, is to cause a spectral broadening of the signal, which can be important for inter-planetary spacecraft when they are occulting the Sun. The charged particles in the atmosphere of the Sun can cause substantially more dispersion than the Earth's ionosphere.

2.2.7 Dispersion

The rate of change of the time delay with frequency, dt/df, is the dispersion of the signal due to time delay. From eqn. 2.10

$$dt/df = -((2{\cdot}68 \times 10^{-7})/f^3) \times \text{TEC} \tag{2.19}$$

If the bandwidth of the signal being transmitted through the ionosphere is df in eqn. 2.19, the difference in the time delay between two signals at the extreme ends of the bandwidth is

$$|\Delta t| = \frac{(2{\cdot}68 \times 10^{-7})}{f^3} \times df \times \text{TEC seconds} \tag{2.20}$$

For an average TEC of 10^{17} electrons/m^2, a frequency of 4 GHz, and a band-width of 36 MHz, Δt is 15 ps. Taking more extreme values of TEC and band-width of 10^{18} and 240 MHz, respectively, Δt is now 1 ns.

The dispersion effects introduced by phase advance are obtained by differentiating eqn. 2.15 with respect to frequency, giving

$$d\phi/df = -\frac{(8{\cdot}44 \times 10^{-7})}{f^2} \times \text{TEC} \tag{2.21}$$

Again, if the bandwidth of the signals being transmitted through the ionosphere is df in eqn. 2.21, the difference in the phase delay between two signals at the extreme ends of the bandwidth is

$$|\Delta\phi| = \frac{(8{\cdot}44 \times 10^{-7})}{f^2} \times df \times \text{TEC radians} \tag{2.22}$$

For the same parameters given in the above two examples of time-delay dis-persion, the corresponding phase-dispersion values are 0·19 and 12·7 rad, respectively. These equate approximately to 11° and 725°, respectively.

The effects of ionospheric dispersion are to reduce the coherence bandwidth. In severe cases, the transmission bandwidth is greatly reduced, which corre-sponds, in a digital system, to imposing an upper limit on the bit rate.

2.3. Ionospheric scintillation

Scintillation of a radio signal is a relatively rapid fluctuation of the signal about a mean level which is either constant or changing much more slowly than the scintillations themselves. A scintillation can be a phase or an amplitude fluctuation, and the amount of scintillation observed is a function of the size of the irregularity causing the signal variation (sometimes called the scale size), the distance between the irregularities and the receiver, and the Fresnel zone size.

2.3.1 Fresnel zone
Figure 2.12 shows two rays, an indirect ray and a direct ray, leaving a transmitter T and arriving at a receiver R.

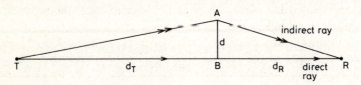

Fig. 2.12 *Illustration of the Fresnel zone*
The distance *d* is a Fresnel zone radius when the difference between the direct and indirect rays is a multiple of a half wavelength.

If the wavelength λ of both signals is the same and

$$\text{distance TAR} = \text{distance TBR} + \lambda/2 \text{ metres} \tag{2.23}$$

then *d* is the first Fresnel zone radius. Components of all rays which pass from T to R through the first Fresnel zone radius will add constructively to some degree. If d_T and d_R, the distances of the transmitter and the receiver from the Fresnel zone, respectively, are very much larger than the Fresnel zone radius *d*, then

$$d = \sqrt{\left(\frac{\lambda \cdot d_T d_R}{d_T + d_R} \right)} \text{ metres} \tag{2.24}$$

For a typical equatorial communications satellite downlink at a frequency of 4 GHz, if the ionospheric scintillation occurs in the E region at a height of 400 km, the following approximate values can be used in eqn. 2.24:

$$\lambda = 0.075 \text{ m}$$

$$d_T = 35\,500 \text{ km}$$

$$d_R = 420 \text{ km}$$

$$\text{yielding } d = 176 \text{ m}$$

The corresponding figure for the uplink Fresnel zone size ($f = 6$ GHz) at the same height in the ionosphere is 144 m. If the active region of the ionosphere is 200 km up, the downlink and uplink first Fresnel zones are 125 m and 102 m, respectively; for 600 km they are 215 m and 175 m, respectively.

The value *d* in eqn. 2.24 is for the first Fresnel zone. The radii of Fresnel zones of higher orders ($n = 2, 3, 4$, etc.) are given by

$$d_n = \sqrt{n} \times d \text{ metres} \tag{2.25}$$

Since only odd-ordered Fresnel zones will add, the scale size of an irregularity which will give the greatest effect is that which is equal in size to the first Fresnel zone size. Below this size, the effect diminishes owing to fewer rays being 'focused'. Above this size and contributions of even-ordered Fresnel zones, beginning with the second, will destructively interfere with the first Fresnel zone rays.

The irregularities are essentially localised variations in electron density which cause small-scale fluctuations in refractive index. These, in turn, will cause a focusing and defocusing effect. Close to the irregularities, the different phase paths through the small-scale fluctuations will give a non-planar phase front similar to the near field of an antenna. Once beyond the Fresnel distance, defined as the square of the scale size divided by the wavelength, the far-field pattern is established and the variations to the signal show up at the receiver essentially as amplitude scintillations. At UHF, these amplitude scintillations are very large, but, as the frequency of the signal goes up, the scintillation amplitude drops off as the inverse of the frequency to the power n, where n lies between 1 and 2. All the ionospheric scintillation models applicable at UHF incorporated this type of scaling with frequency. As a consequence, ionospheric scintillations were not expected to be significant at frequencies above 1 GHz; the gigahertz frequencies.

2.3.2 Observations of gigahertz ionospheric scintillations

In July 1962, Telstar 1 relayed the first live voice and video signals via satellite across the Atlantic Ocean. The transmissions took place between Andover, Maine, in the USA, Goonhilly, in the UK, and Plemeur Bodou, in France; all countries in the temperate regions of the Earth. When the INTELSAT system began operating in early 1965 at 6 and 4 GHz with their first satellite, Early Bird, the first users were again earth stations located in the temperate regions of the Earth, well away from the geomagnetic equator. It was not until the number of earth stations accessing the system grew and there were operators located close to the geomagnetic equator that gigahertz ionospheric scintillations were observed. In October 1969, the Bahrain earth station in the INTELSAT system reported rapid amplitude variations on all received carriers. This was followed in February 1970 with similar reports from both Bahrain and Indonesian earth stations and then by articles in the literature [12, 13]. A world-wide campaign of measurements was instituted by INTELSAT [14] and others [15] from which a global picture of ionospheric scintillation characteristics has been built up [16–18]. Before discussing these, it is useful to outline the various parameters by which ionospheric scintillations are characterised.

2.3.3 Scintillation indices

The amplitude of the received radiowave can fluctuate about the mean level in an apparently random manner. Fig. 2.13 shows the 4 GHz signal measured at the Taipei earth station on 28 April 1977 at an elevation angle of 19° with the antenna directed towards the INTELSAT satellite located in geostationary oribt at 60° *E* over the Indian Ocean.

The period shown in Fig. 2.13 is for approximately 10 min at the peak of that particular period of scintillation. As can be seen, the individual peaks of the scintillation vary from less than 0·5 dB to more than 1·5 dB from the mean. The peak-to-peak fluctuations over the 10 min period exceeded 3 dB but this occurred only once or twice and no excursion of + 1·5 dB was followed immediately by an excursion of − 1·5 dB.

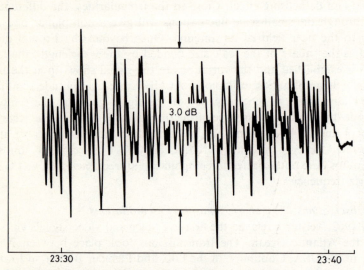

3.0 dB

23:30   23:40

Fig. 2.13 *Expanded samples of the record of ionospheric scintillations on 28–29 April, 1977, at the Taipei Earth Station (from Fig. 4 of Reference 22)*
(Copyright © 1980 AGARD, reproduced with permission)

The peak-to-peak fluctuations can be measured over a given interval (e.g. 10 min) and the peak-to-peak fluctuation statistics prepared for the link in question. Fig. 2.14 shows a collection of such statistical data, each recorded over periods of one year, but in different parts of the sun-spot cycle and over different paths through the ionosphere.

The Sun exhibits a periodicity in its sun-spot activity, peak periods repeating at intervals of approximately 11 years. Fig. 2.15 shows the monthly sun-spot number over a 28-year period.

If the monthly sun-spot number is plotted against the percentage time that the scintillation exceeded 1 dB peak-to-peak over the ensemble of paths, Fig. 2.16 results. There is a definite trend of increased sun-spot number with increased scintillation activity. The problem is to characterise a scintillation phenomenon that is apparently so random in the fine detail on the one hand but generally predictable in the gross features on the other.

To obtain a descriptor that will overcome this apparent randomness in the fine detail of the amplitude of the fluctuations, Briggs and Parkin [19] proposed a series of indices that they termed $S1$, $S2$, $S3$ and $S4$. The one that has stood the test of time is the index $S4$ and this is defined [19] as the standard deviation of received power divided by the mean value of the received power. Mathematically it is

$$S4^2 = (\langle A^2 \rangle - \langle A \rangle^2)/(\langle S \rangle^2) \qquad (2.26)$$

where A is the signal amplitude and $\langle \ \rangle$ signifies the mean has been taken of the enclosed parameter.

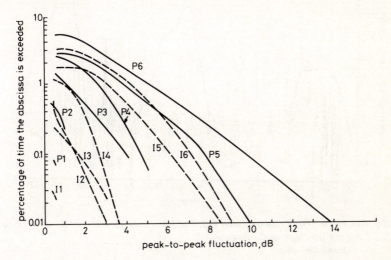

Fig. 2.14 *Cumulative statistics of ionospheric scintillation at a frequency of 4 GHz from solar minimum to solar maximum of sun-spot cycle 21 (from Fig. 2 of Reference 18) The data are for INTELSAT 4 GHz beacons in the Pacific Ocean Region (POR) and the Indian Ocean Region (IOR) with the Sun-Spot Number (SSN) as parameter*

Curves	Period		SSN range
I1, P1	March	1975–1976	10–15
I2, P2	June	1976–1977	12–26
I3, P3	March	1977–1978	20–70
I4, P4	October	1977–1978	45–110
I5, P5	November	1978–1979	110–160
I6, P6	June	1979–1980	153–165

——— POR beacon signal
– – – IOR beacon signal

The parameter $S4$ can be averaged over any appropriate period, 1 min being typical, and plotted against time. Fig. 2.17 shows such a plot for data recorded at the Hong Kong earth station of Cable & Wireless over both the Indian and Pacific Ocean 4 GHz satellite links using INTELSAT satellites.

In the experiment depicted in Fig. 2.17, the elevation angles were 27·8° and 19·2°, respectively. The delay between the onset of the scintillations on the Pacific Ocean link and the Indian Ocean link is of significance to the temporal modelling of the phenomena. Of more immediate significance are the very high values of the $S4$ index which shows that the mechanism producing the scintillation cannot be just a weak scattering phenomen [19].

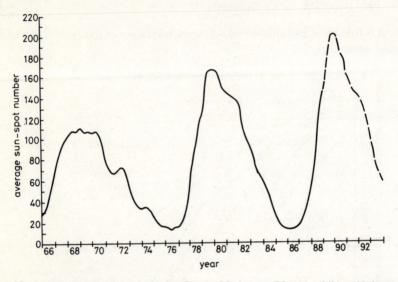

Fig. 2.15 *Average sun-spot number (after Fig. 1 of Reference 23 with additional information from the US Dept. of Commerce)*
———— measured data
– – – predicted data
(Copyright © 1981 American Geophysical Union, reproduced with permission)

A similar descriptor to $S4$ is the Scintillation Index SI [20], where

$$SI = \frac{(P_{max} - P_{min})}{(P_{max} + P_{min})} \tag{2.27}$$

Generally, P_{max} is taken as the third peak down from the highest peak that occured in a given period and P_{min} is similarly the third peak up from the smallest scintillation 'peak' observed in the same period. Occasionally, the fourth peaks are used. Using either prevents one extreme value from biasing the statistics. Because of its simplicity, SI is often used for first-order analysis of scintillation data. $S4$, however, is more accurate as a mathematical description of the phenomenon. The two descriptions have a rough equivalence. Taking the third peaks as the measurement points in a given interval of time, an SI value of 6 dB corresponds approximately to an $S4$ value of 0·3. Similarly, an SI value of 10 dB corresponds approximately to an $S4$ of 0·45. In general, an $S4$ value of 0·5 is taken as the boundary between weak and strong scintillation. Above an $S4$ value of 0·5, saturation effects start to take over and simple scaling laws break down.

2.3.4 Power spectra

In some experiments, a few of the scintillation events were recorded onto magnetic tape and Fourier analysis was conducted to reveal the power spectra of the scintillations [21]. In this way, the frequency components of the scin-

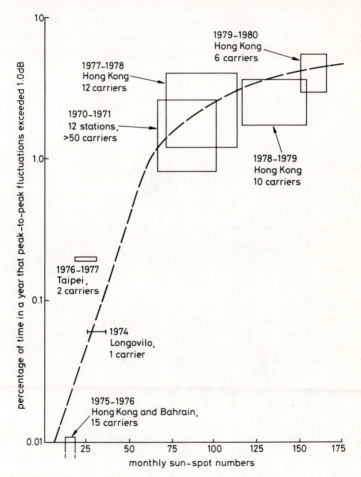

Fig. 2.16 *Dependence of 4 GHz equatorial ionospheric scintillations on monthly sun-spot number (after Fig. 1 of Reference 60)*
(Copyright © 1986 ITU, reproduced with permission)

tillating signal can be identified and an assessment made of their potential impact on a communications channel. The spectra are also invaluable diagnostic tools for investigating the basic physics of the phenomena. Figure 2.18 is one such Fourier analysis from an experiment that took place approximately mid-way between the solar sun-spot maximum and minimum years.

The curves in Fig. 2.18 are interesting for a number of reasons. The roll-off in the power spectrum as f^{-3} is evident at the high-frequency end of the spectrum. This is the part of the spectrum dominated by ionospheric scintillation and it exhibits the anticipated slope for log-amplitude fluctuations [22]. The transition between the low-frequncy and high-frequency slopes is complex, however, and is called the Fresnel frequency or 'corner frequency'. Usually, this transition takes place at a clear corner frequency, but in Fig. 2.18 there appear to be one or two such corner frequencies. The slope of one section is also $f^{-8/3}$ which is the roll-off expected for tropospheric scintillations (see Section 3.5).

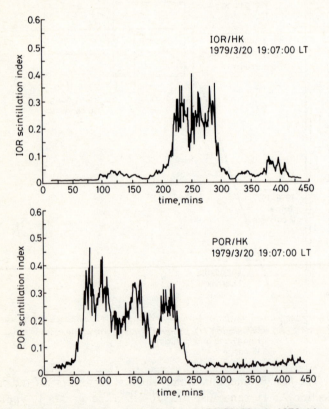

Fig. 2.17 *S4 scintillation statistics for an event on 20/21 March 1979 (from Fig. 7 of Reference 18)* The Local Time (LT) given is the start time of the data.
(Copyright © 1984 American Geophysical Union reproduced with permission)

The uplink for this experiment was at an elevation angle of 15° and it is probable that the tropospheric scintillations that occurred on the uplink were simply transferred to the downlink by the linear transponder action of the satellite.

In many cases, the slopes of the spectra are difficult to interpret owing to a number of factors. The signal will be inherently noisy and quite often the data have to be smoothed to enable the roll-off slope to be gauged accurately. Even with this smoothing, which has been applied in the case of Fig. 2.19, two aspects can still colour the interpretation. The first is spin modulation. If the satellite is spin stabilised, enhanced noise spikes will occur at harmonics of the spin frequency. The second aspect is due to the Fresnel filtering action of the

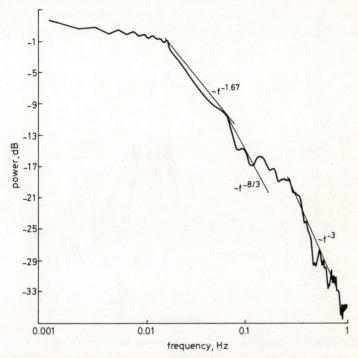

Fig. 2.18 *Power spectrum of combined tropospheric and ionospheric scintillation as observed at Hong Kong earth station (D. J. Fang, private communication)*

ionosphere. The first Fresnel minimum, which is the inverse of the coherence time (see Section 2.5.1), can cause the roll-off in the spectra to appear to be steeper than it actually is. This steeper-than-expected roll-off can be seen in Fig. 2.19. Great care must therefore be taken in setting the processing parameters of the data analysis in order to produce the correct roll-off slope.

Spectra of ionospheric scintillations taken at or near the solar sun-spot maximum are shown in Fig. 2.20. In these spectra [23], there are clear corner frequencies but the slopes vary from -2.3 to -3.7. At times the peak-to-peak scintillation was 9 dB.

The negative increase in the slope from −3 and, in some cases, the slight increase in the corner frequency both point to some other mechanism than weak scattering being present. The latter increase is sometimes a suggestion of multiple scattering [24, 32, 33]. Whether or not weak scattering is the dominant scintillation mechanism is crucial to the modelling of the phenomena. Before looking at this aspect, it is useful to summarise the current state of experimental evidence and identify the correlations noted for gigahertz ionospheric scintillation.

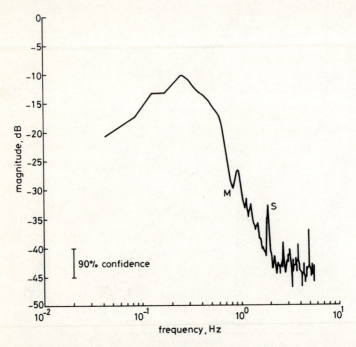

Fig. 2.19 *Temporal amplitude spectra, after smoothing, of the 4 GHz beacon on the Marisat satellite. The Fresnel minimum is shown as M and the spin modulation as S. (C. H. Liu, private communication)*
1981/1/30 22:22:59
S4 index: 0·1749
Sample length: 178·2 s

2.4. Ionospheric scintillation characteristics

The ionospheric scintillation characteristics and correlations can be summarised as follows [22].

Sun-spot number dependence
● No strong correlation between individual scintillation event occurrences and daily sun-spot number

- Stong correlation between annual scintillation occurrence and the annual sun-spot number
- Strong correlation between the amplitude of the scintillations and the monthly sun-spot number

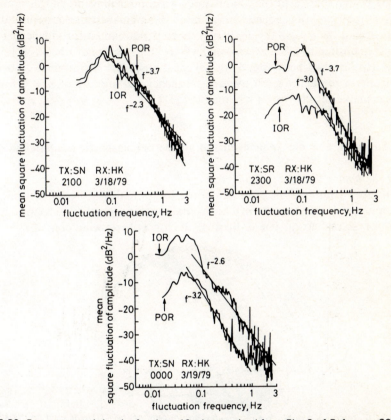

Fig. 2.20 *Power spectral density for three 10 min samples (from Fig. 8 of Reference 23)*
(Copyright © 1981 American Geophysical Union, reproduced with permission)

Temporal dependence
- Annual scintillation activity varies in an 11-year cycle in concert with solar sun-spot cycle
- Peak annual scintillation activity occurs at or just after the equinox periods
- Peak daily scintillation activity occurs approximately one hour after sunset at the ionospheric height

Geographic dependence
- Gigahertz ionospheric scintillations of any significant amplitude only occur within approximately ± 30° of the magnetic equator for geostationary communications satellite links

Frequency dependence
- The scintillation frequency or fading rate is below 1 Hz with a corner frequency of about 0·1 Hz
- The period of the scintillations are generally less than 15 s
- The power spectra of the scintillations generally roll off as f^{-3}
- The frequency dependence varies as f^{-n} for gigahertz frequencies. Weak scattering theory [4, 19] predicts n to be 2 but experimental evidence has shown a value of 1·5 between 4 and 6 GHz [22]. A value of 1 between 1·5 and 4 GHz [25] has been found for severe ionospheric disturbances, but, for gigahertz frequencies, an average frequency scaling law of $n = 1·5$ gives reasonable results for $S4$ less than 0·4 [51]

2.5 Theory and predictive modelling of gigahertz ionospheric scintillations

2.5.1 Summary of current theories
Owing to a combination of gravitational forces and heating by the Sun, the ionosphere will tend to move in a horizontal direction, much like the wind in the troposphere and stratosphere. Because the ionospheric wind contains electric-

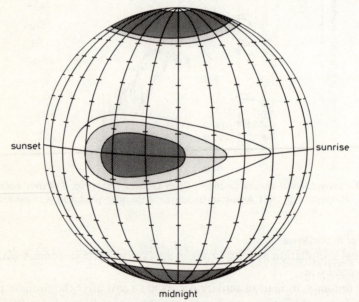

Fig. 2.21 *Night-time picture of scintillation occurrence (from Fig. 2 of Reference 60)*
The density of hatching is proportional to the occurrence of deep fading
(Copyright © 1986 ITU, reproduced with permission)

ally charged particles and it is cutting the magnetic field lines of the Earth, a current will be generated. The movement of the current is west to east at the geomagnetic equator, where the current is at a maximum. This peak current is referred to as the equatorial electrojet and it flows in the E region at a height of about 110 km [26, 27].

The generation of the current is at a maximum on the side of the Earth exposed to the Sun. The diurnal characteristics of ionospheric scintillation have been established and an approximate set of contours can be laid over the Earth to show where and when ionospheric scintillation is most likely to occur. Fig. 2.21 shows such a set of contours under which the Earth can be considered to rotate from left to right (sunset to sunrise).

Note that as an equatorial point on the surface of the Earth moves from the sunset rim, there is a rapid increase in the likelihood of intense scintillation occurring followed by a gradual decrease in both the probability and the severity of the scintillation. Another method of using contours to display such characteristics is illustrated in Fig. 2.22 [28]. Here the contours enclose equiprobable points for a given level of amplitude scintillations, in this case 2 dB.

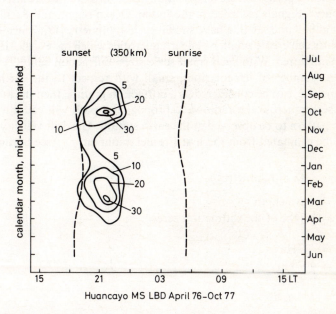

Fig. 2.22 *Monthly percentage occurrence of 1·54 GHz scintillations* ⩾ 2 dB *at Huancayo*
(*from Reference: 28*)
S4 = 0·13
(Copyright © 1980 American Geophysical Union, reproduced with permission)

Again, the higher probabilities occur closer to sunset than to sunrise; the equinoctal characteristics are also clearly brought out in such a presentation. The scintillation phenomenon is therefore induced by causes that have a higher probability of occurrence at equinoctal periods and close to sunsets. The reason for the former is straightforward since this a period of enhanced solar flux at the equator: the greater the flux the greater the TEC. The reason for the latter was difficult to explain until large, plume-like structures were observed by the Ticamarca Radar Observatory near Lima, Peru, to be rising in the ionosphere after dusk [29].

The plume-like structures, or funnels [18], contain rising bubbles of ionised plasma with electron densities one or two orders of magnitude less than the surrounding plamsa. The size of the bubbles can be in excess of 100 km, well in excess of the Fresnel zone size as observed by earth stations. With these sizes, refraction of incident radiowaves can occur, rather than diffraction which requires the irregularities to be equivalent in size to the Fresnel zone size, or smaller. Refractive effects can be severe, occasionally causing the $S4$ index to exceed unity [25], indicating the presence of focusing and defocusing.

The movement and displacement of the plume and bubble(s) with respect to the radiowave propagation direction can explain the asymmetry in scintillation data between the paths that travel essentially eastwards and westwards through the ionosphere. Signals recorded in the Indian Ocean region and Pacific Ocean region using the same satellite have shown such asymmetry [18]. The movement of the bubbles can create smaller-scale irregularities which have scale sizes down to a few centimetres. With such small scale sizes, diffraction can occur.

If a large number of irregularities, small with respect to the earth-station Fresnel zone size, can be considered as a diffraction grating, then the movement of such a screen across the radiowave propagation path will cause a varying diffraction pattern to be observed at the earth station. The drift velocity v of the screen can be estimated from the first Fresnel minimum f_{min} (see Section 2.3.4) as

$$v = \sqrt{(\lambda z)}\, f_{min} \text{ metres/s} \tag{2.28}$$

where λ = wavelength
        z = height of the diffraction screen

If $f = 4\,\text{GHz}$
        $z = 400\,\text{km}$
        $f_{min} = 0.84\,\text{Hz}$
then $v = 145\,\text{m/s}.$

This order of drift velocity has been confirmed in another experiment using a different technique [45] and the diffraction screen model [30] appears to explain the weak scattering phenomena. The sum of the plane waves produces an interference pattern which in turn produces the amplitude scintillations [31].

If the scintillations resulting from this diffraction pattern are digitised and submitted to an autocorrelation analysis, a correlation interval τ can be obtained, where τ is the time lag for which the level of correlation has decreased by 50% [34]. The data in Fig. 2.17 were subjected to such an analysis [18] and the correlation interval was found to have a mean of 1.5–2 s. The correlation interval is the inverse of the first Fresnel mimimum.

In Fig. 2.19, the first Fresnel minimum was at 0·84 Hz, approximately, which gives a correlation interval of 1·19 s in that instance. The more severe the scintillation, the higher is the Fresnel frequency (the 'corner frequency') and also the first Fresnel minimum. This is referred to as spectrum broadening and results in a shorter correlation time and increased dispersion. In general, the onset of ionospheric scintillation shows a more intense characteristic than the end of the scintillation activity, which tends to tail off gradually. The data analysed in Fig. 2.17 followed this trend, with the correlation interval being noticeably longer after midnight than before midnight, increasing to between 2·3 and 3·5 s [18]. The Indian Ocean links showed longer correlation intervals than the Pacific Ocean links in both instances [18].

Note that, if λ and z are constant in eqn. 2.23, as v increases so does f_{min}. Conversely, as v increases, the correlation interval decreases and the ionospheric impairments become worse. While the difference in correlation interval before and after midnight can be explained by a simple drop in drift velocity [45], the difference between the correlation intervals in the Pacific and Indian Oceans is probably not a simple difference in drift velocity. The explanation for this difference is probably due to a complex interaction of scale size, screen height, propagation direction and magnetic-field interactions.

The above results seem to indicate a non-stationarity of the irregularities, a non-constant drift velocity of the ionosphere across paths, and that weak scattering and multiple scattering could both be occurring. The problem arises when the boundaries between weak and strong scattering on the one hand and scattering and refraction on the other have to be merged into a unified theory to explain gigahertz ionospheric scintillations. To date, this has not been done successfully, largely because of the lack of data with which accurate models can be evolved. The approach to predictive modelling, at least in regard to geostationary communications satellite links, has been to resort to empirical models with suitable scaling applied.

2.5.2 An approach to an engineering model

The largest bodies of measured data that exist at present for gigahertz ionospheric scintillation on geostationary communications satellite links are those due to the INTELSAT measurement programm [23] and experiments conducted with the maritime mobile satellites of INMARSAT [e.g. Reference 38]. A large proportion of the INTELSAT data were recorded at the Cable & Wireless Stanley earth station in Hong Kong. The Hong Kong data are therefore used by INTELSAT as a reference data base and all predictions are scaled temporally and spatially therefrom. The flow diagram of the INTELSAT prediction model is shown in Fig. 2.23.

2.6 System impact

Ionospheric induced perturbations will have differing effects on different systems. One system, such as a synthetic-aperture radar operating at 1·275 GHz

[42], will require accurate time information as well as coherent phase over the pulse interval, so that range and phase errors are of paramount importance. For those systems, an accurate knowledge of group delay and relative phase advance provided by the Global Positioning System (GPS) of satellites [37] could be crucial. In other systems, where voice and/or video information is being transmitted, the modulation and bandwidth utilised are important since the variations in amplitude and phase induced by the ionosphere could reduce the quality of the transmissions.

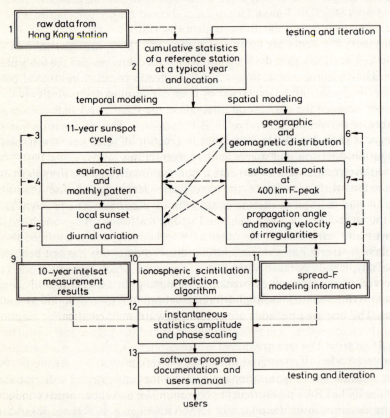

Fig. 2.23 *INTELSAT ionospheric predictive model flow diagram(from Fig. 6 of Reference 62)*
→ Modelling sequence
--→ Database input path
(Copyright © 1983 Communications Satellite Corporation, reproduced with permission)

Another aspect of importance when considering system impact is the reliability and operating mode of the system. A meterological data transmission circuit which is only utilised sporadically, such as the NOAA 7 and NOAA 8 satellite links to Greenland [44] or a data relay satellite [56], can 'work around' the random fluctuations induced by the ionosphere on the radiowave signals. On the other hand, a commercial communications satellite system [43] requires continuous operation at a reliability level well in excess of 99%. Such a service has to be designed to work through the impaired interval and a knowledge of the additional margin required is important in the economic design of such communications satellite systems. To assess the various elements of system impact, the impairements have been broadly subdivided into amplitude effects, phase effects and system effects.

2.6.1 Amplitude effects

Table 2.1 indicated that the absorption of the ionosphere at frequencies above 1 GHz is very low, certainly below the resolution of most communications links to detect. The signal level does vary significantly, however, due to scintillation effects. The amplitude effects induced by the ionosphere fall into three categories; decrease in power (fading), increase in power, and differential amplitude.

a) Decrease in power

At frequencies above 1 GHz, variations in the amplitude of the received signal due to ionospheric disturbances tend to dominate other varying parameters such as differential phase effects, ranging errors etc. Fig. 2.13 shows the variations in the amplitude of the received signal at a frequency of 4 GHz measured at the Taipei earth station. The problem in assessing and predicting the effect of amplitude scintillations is to know how to quantify an instantaneous parameter, such as peak-to-peak fluctuations, in terms of a statistical quantity, such as average fade level. Fade level here means the drop in the average signal power, in decibels, below the mean signal level.

There are four methods for decribing the instantaneous amplitude fluctuations in terms of statistics: log normal, gaussian, Nakagami-*m*, and two component. The first two methods are straightforward mathematical descriptions, the third is a semi-empirical procedure [46] applied to ionospheric scintillations [32, 47], and the last is a treatment of ionospheric scintillation as a product of a diffractively scattered component and a refractively focused component [48]. An examination of the four methods as they are applied to ionospheric scintillations has shown [38] that the Nakagami-*m* distribution gives the most accurate description of the phenomenon on average. The Nakagami-*m* distribution is very similar to the Nakagami–Rice distribution for $m > 1$ and the Nakagami–Rice distribution is therefore used to convert peak-to-peak measurements to average fade statistics [49]. An example [50] of such a conversion is shown in Fig. 2.24.

Given a peak-to-peak scintillation of 4 dB for instance, Fig. 2.24 indicates that, for 11% of the time, the signal is 1 dB below the average value. If the 4 dB peak-to-peak scintillations occur for 1% of the total time, the signal will experience an equivalent fade of 1 dB for 11% of 1% which is 0·11% of the total time. A practical example is indicated in Figs. 2.25 and 2.26 [38]. Fig. 2.25 gives

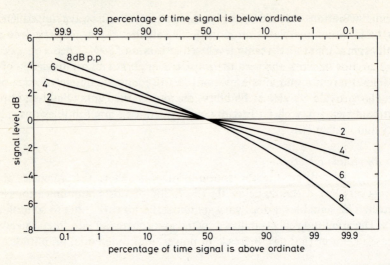

Fig. 2.24 *Cumulative distribution of signal amplitude for different peak-to-peak amplitudes (in dB) of scintillations using the Nakagami–Rice distribution (from Reference 50)*

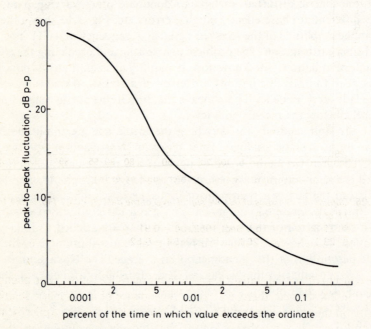

Fig. 2.25 *Cumulative statistics of peak-to-peak scintillations at a frequency of 1.542. GHz (from Fig. 10 of Reference 38)*
29 April 1982–26 May 1983
Elevation angle: 17·3°
Frequency: 1541·5 MHz
(Copyright © 1985 American Geophysical Union, Reproduced with permission)

the cumulative statistics of ionospheric scintillations measured at a frequency of 1·542 GHz with a path elevation of 17·3°. The data are expressed as peak-to-peak scintillations. Using a Nakagami–Rice distribution, Fig. 2.25 is converted to Fig. 2.26. For this link and for the measurement period in question, the relative signal level stayed above a fade level of − 10 dB for approximately 99·9% of the time. To provide 99·9% availability, therefore, this link would require an additional fade margin of 10 dB to counteract amplitude scintillations.

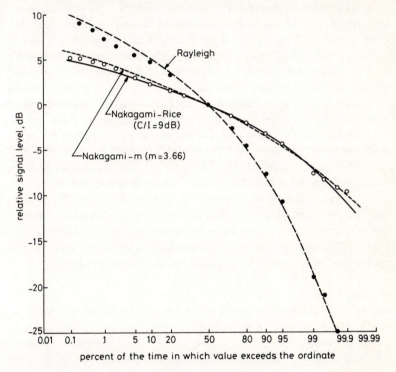

Fig. 2.26 *Cumulative statistics of relative signal level of the data shown in Fig. 2.25 (from Fig. 11 of Reference 38)*

● 21·35 ~ 21·43 h 6 Sept 1982: S4 = 0.91
○ 23.31 ~ 0·06 h 26 March 1983: S4 = 0·52

The concept of annual statistics is of dubious merit for ionospheric phenomena. As Section 2.4 showed, there is a cyclic dependence of ionospheric scintillation intensity on the equinoxes and the 11-year sun-spot cycle. In a 'quiet' year, with the monthly sun-spot number below 30, the ionospheric scintillation effects can almost disappear at gigahertz frequencies. On the other hand, in peak sun-spot years, the intensity of ionospheric scintillation can be extraordinarily severe. At Ascension Island, for instance, a 1·542 GHz maritime satellite link operating at an elevation angle of 81° experienced ionospheric scintillations with $S_T = 29$ dB for more than 10% of the time between September 1980 and March 1981 [52].

This *SI* is roughly equivalent to a peak fade level of 20 dB [52]. If the scintillations are approximately symmetrical about the mean level, the average fade level would be between 5 and 7 dB for 10% of the time, depending on the degree of saturation and the exact Nakagami–Rice presentation used. With these scintillations, a maritime satellite link with a fade margin of 5 to 7 dB would have been out of service for 10% of that period (see Figs. 2.24, 2.25 and 2.26 for the method of conversion to equivalent fade level).

Another extreme phenomenon, connected mainly with the years of high sun-spot activity, is the occurrence of isolated 'spikes' of fading [38]. In these cases, the signal level can drop by up to 25 dB at an operating frequency of 1·542 GHz for periods of several seconds. Such spikes are usually more severe for narrow-band systems such as maritime and aeronautical mobile services. The isolated spikes have been explained in terms of quasiperiodic scintillations [53, 54] and their severity can adversely bias both seasonal and annual statistics. For system purposes, therefore, ionospheric scintillation data are usually presented as monthly statistics, with a worst month figure quoted for equinoctal and maximum sun-spot periods. For example, a 4 GHz communications satellite link operating in a moderate sun-spot year (monthly sun-spot number about 80) will experience the equivalent of a 2 dB drop in signal level for an average of 4 min per day in the worst month. This is equivalent to 0·3% of the worst month. The designer of the satellite communications channel must decide whether is is necessary to ensure transmissions of acceptable quality at all times or whether to design the system for an 'average' scintillation level. Techniques for combatting scintillation impairments are discussed in Chapter 7.

b) Increase in power

At first sight, an increase in signal power should be beneficial. Most satellite systems, however, operate with more than one channel in a transponder [43]. Successful operation in such a multi-carrier system requires not only close control of the carrier frequencies but accurate power-level monitoring. An increase in power of one carrier with respect to another in the same transponder can cause intermodulation. In extreme cases, the carrier can increase to such a high level that it pushes the transponder into saturation. If the satellite is additionally operating in a dual-polarised mode, with a second transponder centred on the same frequency but using an orthogonal polarisation, increasing the power of the carrier in one transponder will cause increased interference into the co-channel signals in the other transponder. An increase in power of more than a decibel or so in one signal can therefore cause a degradation in other channels for a number of reasons.

The magnitude of the degradation will depend on whether the modulation is analogue or digital and whether the access to the satellite is Frequency Division Multiple Access (FDMA) or Time Division Multiple Access (TDMA). FDMA utilising Frequency Modulated (FM) carriers needs a more linear amplifier than TDMA. The operating point of a Travelling Wave Tube Amplifier (TWTA) is therefore reduced so that a 1 dB change in the input to the TWTA results in a change of 1 dB on the output. A large increase in signal into a TWTA will therefore drive the TWTA into the non-linear portion of its characteristics. For TDMA operations, a 1 dB change in input power results in only an 0·6 dB change in output power as the TWTA is being operated close to saturation. A moderate amount of such degrations are taken into account in the link budgets [43], but, again, extreme conditions can cause unacceptable interference.

c) Differential amplitude
Differential amplitude effects across a typical frequency band are minimal. The quasiperiodic scintillations, however, tend to be a narrow-band effect, very similar to in-band distortion produced by multipath activity in terrestrial systems [35, 55], and this frequency-selective scintillation can cause apparent differential amplitude effects. It is more serious for narrow-band Single Channel Per Carrier (SCPC) or similar services; TDMA systems which utilise a full transponder will tend to average out the effect of quasi-periodic scintillations.

2·6·2 Phase effects
Time, phase and frequency are all inter-related in their effect on a radiowave. The rate of change of phase is frequency, so that any disruption to one of the parameters − phase, frequency or time − will cause a corresponding disruption to the other two. The overall effect is essentially group delay (see Section 2.2.4) from which all the other parameters can be derived. The effect of the ionosphere on radiowaves decreases with increasing frequency and so only examples below 10 GHz will be given. Two types of communications satellite systems will be considered: maritime communications satellites which operate to ships on links at a frequency of approximately 1·6 GHz and communications satellites in the Fixed Satellite Service [43] which have downlinks in the region of 4 GHz. A third category, relating to radars, will also be discussed.

c) Maritime mobile links
The uplink and downlink frequencies to the mobile stations are centred on 1653 MHz and 1552 MHz, respectively, with a total bandwidth of 8 MHz. Individual voice and data channels use a much smaller bandwidth than 8 MHz but, for these examples, an 8 MHz bandwidth will be assumed. Since the system operates with circularly polarised signals, Faraday rotation will not cause a problem.

Phase advance: Phase advance $\Delta\phi$ is given by eqn. 2.15, for a frequency of 1·6 GHz and a TEC of 10^{17} electrons/m², as

$$\Delta\phi = 52\cdot75 \text{ rad}$$

$$= 3022°$$

More important than this phase advance is the relative change in phase advance.

If the TEC changes by 1% from an average value of 10^{17} electrons/m^2, the relative change in phase advance $\delta\phi$ is

$$\delta\phi = 0\cdot5275 \text{ rad}$$

$$= 30\cdot22°$$

The modulation/demodulation scheme must therefore be capable of adjusting to changes in phase of this order over the period of time that the changes take place. The latter typically varies from about 0·5 to 5 s in a worst-case situation, and most demodulation schemes will cope with this rate of change of phase. The change due to Doppler frequency will be much smaller than this and can be ignored (see Section 2.2.6).

Dispersion: Eqns. 2.20 and 2.22 give the time-delay and phase-advance dispersion effects, respectively. For the time delay component Δt, given that

$$f = 1\cdot6 \text{ GHz}$$

$$df = 8 \text{ MHz}$$

$$\text{TEC} = 10^{17} \text{ electrons/m}^2$$

Then:

$$|\Delta t| = 52\cdot344 \text{ ps}$$

The bandwidth of 8 MHz, if occupied completely by one signal, can support a pulse length τ_I equal to the inverse of the instantaneous bandwidth [5]. Hence

$$\tau_I = 1/(8 \times 10^6) \text{ s}$$

$$= 0\cdot25 \,\mu\text{s}$$

The time-delay dispersion effects are therefore much smaller than the maximum pulse length and so time delay dispersion will not have a significant impact. The phase advance component, from eqn. 2.22, using the same parameters as above, results in

$$\Delta\phi = 0\cdot2638 \text{ rad}$$

$$= 15\cdot11°$$

This is well within the range most modulation schemes can cope with. If the TEC increases by an order of magnitude, $\Delta\phi$ can become quite large over the 8 MHz bandwidth, but most of the maritime transmissions utilise bandwidths of a few kilohertz and so the phase dispersion effects will be quite small.

b) Fixed satellite systems

Phase advance, relative change in phase advance, time delay dispersion, and phase advance dispersion, calculated for a frequency of 4 GHz and a TEC of 10^{17} electrons/m^2, are given for two bandwidths, 36 and 240 MHz, as follows:

Phase advance
36 MHz and 240 MHz:

$$\Delta\phi = 21 \cdot 1 \text{ rad}$$

$$= 1208 \cdot 9°$$

Given a 1% change in TEC:

$$\delta\phi = 0 \cdot 211 \text{ rad}$$

$$= 12 \cdot 1°$$

Dispersion
36 MHz:

$$\Delta t = 15 \cdot 075 \text{ ps}$$

$$\tau_l = 27 \cdot 78 \text{ ns}$$

$$\Delta\phi = 0 \cdot 1899 \text{ rad}$$

$$= 10 \cdot 88°$$

240 MHz:

$$\Delta t = 100 \cdot 5 \text{ ps}$$

$$\tau_l = 4 \cdot 17 \text{ ns}$$

$$\Delta\phi = 1 \cdot 266 \text{ rad}$$

$$= 72 \cdot 54°$$

The large phase advance value of 72·54° across the 240 MHz bandwidth can be compensated for by phase pre-emphasis since this phase advance value is relatively constant. The differential effect due to variations in the TEC will be of the order of 1% of this value which is small enough to be coped with by the modulator/demodulator. In general, coherence is usually very high across bandwidths up to about 2% of the carrier frequency [39] and it is usually acceptable for most of the time to almost 10% instantaneous bandwidths.

c) Synthetic aperture radars
Synthetic Aperture Radars (SAR), operating at a frequency of about 1275 MHz (42, 57) and in a low earth orbit approximately 600 km up, have as one of there objectives, a resolution of 25 × 25 m on the surface of the Earth. The bandwidth of the radar signal is about 12 MHz and so coherence is not a serious problem over an individual pulse. The radar image is made up of sequences of pictures of overlapping regions which are recorded as the satellite moves in its orbit. The synthetic aperture of the radar is therefore much larger than its physical dimensions, in much the same way that two radio telescopes can increase their effective size by operating as an interferometer. Time and range errors are therefore important to an SAR.

The range error ΔR, is given by eqn. 2.8 and, for $f = 1.275\,\text{GHz}$ and a TEC value of 10^{17} electrons/m², is

$$\Delta R = 2.48\,\text{m}$$

giving

$$\Delta t = 8.27\,\text{ns}$$

These are the one-way errors. For a radar, the errors will be doubled since they include the return path as well. Clearly, in periods of really severe ionospheric disturbances, the range errors will exceed 50 m and so the required resolution will not be achieved. At most other times, and at mid-latitudes for nearly all the time, the ionosphere will not prevent an SAR from achieving the required resolution.

2.6.3 System effects

System effects are defined here as those effects that are not uniquely amplitude or phase effects, but which rely on aspects of the hardware to manifest the phenomenon. These can be divided into two classes: on-axis and off-axis.

A receiving antenna is usually designed to achieve its maximum performance for signals that arrive on-axis. If the signal arrives slightly off-axis, the performance of the antenna will drop accordingly. Such off-axis effects can be produced by ray bending; the geometric path to the satellite being distorted by refractive effects. If the refractive effects vary with time, so will the angle of arrival of the signals. The result will be a drop in the received co-polar (wanted) signal and an increase in the cross-polarised (unwanted) signal. The ratio of the wanted signal to the unwanted signal, the cross-polarisation discrimination (XPD), gives a measure of the isolation performance of the antenna. At frequences above 1 GHz, angle-of-arrival problems induced by the ionosphere are insignificant and hence XPD degradations induced by off-axis arrival of the signal are equally insignificant.

On-axis reduction in XPD can be caused by Faraday rotation effects. Simple geometry shows that the rotation of the linearly polarised electric vector $\Delta\theta$ away from the required orientation will give an XPD of

$$XPD = -20 \log_{10} \tan(\Delta\theta) \text{ decibels} \tag{2.31}$$

A rotation of $1°$ yields an XPD of 35 dB, which is better than the usual system requirement of 27–30 dB [43]. A $3°$ rotation, however, reduces the XPD to 26 dB. The linearly polarised systems at 4 and 6 GHz can be severely limited by this reduction in XPD [9] unless steps are taken to reduce the impact [10]. Severe amplitude scintillations do not of themselves produce depolarisation, i.e. the XPD does not fall below 30 dB [38], since the amplitude scintillations are an on-axis effect.

2.7 References

1 ECKERSLEY, T. L.: 'An investigation of short waves', *J. Inst. Elec. Engrs.*, 1929, **67**, pp. 992–1032
2 ECKERSLEY, T. L.: 'Studies in radio transmission', *ibid.* 1932, **71**, pp. 405–454
3 LOVELL, A. C. B., and CLEGG, J. A.: 'Radio astronomy' (Chapman and Hall, 1952)
4 BROOKER, H. G., and GORDON, W. E.: 'A theory of radio scattering in the troposphere', *Proc. IRE*, 1950, **38**, pp. 401–412
5 FLOCK, W. L.: 'Propagation effects on satellite systems at frequencies below 10 GHz'. NASA Reference Publication 1108, Dec. 1983: Fig. 2.2 abstracted from SMITH, E. K.: 'A study of ionospheric scintillation as it affects satellite communications'. Office of Telecommunications, US Dept. of Commerce, Technical Memorandum 74–186, Nov. 1974
6 RATCLIFFE, J. A.: 'The magnetoionic theory' (Cambridge University Press, 1959)
7 FLOCK, W. L.: 'Electromagnetics and the environment' (Prentice–Hall, 1969)
8 DAVIES, K.: 'Ionospheric radio waves' (Blaisdell Publishing Co., 1969)
9 WOLFF, R. S.: 'The variability of the ionosphere total electron content and its effect on satellite microwave communications,' *Int. J. Satellite Commun.*, 1985, **3**, pp. 237–243
10 WOLFF, R. S.: 'Minimisation of Faraday depolarisation effects on satellite communications systems at 6/4 GHz', *idid.* 1985, **3**, pp. 275–286
11 CLARKE, A. C.: 'Extraterrestrial relays', *Wireless World*, 1945, pp. 305–308
12 CHRISTIANSEN, R. M.: 'Preliminary report of S-Band propagation disturbances during ALSEP mission support (19 Nov. 1969–30 June 1970), Goddard Space Flight Center', NASA X-861-71-239, June 1971
13 SKINNER, N. J., KELLEHER, R. F., HACKING, J. B., and BENSON, C. W.: 'Scintillation fading of signals in the SHF band', *Nature Phys. Science*, 1971, **232**, pp. 19–21
14 ALLNUTT, J. E.: 'The INTELSAT propagation measurements programme', ICAP '81, IEE Conf. Publ. No. 195, 1981, Part 2, pp. 46–53
15 International Symposium on Beacon Satellite Studies of the Earth's Environment, New Delhi, India, 3–4 February 1983, and Workshop on Beacon Techniques and Applications, New Delhi, India, 3–4 February 1983. Proceedings (A85-27551 11–46), New Delhi, National Physical Laboratory of India, 1984
16 AARONS, J.: 'Global morphology of ionospheric scintillations', *Proc. IEEE*, 1982, **70**, pp. 360–378
17 BASU, S., and BASU, S.: 'Equatorial scintillations – A review', *J. Atmos. Terr. Phys.*, 1981, **43**, pp. 473 *et seq.*
18 FANG, D. J. and LIU, C. H.: 'Statistical characterisation of equatorial scintillation in the Asian region', *Radio Science*, 1984, **19**, pp. 345–358
19 BRIGGS, B. H., and PARKINS, I. A.: 'On the variation of radio star and satellite scintillations with zenith angle', *J. Atmos. Terr. Phys.*, 1963, **25**, pp. 334–365
20 WHITNEY, H. E., AARONS, J., and MALIK, C.: 'A proposed index of measuring ionospheric scintillation', *Planet. Space Science*, 1969, **7**, pp. 1069–1073
21 CRANE, R. K.: 'Spectra of ionospheric scintillations', *J. Geophys. Res.*, 1976, **81**, pp. 2041–2050
22 FANG, D. J.: '4/6 GHz ionospheric scintillation measurements'. AGARD Conf. Proc. 284, Propagation Effects in Space/Earth Paths, 1980, pp. 33–1 to 33–12
23 FANG, D. J.: 'C-Band ionospheric scintillation measurements at Hong Kong earth station during the peak of solar activities in sunspot cycle 21', Proceedings of the 3rd. Ionospheric Effects Symposium, 1981, Alexandria, Va, USA, pp. 3–1 to 3–12
24 UMEKI, R., LIU, C. H., and YEH, K. C.: 'Multifrequency spectra of ionospheric amplitude scintillations', *J. Geophs. Res.*, 1977, **82**, pp. 2752–2770
25 OGAWA, T., SINNO, K., FUJITA, M., and AWAKA, J.: 'Severe disturbances of UHF and GHz waves from geostationary satellites during a magnetic storm', *J. Atmos. Terr. Phys.*, 1980, **42**, pp. 637–644

26   EVANS, J. V.: 'Theory and practice of ionospheric study by Thompson scatter radar', *Proc. IEEE*, 1969, **57**, pp. 496–530

27   BALSLEY, B. B.: 'Some characteristics of non-two-stream irregularities in the equatorial electrojet', *J. Geophys. Res.*, 1969, **74**, pp. 2333–2347

28   BASU, S., BASU, D., MULLEN, J. P., and BUSHBY, S.: 'Long-term 1·5 GHz amplitude scintillation measurements at the magnetic equator', *Geophys. Res. Lett.*, 1980, **7**, pp. 259–262

29   WOODMAN, R. F., and LA HOZ, C.: 'Radar observations of F-region equatorial irregularities', *J. Geophys. Res.*, 1976, **81**, pp. 5447–5466.

30   CRONYN, W. M.: 'The analysis of radio scattering and space-probe observations of small-scale structures in the interplanetary medium', *Astrophys. J.*, 1970, **161**, pp. 755–763

31   COLES, W. A.: 'Interplanetary scintillations', *Space Sci. Rev.*, 1978, **21**, 411–425

32   YEH, K. C., and LIU, C. H.: 'Radio wave scintillations in the ionosphere', *Proc. IEEE*, 1982, **70**, pp. 324–360

33   RINO, C. L., and OWEN, J.: 'On the temporal coherence loss of strongly scintillating signals', *Radio Science*, 1981, **16**, pp. 31–33

34   WHITNEY, H. E., and BASU, S.: 'The effects of ionospheric scintillation on VHF/UHF satellite communications', *Radio Science*, 1977, **12**, pp. 123–133

35   BOITHIAS, L., 'Propagation des ondes radioélelectrique dans l'environment terrestre', CNET, Collection technique et scientifique des télécommunications, Paris, 1983. The 2nd. French edition was published in 1984 and an English language version was published in 1987 by Philip Kogan, London, U.K

36   SOICHER, H., and GORMON, F. J.: 'Seasonal and day-to-day variability of total electron content at mid-latitudes near solar maximum', *Radio Science*, 1985, **20**, pp. 383–387

37   BISHOP, G. J., KLOBUCHAR, J. A., and DOHERTY, P. H.: 'Multipath effects on the determination of absolute ionospheric time delay from GPS signals', *Radio Science*, 1985, **20**, pp. 388–396

38   KARASAWA, Y., YASUKAWA, K., and YAMADA, M.: 'Ionospheric scintillation measurements at 1·5 GHz in mid-latitude regions', *Radio Science*, 1985, **20**, pp. 643–651

39   RUFENACH, C. L.: 'Coherence properties of wideband satellite signals caused by ionospheric scintillation', *Radio Science*, 1975, **10**, pp. 973 *et seq.*

40   RATCLIFFE, J. A.: 'An introduction to the ionosphere and magnetosphere' (Cambridge University Press, 1972)

41   RISBETH, H., and GARRIOTT, O. K.: 'Introduction to ionospheric physics' (Academic Press, 1969)

42   AKAISHI, A., IMURA, N., MISUTAMARI, H., OGATA, Y., HISADA, Y., and ITOH, Y.: 'Research and development of a synthetic aperture radar antenna'. Proceedings 1985 Intl. Symp. on Antennas and Propagation, 1985, II, paper 151–2, pp. 639–642

43   MIYA, K. (Ed.): 'Satellite communications technology' (Institute of Electrical and Communications Engineers of Japan, Tokoy, 1985, 2nd Edn.)

44   JOHNSON, A., and TAAGHOLT, J.: 'Ionospheric effects on C^3 I satellite communications systems in Greenland', *Radio Science*, 1985, **20**, pp. 339–346

45   MENDILLO, M., and BAUMGARDNER, J.: 'Airglow characteristics of equatorial plasma depletions', *J. Geophys. Res.*, 1982, **87**, pp. 7641 *et seq.*

46   NAKAGAMI, M., in HOFFMAN, W. C. (Ed.): 'The *m*-distribution: A general formula of intensity distribution of rapid fading' (Pergamon, New York, 1960)

47   CRANE, R. K.: 'Ionospheric scintillation', *Proc. IEEE*, 1977, **65**, pp. 180–199

48   FREMOUW, E. J., LIVINGSTONE, R. C., and MILLER, D. A.: 'On the statistics of scintillation signals', *J. Atmos. Terr. Phys.*, 1980, **42**, pp. 717–731

49   AARONS, J., WHITNEY, H. E., and ALLEN, R. S.: 'Global morphology of ionospheric scintillations', *Proc. IEEE*, 1971, **59**, pp. 159–172

50   MAAS, J.: Private communication, May 1986

51   FREMOUW, E. J., LIVINGSTONE, R. C., RINO, C. L., COUSINS, M., FAIR, B. C., and LEADBRAND, R. L.: 'Complex signal scintillations — Early results from DNA — 002 coherent beacon', *Radio Science*, 1978, **13**, pp. 167 *et seq.*

52   MULLEN, J. P., MACKENZIE, E., BASU, S., and WHITNEY, H.: 'UHF/GHz scintillation observed at Ascension Island from 1980 through 1982', *Radio Science*, 1985, **20**, pp. 357–365

53   SLACK, F. F.: 'Quasiperiodic scintillation in the ionosphere', *J. Atmos. Terr. Phys.*, 1972, **34**, pp. 927 *et seq.*

54   HAIJKOWICZ, L. A., BRAMLEY, E. N., and BROWNING, R.: 'Drift analysis of random and quasiperiodic scintillations in the ionosphere', *J. Atmos. Terr. Phys.*, 1981, **43**, pp. 723 *et seq.*

55   HALL, M. P. M.: 'Effects of the troposphere on radio communications' (Peter Perigrinus, 1979)

56   TESHIROGI, T., CHUJO, W., KOMORU, H., AKAISHI, A., and HIROSI, H.: 'Development of 19-multibeam array antenna for data relay satellite'. Proceedings of ISAP '85, 1985, II, Paper 112-4, pp. 381–384

57   ITOH, Y., and HISHADA, Y.: 'Research and development on synthetic aperture radar'. Proceedings of ISAP '85, 1985, II, Paper 151-1, pp. 635–638

58   Report 340-5: 'CCIR atlas of ionospheric characteristics', 1988, ITU, 2 Rue Varembé 1211, Geneva 20, Switzerland

59   Recommendations and Reports of the CCIR, XVIth Plenary Assembly, Dubrovnik, 1986, Volume VI (Propagation in ionized media); Report 725-1: 'Ionospheric properties'.

60   Ibidem; Report 263-5: 'Ionospheric effects upon Earth-space propagation'.

61   WILLIAMS, P.: 'European radar unscrambles the ionosphere', *New Scientist*, 1985, pp. 46–52

62   FANG, D. J.: 'Final Report (INTEL 222/RAE-104, Milestone No. 4); COMSAT Laboratories Report to INTELSAT, Task No. 157-6117, 10 December 1983.

Chapter 3

Clear-air effects

3.1 Introduction

Despite the fact that Guglielmo Marconi had succeeded in transmitting signals across the Atlantic in 1901, no real investigations were conducted on radiowave propagation in the atmosphere until after 1930. At frequencies above 30 MHz, it was generally believed that radiowaves propagated in geometric straight lines with only the inverse square law causing their intensity to diminish with distance. Marconi proved this belief to be untrue in 1932 when he succeeded in transmitting radiowaves at frequencies above 30 MHz for distances many times the optical range [1].

A radio signal that is transmitted approximately horizontally will consist of two basic components: a space wave (or direct wave) and a ground wave (or reflected wave). The constructive and destructive interference of these two signals continues alternately with distance from the antenna, giving rise to a rippling pattern of amplitude fluctuations centred about the inverse square loss value. At the optical range from the transmitter, where grazing incidence occurs, some components of the signal will be diffracted; the smooth ripple pattern is destroyed and the loss in signal strength starts to exceed the inverse square law loss. At further distances still, scattering of the signal energy occurs from the non-uniform structure − or turbulence − of the atmosphere. This phenomenon is called tropospheric scatter [2, 3] and is of little direct relevance to satellite-to-ground propagation except for elevation angles below about 1°. Fig. 3.1 gives a schematic of the three propagation ranges.

As well as the scattering phenomenon, which was invoked [4] to explain the reliable reception of radiowaves at distances well beyond the horizon, it was apparent that the radiowaves were bent, or refracted, as they passed through the atmosphere. The development of transmitting devices in the gigahertz range also highlighted the absorptive effect of the atmosphere, both in rain and in apparently clear-sky conditions. Later Chapters will deal with the effect of rain and other particulates on radiowaves. In this Chapter, only the effects of an apparently clear atmosphere on radiowave propagation are discussed.

3.2 Refractive effects

3.2.1 Refractive index

The ratio of the speed of radiowaves *in vacuo* to the speed in the medium under consideration is called the refractive index *n* of that medium. For clean dry air in the lower atmosphere, the refractive index is given by

$$n(\text{dry}) = 1 + 77 \cdot 6(P/T) \times 10^{-6} \qquad (3.1)$$

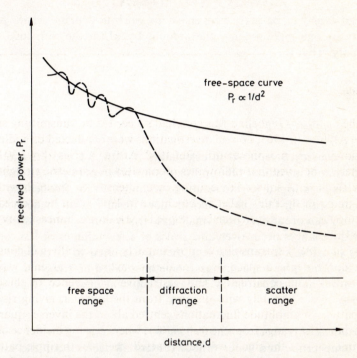

Fig. 3.1 *Schematic of the three propagation ranges (from Fig. 11.5 of Reference 1)*
Direct and reflected rays interfere in the free-space range.
Edge diffraction over obstacles on the ground dominates the diffraction range.
Refractive effects are the major phenomenon in the scatter range.
(Copyright © 1983 Butterworths, reproduced with permission)

where *P* is the pressure in millibars (mb) and *T* is the absolute temperature in kelvins. The gaseous composition of the atmosphere is fairly constant with geographic location and with height, at least up to 50 km. Eqn. 3.1 is therefore generally applicable for a clean dry atmosphere anywhere on Earth. The presence of water vapour, however, will modify the refractive index considerably. The 'wet' refractive index is

$$n(\text{wet}) = 375\,000(e/T^2) - 5 \cdot 6(e/T) \qquad (3.2)$$

where e is the water vapour pressure in millibars. The two terms in eqn. 3.2 can be combined to give an approximate expression

$$n(\text{wet}) = 373\,000(e/T^2) \tag{3.3}$$

The 'wet' and 'dry' indices can be summed arithmetically to give the total refractive index n. Re-arranging and combining the two indices gives

$$n - 1 = (77 \cdot 6/T) \times (P + 4810(e/T)) \times 10^{-6} \tag{3.4}$$

It has been found to be more convenient to work in N units, where N, the refractivity, is one million times the amount by which the refractive index exceeds unity. That is

$$n = 1 + N \times 10^{-6} \tag{3.5}$$

or

$$N = (77 \cdot 6/T) \times (P + 4810(e/T)) \tag{3.6}$$

If extreme accuracy is required in calculating N, eqn. 3.1 and 3.2 should be combined without introducing the approximation inherent in eqn. 3.3. Eqn. 3.6, however, has been found to give results to within 0·5% for frequencies below 100 GHz and can, in fact, be used with good results at all frequencies [5]. Changes in N can occur over both small and large volumes, and with different speeds. If the changes occur over time scales of a few minutes or less, or over scale sizes of a few kilometres or less, they are referred to as turbulent fluctuations. Both the large scale changes, sometimes called macroscopic changes, and the turbulent fluctuations in N have significant effects on Earth–space propagation.

3.2.2 Variations of refractivity with height
Pressure, temperature and water vapour content all decrease, on the average, with height in the troposphere. The contributions due to the drop in pressure and humidity with height outweigh those due to the drop in temperature in eqn. 3.6 leading to an overall drop in refractivity, on the average, with height. The variations of N with height are usually much larger and more rapid than those which occur on a horizontal plane except, perhaps, in very windy conditions. The most rapid variations with height, and consequently the most important ones, are those that occur in the lowest levels of the troposphere. Fig. 3.2 illustrates the radio refractivity profiles of the lower atmosphere [5].

As can be seen in Fig. 3.2, refractivity decreases approximately exponentially with height. If N is the refractivity at a height h above a level where the refractivity is N_s, then the exponential decay of the refractivity can be expressed as

$$N = N_s \times e^{-(h/H)} \tag{3.7}$$

where H is an appropriate 'scale height'. Generally, N_s is taken as the refractivity at the Earth's surface, and the lower the scale height, the more rapid is the decay in the value of N. Measurements have shown [6], however, that the median rate of change in N is $-40\,N/\text{km}$, independent of the scale height chosen.

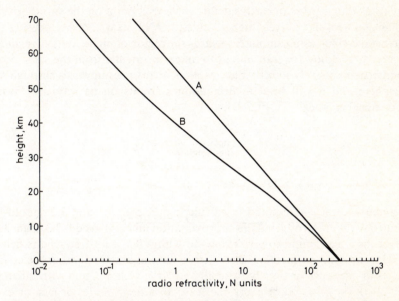

Fig. 3.2 *Radio-refractivity profiles for model atmospheres (from Fig. 1 of Reference 5)*
A: Average exponential model
B: Mid-latitude model (dry atmosphere)
(Copyright © 1986 ITU, reproduced with permission)

The CCIR defines [5] an average exponential atmosphere as one in which the scale height is $7 \cdot 36\,\text{km}$, the surface refractivity is 315, and the mean refractive index gradient in the first kilometre is $-40\,N/\text{km}$. For an average exponential atmosphere, eqn. 3.7 reduces to

$$N = 315 \times e^{-(h/7\cdot36)} \tag{3.8}$$

The world maps of refractive index are usually prepared in terms of the refractivity at sea level, N_0. To convert the sea level values to those that apply to the surface of the land, N_s, an equation similar to eqn. 3.7 can be used, namely

$$N_s = N_0 \times e^{-(h_s/7.36)} \tag{3.9}$$

where h_s is the height of the surface above mean sea level. In a like manner, information on humidity is usually provided in terms of ϱ, the absolute humidity or water vapour density. Water vapour pressure e can be related to ϱ by [7]

$$\varrho = 216 \cdot 5 \times (e/T) \tag{3.10}$$

As a consequence of the variation of refractivity with height, a radiowave transmitted from the surface of the Earth towards an earth satellite will not travel in a straight line. This is called ray bending.

3.2.3 Ray bending

A ray transmitted from an earth satellite towards a point on the surface of the Earth encounters successively higher values of N on its way down, causing the ray to bend by increasing amounts towards the region of higher refractive index. Conversely, a radiowave transmitted to an earth satellite from the same point encounters successively lower values of N as it traverses upwards through the atmosphere and it will bend in decreasing amounts on its way up. This is illustrated schematically in Fig. 3.3.

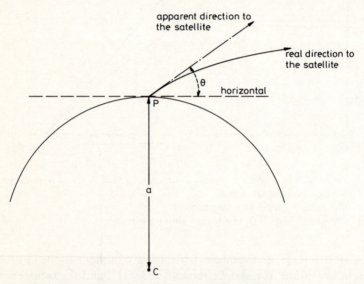

Fig. 3.3 *Apparent and real directions to an earth satellite*
An earth station at point P on the surface of the earth transmits a signal with an initial elevation angle θ in the apparent direction to an earth satellite. The varying refractive index with height causes the ray to bend. C is the centre of the Earth of radius a. (See Appendix A2.2 for the calculation of the geometric (straight line) and real (refracted) elevation angles)

For a vertical gradient of refractive index dn/dh, the radius of curvature r is given by [8]

$$1/r = -(1/n) \times (dn/dh) \times \cos\theta \tag{3.11}$$

where θ is the initial elevation angle of the ray at the transmitter with respect to the local horizontal. The negative sign indicates a decrease in refractive index with height. For terrestrial propagation, where θ is generally close to zero, eqn. 3.11 simplifies to

$$1/r = -dn/dh \tag{3.12}$$

with n assumed to be unity in eqn. 3.11. The difference in curvature between this ray and the curvature of the Earth is given by

$$(1/a) - (1/r) = (1/a) + (dn/dh) \tag{3.13}$$

where a is the radius of the Earth. To simplify the tracing of the curved ray paths in the planning of terrestrial microwave systems, it is customary to assume that the Earth has a bigger curvature than normal, thus permitting the ray paths to be drawn as straight lines. Fig. 3.4 illustrates this technique and Table 3.1 from Reference 7 gives the value of k corresponding to the various gradients of refractivity.

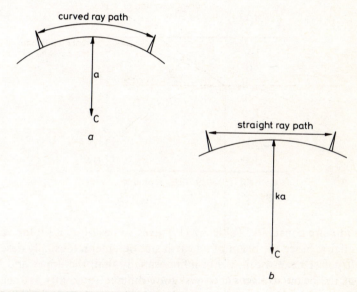

Fig. 3.4 *Illustration of two methods of ray tracing*
a Real radius *a* and a curved ray path.
b Effective radius *ka* and a geometric ray path.
In both cases, the two terrestrial antennas are the same distance apart. In (*a*), the antennas are pointed with a slight elevation angle to compensate for the curved ray path due to refraction. In (*b*), the curved ray path is compensated for by decreasing the curvature of the earth by increasing the radius by a factor *k*.

The term k is the factor which converts the real radius of the Earth to an effective radius ka. A value of $k = \frac{4}{3}$, corresponding to a $-40\,N$/km gradient, is usually used to derive the normal effective Earth's radius for terrestrial systems design. A value of $k = \infty$ corresponds to a gradient of $-157\,N$/km and results in a radius of curvature for the ray that is equal to the radius of the Earth. In this case, the ray propagates parallel to the surface of the Earth by a mechanism called 'ducting' (see Section 2.5 of Reference 9). For most satellite-to-ground applications, ducting will not occur unless the elevation angle is at, or below, 1°. Significant ray bending can occur, however, for elevation angles below 10°. Since the bending is due to the refractive index changes and the refractive index is relatively insensitive to frequency, ray bending is substantially independent of frequency. A relationship between ray bending τ, and surface refractivity N_s has been developed [10] giving

$$\tau = a + (b \times N_s) \text{ millidegrees} \tag{3.14}$$

Table 3.1 *Corresponding values of dN/dh and k (from Table 3.2 of Reference 6)*

$\dfrac{dN}{dh}$ (N/km)	k
157	0·5
78	$\frac{2}{3}$
0	1
− 40	$\frac{4}{3}$
− 100	2·75
− 157	∞
− 200	− 3·65
− 300	− 1·09

(Copyright © 1969 IEE, reproduced with permission)

where a and b are constants. Table 3.2 [13] gives values of a and b for various elevation angles. Since the beam of an earth station antenna, usually described by the half-power beamwidth, is not infinitesimally thin, the upper and lower elements of the beam will be bent by slightly different amounts as they pass through the atmosphere. The result is that the beamwidth is increased or, conversley, the beam becomes defocused.

3.2.4 Defocusing

The need to define accurately the defocusing loss of the atmosphere when using radio stars to calibrate the gain of large earth stations lead to an initial investigation of atmospheric defocusing loss [11]. Later calculations [12] from refractive index profiles resulted in Fig. 3.5.

These results apply to the Albany, New York, area. Coastal areas would generally exhibit a high loss owing to their generally higher humidity, and dry climates a lower loss. The results are independent of frequency between 1 and 100 GHz. Above 100 GHz, it is expected that the effects would be less [13].

The ray bending is due to the large-scale refractive index variations in the

Table 3.2 *Regression parameters for estimating the bending angle through the atmosphere given surface refractivity (from Table 2 of Reference 13)*

Elevation angle (deg)	a (mdeg)	b (mdeg/N unit)	Correlation coefficient	RMS error (mdeg)	95% deviation (mdeg)
0·1	−1112·8	5·778	0·81	89·0	151·0
0·2	−889·2	4·951	0·85	64·0	119·0
0·5	−512·3	3·473	0·94	26·0	58·0
1·0	−268·3	2·372	0·97	12·0	28·0
2·0	−95·9	1·409	0·99	5·0	11·0
3·0	−41·0	0·985	0·99	3·1	6·6
5·0	−10·2	0·610	0·99	1·9	3·8
10·0	−0·3	0·309	0·99	0·99	1·8
20·0	+0·6	0·151	0·99	0·49	0·88
50·0	+0·2	0·046	0·99	0·15	0·27

atmosphere. These are rarely stationary processes and large scale fluctuations will cause variations in the amount of ray bending. At the receiving antenna, these will appear as apparent changes in the angle of arrival of the signal and, in extreme cases, in multipath phenomena.

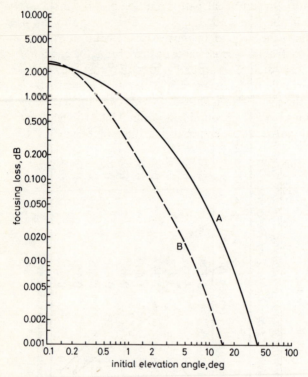

Fig. 3.5 *Focusing loss and standard deviation about the average (Fig. 1 from Reference 13)*
A: average loss
B: standard deviation
(Copyright © 1982 ITU, reproduced with permission)

3.2.5 Angle of arrival and multipath effects

In Table 3.2, the RMS error in the elevation angle due to the refractive effects was given. This error constitutes a variation in the angle of arrival of the signal. The fluctuations in elevation angle are about an order of magnitude greater than those that occur in the azimuth angle. Fig. 3.6 [14] shows the median standard deviation in angle of arrival fluctuations. The fluctuations are higher in the summer, consistent with the increase in surface refractivity that normally occurs in this period. Angle of arrival fluctuations, like ray bending in general, are independent of frequency between 1 and 100 GHz.

Angle of arrival fluctuations can be considered to be a single ray that is being deviated from its normal path. In some situations, however, several possible paths can exist simultaneously through the atmosphere between the transmitter and the receiver. The rays travelling the various paths arrive at the receiver with different amplitudes and phases, and interference results. This phenomenon is called multipath. On terrestrial paths, multipath is the most common propagation outage in the frequency range 1–10 GHz owing to the proximity of a reflecting surface, the ground in most cases, to the ray path. Only a slight modification in k from the usual value of $\frac{4}{3}$ can bring the ground within the first Fresnel zone. On satellite-to-ground paths above an elevation angle of 10°, multipath is virtually non-existant. If the elevation angle is low enough or the beamwidth of the earth station antenna is wide enough, destructive interference due to reflections from the ground can occur. This problem is expanded on in Chapter 6 with particular reference to maritime mobile systems.

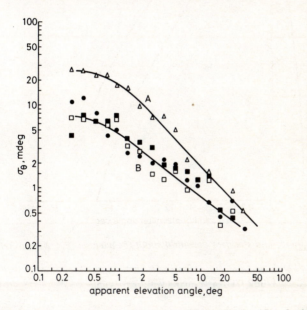

Fig. 3.6 *Median standard deviation in elevation angle scintillation (Fig. 3 of Reference 28)*

● spring   A: summer
△ summer   B: rest of year
□ fall $f = 7.3\,\text{GHz}$
■ winter $D = 36.6\,\text{m}$

(Copyright © 1986 ITU, reproduced with permission)

Multipath, especially that due to a tilted or elevated duct or to a smooth sea surface, generally causes relatively long periods of signal fading ranging from several seconds to many minutes. This is because of the large scale, stable nature of the atmosphere or sea required to produce the conditions condusive to multipath. Although the degree of destructive interference (fading) that occurs is not of itself frequency sensitive, an individual multipath effect is frequency selective since generally only one particular frequency will have two phase paths that cause exact cancellation at the receiver. Terrestrial microwave engineers refer to this as in-band distortion. Over a bandwidth of 50 MHz, say, multipath fading will only effect a few megahertz of the instantaneous bandwidth at

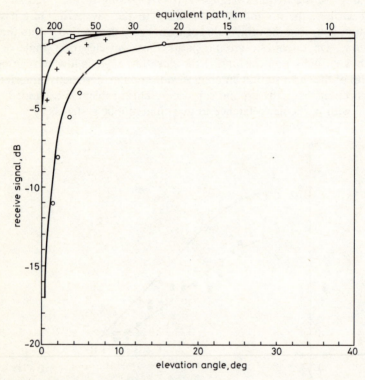

Fig. 3.7 *Predicted and measured signal level as a function of elevation angle with frequency as parameter (from Fig. 12 of Reference 15)*
—— Model for *R* plus atmospheric gas loss
Measured
□ 2 GHz OSU
+ 7·3 GHz McCormick & Maynard
o 30 GHz OSU
(Copyright © 1978 Ohio State University, reproduced with permission)

any instant. If the atmosphere becomes mixed owing to moderate wind or rainfall being present in the ray path, the likelihood of multipath is very much reduced. The turbulence, however, will cause a degree of phase incoherence across the aperture of the receiving antenna, thereby producing an apparent gain reduction.

3.2.6 *Antenna gain reduction*

The apparent reduction in gain of an antenna increases as the effective aperture increases. For a given antenna size, therefore, the effect becomes larger with frequency. It also increases as the elevation angle decreases. Using low elevation angle data acquired from measuring beacon signals transmitted by ATS–6 at a frequency of 2 and 30 GHz [15] and by TACSATCOM–1 at a frequency of 7·3 GHz [16], Theobald and Hodge derived an empirical model [15] for antenna gain reduction R that is a complicated mix of signal amplitude and angle-of-arrival variances, path length and antenna beamwidth. Fig. 3.7 shows the predicted signal levels using the Theobald and Hodge model for 2, 7·3, and 30 GHz for antenna beamwidths of 1·8, 0·3 and 0·15° respectively, against measurements.

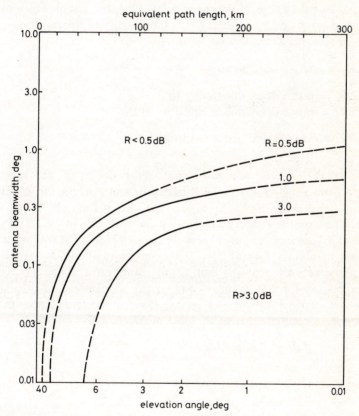

Fig. 3.8 *Gain degradation as a function of beamwidth and elevation angle (Fig. 6.6.14 of Reference 17)*
R = antenna gain degradation

With antenna degradation R as parameter, Fig. 3.8 was derived [17] from the same data. It can be seen from Fig. 3.8 that, if the elevation angle is above 5° and the antenna beamwidth is larger than 0·3°, antenna gain degradation is less than 0·5 dB. At elevation angles and antenna beamwidths below these values, however, antenna gain degradation increases appreciably. The variation in refractive index along the path can also introduce both an absolute and a relative phase error with respect to a single frequency and a signal bandwidth. The latter will be of relevance in calculating the coherence bandwidth due to dispersion across the instantaneous communications bandwidth. This is discussed in Chapter 4. The former is only of importance when accurate ranging is required.

3.2.7 *Phase advance*

Exactly as in the case of the ionosphere, the effect of a non-unity refractive index is to delay the signal, causing an apparent phase advance or, conversely, an over-estimate of the range. If the range delay ΔR is the distance by which the range is over-estimated by assuming that the velocity of the radiowave in the atmosphere was the same as *in vacuo*, then

$$\Delta R = \Delta R_d + \Delta R_w \text{ metres} \tag{3.15}$$

where ΔR_d = range delay due to dry air
ΔR_w = range delay due to moisture in the air

It can be shown [17] that, for a zenith path,

$$\Delta R_d = 2\cdot2757 \times 10^{-3} \times P_d \text{ metres} \tag{3.16}$$

where ΔR_d is the range delay due only to dry air and P_d is the surface pressure for dry air, in millibars. For a dry surface pressure of 1000 mb, $\Delta R_d = 2\cdot28$ m at zenith. The important aspect, however, is the variability of ΔR_d. The general variation of dry surface pressure is between about 975 and 1025 mb which gives a variation of about 11·4 cm. Careful measurement of the dry atmospheric pressure, however, will probably limit the dry range error to about 0·5 cm [17].

The expression for ΔR_w is more complicated and, instead of calculating a value for ΔR_d and ΔR_w independently, it is easier to use two composite values ΔR_1 and ΔR_2 to compute the zenith value of ΔR, namely

$$\Delta R = \Delta R_1 + \Delta R_2 \text{ metres} \tag{3.17}$$

where $\Delta R_1 = 2\cdot2757 \times 10^{-3} \times p$
$\Delta R_2 = 1\cdot7310 \times 10^{-3} \int (\varrho/T) \, dl$
p = total surface pressure,
ϱ = water vapour density,
T = temperature in kelvins,
dl = incremental distance along the path

Using normal values for ϱ, T, and the scale height of 7·5 g/m³, 280 K, and 2 km, respectively, $\Delta R_2 = 9\cdot23$ cm. It is worth noting that only slight variations in ϱ, T and the scale height can double this value. The variations in ΔR due to the moist component of the atmosphere are therefore larger, in general, than those due to the dry component. Crane [12] has calculated ΔR for elevation angles of 0, 5° and 50° for a standard atmosphere. These are shown in Table 3.3. The ray paths extend to the heights shown. For geostationary communications satellites,

Table 3.3 *Ray parameters for a standard atmosphere (from Table 3.3 of Reference 7)*

Initial elevation angle (deg)	Height (km)	Range (km)	Bending (mdeg.)	Elevation angle error (mdeg.)	Range error (m)
0·0	0·1	41·2	97·2	48·5	12·63
	1·0	131·1	297·9	152·8	38·79
	5·0	289·3	551·2	310·1	74·17
	25·0	623·2	719·5	498·4	101·1
	80·0	1081·1	725·4	594·2	103·8
5·0	0·1	1·1	2·6	1·3	0·34
	1·0	11·4	25·1	12·9	3·28
	5·0	55·2	91·7	52·4	12·51
	25·0	241·1	176·7	126·3	24·41
	80·0	609·0	181·0	159·0	24·96
50·0	0·1	0·1	0·2	0·1	0·04
	1·0	1·3	1·9	1·0	0·38
	5·0	6·5	7·0	4·0	1·47
	25·0	32·6	14·3	10·3	3·05
	80·0	104·0	14·8	13·4	3·13

the height is approximately 36 600 km, but the additional range error beyond a height of 80 km is only about 2% of the 80 km value. The values in Table 3.3, while applicable only to a standard atmosphere, are fairly representative. To interpolate to angles above 5°, an equation of the form

$$\Delta R(\theta) = \Delta R / \sin \theta \text{ metres} \qquad (3.18)$$

can be used where $\Delta R(\theta)$ is the range error at the required elevation angle θ in degrees.

3.3 Reflective effects

In Section 3.2.5, multipath effects were considered to have occurred from the presence of multiple paths that arose owing to the variation of the refractive index in the atmosphere. Such multipath effects are generally much smaller than those that arise owing to reflection from smooth or almost smooth surfaces.

These surfaces are usually the boundaries between two media of very different refractive indices. Occasionally, such boundaries can be formed by inversion layers in the lower atmosphere creating ducts [9], but more usually the boundary is between the air and the sea or the air and the ground.

For significant multipath effects to occur owing to ducting or surface reflection, small elevation angles are necessary, typically less than 3° or half of the beamwidth of the antenna on the surface of the Earth, whichever is the greater. The power and phase of the reflected wave depend a great deal on the mechanical and electrical properties of the reflecting surfaces. To produce a significant degree of power in the reflected wave, the reflected components should remain coherent; i.e. their phases must be non-random and the elements, when vertorially summed, produce a plane wavefront. To achieve this, the reflecting surface must be smooth. If the surface is not smooth, the reflected elements will tend to have random directions and phases, i.e. they will be incoherent. For this reason, reflection from a smooth surface is referred to as specular reflection while that from a rough surface is referred to as diffuse reflections.

3.3.1 Reflection from a smooth surface
The reflection coefficient ϱ is defined as

$$\varrho = E_r/E_i \tag{3.19}$$

where E_r and E_i are the reflected and incident fields, respectively. Two values of ϱ will exist for a surface that is partially conducting and these will depend on the polarisation of the incoming wave. The two polarisations considered are horizontal and vertical. In this case, horizontal polarisation is defined as the polarisation with the electric field both perpendicular to the plane of incidence and parallel to the reflecting surface. Vertical polarisation is defined as the polarisation with the magnetic field parallel to the reflecting surface. Fig. 3.9 illustrates the two cases. Note that the definition for vertical polarisation seems to break down when the elevation angle is 90° [7].

The reflection coefficient for vertical polarisation ϱ_v is given by [7]

$$\varrho_v = \frac{[k - j\sigma/\omega\varepsilon_0]\sin\theta - [k - j\sigma/\omega\varepsilon_0 - \cos^2\theta]^{1/2}}{[k - j\sigma/\omega\varepsilon_0]\sin\theta + [k - j\sigma/\omega\varepsilon_0 - \cos^2\theta]^{1/2}} \tag{3.20}$$

where k = relative dielectric constant
σ = conductivity, *mhos/m*
θ = elevation angle
ω = angular frequency $(2\pi/f)$
ε_0 = permittivity of free space $(8\cdot854 \times 10^{-12})$, F/m

The relative dielectric constant k is also known as the relative permittivity ε_r. For any medium that is partially conducting, the permittivity is complex and is given by

$$\varepsilon^* = \varepsilon' - j\varepsilon'' \tag{3.21}$$

where ε' and ε'' are the real and imaginary parts of the complex permittivity ε^*. ε' is referred to as the dielectric constant and ε'' as the loss component. By substituting for ε''

$$\varepsilon^* = \varepsilon' - j\frac{\sigma}{\omega} \tag{3.22}$$

giving

$$\varepsilon^*/\varepsilon_0 = \varepsilon_r - j\frac{\sigma}{\omega\varepsilon_0}$$

or

$$= k - j\frac{\sigma}{\omega\varepsilon_0} \tag{3.23}$$

but

$$\varepsilon^*/\varepsilon_0 = n^2 \tag{3.24}$$

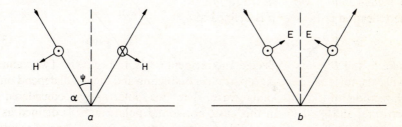

Fig. 3.9 *Phase reversal on reflection at a conducting surface (Fig. 4.8 of Reference 9)*
a Horizontal polarisation: direction of E vector changes from out of reflecting plane ⊙ to into reflecting plane ⊗. Direction of H vector remains constant. Nominal phase change ϕ = 180°
b Vertical polarisation: direction of E vector remains constant. Direction of H vector remains constant. Nominal phase change ϕ_ϱ = 0°
(Copyright © 1979, IEE, reproduced with permission)

where n is the refractive index. Eqn. 3.20 can therefore be written as [9]

$$\varrho_V = \frac{n^2 \sin\theta - (n^2 - \cos^2\theta)^{1/2}}{n^2 \sin\theta + (n^2 - \cos^2\theta)^{1/2}} \tag{3.25}$$

In a like manner, the reflection coefficient for horizontal polarisation ϱ_H can be given by [7]

$$\varrho_H = \frac{\sin\theta - \left(k - j\dfrac{\sigma}{\omega\varepsilon_0} - \cos^2\theta\right)^{1/2}}{\sin\theta + \left(k - j\dfrac{\sigma}{\omega\varepsilon_0} - \cos^2\theta\right)^{1/2}} \tag{3.26}$$

or by [9]

$$\varrho_H = \frac{\sin \theta - (n^2 - \cos^2 \theta)^{1/2}}{\sin \theta + (n^2 - \cos^2 \theta)^{1/2}} \qquad (3.27)$$

For a vertically polarised wave, if the conductivity is zero, there is a value of θ for which ϱ_V is zero. This is called the Brewster angle, θ_B. If the conductivity is finite, the value of ϱ_V still approaches a minimum at the Brewster angle. For a wave going from medium 1 to medium 2, the Brewster angle, θ_B, is given by [7]

$$\theta_B = \tan^{-1} \sqrt{\frac{k_1}{k_2}} \text{ degrees} \qquad (3.28)$$

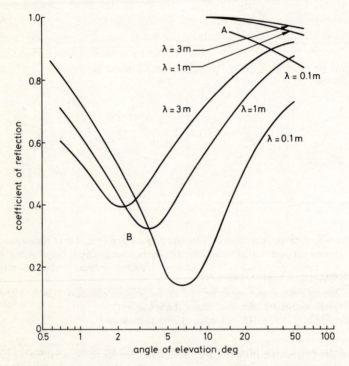

Fig. 3.10 *Reflection coefficients for a smooth, plane sea (Fig. 6.12 of Reference 7)*
A: horizontal polarisation
B: vertical polarisation

If medium 1 is air, then

$$\theta_B = \tan^{-1} \sqrt{\frac{1}{k_2}} \text{ degrees} \qquad (3.29)$$

The relative dielectric constant k (or ε_r) varies with frequency. At 1 GHz, k for sea water is 80 but this value has fallen to 65 at a frequency of 10 GHz [7].

The Brewster angle for an air/sea boundary therefore increases slightly with frequency and also becomes more pronounced. Fig. 3.10 illustrates this [7].

3.3.2 Reflection from rough surfaces

The reflected signal from a smooth surface is made up of elements that are coherent in phase and which do not vary much in amplitude. A small antenna used to sample such a signal would detect gradual and predictable changes in amplitude as the antenna was swept through the area. If a surface is not smooth, the reflected signals from elements are no longer mutually parallel. The reflection coefficient will therefore be reduced in the 'smooth reflection' direction. As the surface roughness increases, so the degree of incoherence increases. The reflected signal from a rough surface will be made up of elements that are generally incoherent in phase and which vary greatly in amplitude. A small antenna used to sample such reflections would detect large and seemingly random changes in amplitude as the antenna swept the area. These random changes can be described by a Rayleigh distribution [19].

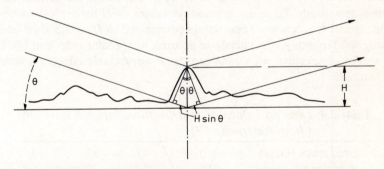

Fig. 3.11 *Schematic of surface roughness*
Two parallel rays at an elevation angle θ to the horizontal impinge upon a rough point of height H. The difference in the path lengths of the two parallel reflected rays is $2 H \sin \theta$.

Fig. 3.11 shows a random surface from which a signal is reflected from two locations with a height difference H. The difference in the two path lengths Δl is

$$\Delta l = 2H \sin \theta \text{ metres} \tag{3.30}$$

giving a phase difference $\Delta\phi$ of

$$\Delta\phi = (4\pi/\lambda) H \sin \theta \text{ degrees} \tag{3.31}$$

For destructive interference, $\Delta\phi = \pi$. Substituting this value in eqn. 3.31 and re-arranging gives the destructive interference value of H as

$$H = \lambda/4 \sin \theta \text{ metres} \tag{3.32}$$

If H is much smaller than this value, the surface can be considered to be smooth. Suggested 'smooth' values of H range from a quarter [7], through an eighth [9, 18], to a sixteenth [9] of a wavelength, giving values of H of $\lambda/8 \sin \theta$, $\lambda/16 \sin \theta$, and $\lambda/32 \sin \theta$, respectively. An approach that corresponds to the more relaxed of these three path length differences introduces a roughness factor, $f(\sigma)$, and a Rayleigh criterion [19]. The variable σ is the standard deviation of the variation in H and the roughness factor is given by [19]

$$f(\sigma) \;=\; \exp\left(-1/2[4\pi\sigma(\sin \theta)/\lambda]^2\right) \tag{3.33}$$

The roughness factor is used to multiply the smooth earth reflection coefficient to arrive at an overall specular reflection coefficient. If the surface is very rough, the value of σ is large and $f(\sigma)$ tends to zero. The specular reflection coefficient therefore tends to zero leaving only the diffuse reflection. For a smooth surface, $f(\sigma) = 1$ and the specular reflection is the same as that for a smooth earth. The Rayleigh criterion, given by [19]

$$H \;=\; 7 \cdot 2\lambda/\theta \tag{3.34}$$

with θ in degrees, gives a limiting value of H for which the surface can be described as smooth. Table 3.4 gives some values of H for typical communications satellite frequencies. These values correspond to $H = \lambda/8 \sin \theta$ and are probably too large to give an accurate account of specular reflection [9]. Even small values of specular reflections can cause appreciable effects on received signal strength.

Table 3.4 *Limiting values of H using the Rayleigh criterion (from Reference 19)*

Frequency (GHz)	1·5	4	11	20
wavelength (cm)	20	7·5	2·7	1·5
$\theta = 10°$, H (cm) =	14·4	5·4	1·9	1·1
$\theta = 5°$, H (cm) =	29	10·8	3·9	2·2
$\theta = 1°$, H (cm) =	144	54	19·4	10·8

(Reproduced by permission of the authors)

3.4 Absorptive effects

The effect a dielectric and, to some extent, a partially conducting medium have on a radiowave passing through them is described by their complex permittivity ε^*. If the medium is low-loss, the imaginary part of the complex permittivity ε'' is practically non-existent and the dielectric constant, the real part of the complex permittivity, can be used alone. Low-loss media are generally those that exhibit a symmetry in their atomic or molecular make up.

When a molecule is non-symmetrical, it has a preferred orientation if placed in an electric field and the molecule is said to be polar. Polar molecules exhibit significantly more loss than other molecules [20] owing to a general relaxation of the dipoles [21] within the medium when they are made to move by changing the external field. This loss has been found to depend markedly on the frequency of measurement.

The principal constituents of the lower atmosphere, oxygen and nitrogen, are both electrically non-polar and no absorption occurs due to electric dipole resonance. Oxygen, however, is a paramagnetic molecule with a permanent magnetic moment [18] which causes resonant absorption at particular frequencies. Water, and hence water vapour, is a polar molecule and so absorption occurs due to electric dipole resonance at critical frequencies. If ε'' and ε' are plotted against frequency for a polar molecule, the critical frequency or frequencies show up as peaks in ε''.

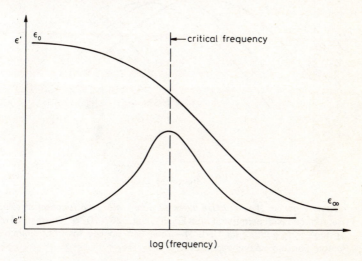

Fig. 3.12 ε' *and* ε'' *plotted against log(frequency) for a substance that has one critical frequency*

Fig. 3.12 illustrates this effect for a polar molecule with one critical frequency. The precise frequency at which the absorption peaks can vary with temperature. Fig. 3.13 illustrates this effect for water [71]. The radiowave absorption effects of CO, NO, N_2O, NO_2, SO_2, O_3 and other polar molecules in the atmosphere are negligible compared to the absorption of oxygen, water and water vapour [22] but, at frequencies above 70 GHz, these and other trace gases can contribute significant attenuation in the absence of water vapour [23].

3.4.1 Oxygen and water vapour resonance lines
Oxygen and water vapour are the principal absorbers of radiowave energy. As well as causing an increasing background level of attenuation with an increase

in frequency, there are a number of critical frequencies where resonant absorption takes place. Below a frequency of 350 GHz, water vapour has three resonant absorption lines at 22·3, 183·3 and 323·8 GHz. Oxygen has an isolated absorption line at 118·74 GHz and a host of absorption lines around approximately 60 GHz. The latter broad absorption spectrum can be split up into the individual lines [24] which become more distinct as the pressure is reduced. This has led to the term 'pressure broadening' being used to describe the smearing of

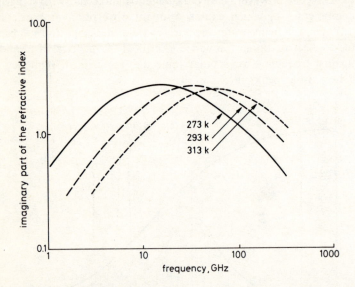

Fig. 3.13 *Imaginary part of the refractive index of pure water as a function of temperature* (Reproduced with permission from Fig. 5–32 pp. 207–208 of 'The manual of remote sensing', First edition, © 1975 The American Society for Photogrammetry and Remote Sensing, see reference 71)

the multitude of individual resonance lines into one broad absorption band. Fig. 3.14 plots the specific attenuation due to water vapour and oxygen at sea level on a horizontal path with the given pressure and temperature. This corresponds to 1% humidity (i.e. 1% of water vapour molecules mixed with 99% of dry air molecules). This rapidly increases to 75% humidity if the temperature falls to 10° Celsius [23].

Complex analytical calculations of attenuation due to moist air up to a frequency of 1 THz can be made [24] and are readily available in FORTRAN IV computer code [60]. An approximate, but widely used, approach for calculating gaseous absorption [23] is given below.

3.4.2 Gaseous absorption
For dry air, the specific attenuation at a pressure of 1013 mb and a temperature

of 15° Celsius is mainly due to oxygen and is given by [23 from 72]

$$\gamma_0 = \left[7 \cdot 19 \times 10^{-3} + \frac{6 \cdot 09}{f^2 + 0 \cdot 227} + \frac{4 \cdot 81}{(f - 57)^2 + 1 \cdot 50} \right]$$
$$\times f^2 \times 10^{-3} \text{ decibels/km} \qquad \text{for } f < 57 \text{ GHz} \qquad (3.35a)$$

$$\gamma_0 = \left[3 \cdot 79 \times 10^{-7} f + \frac{0 \cdot 265}{(f - 63)^2 + 1 \cdot 59} + \frac{0 \cdot 028}{(f - 118)^2 + 1 \cdot 47} \right]$$
$$\times (f + 198)^2 \times 10^{-3} \text{ decibels/km} \qquad \text{for } f > 63 \text{ GHz} \qquad (3.35b)$$

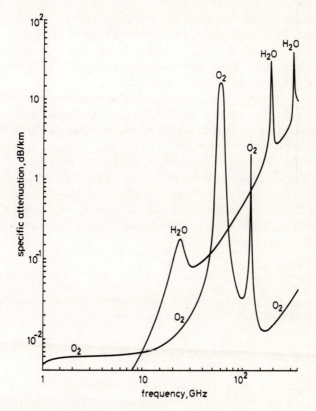

Fig. 3.14 *Specific attenuation due to atmospheric gases (Fig. 2 of Reference 23)*
   Pressure: 1013 mb
   Temperature: 15° C
   Water vapour: 7·5 g/m³
   (Copyright © 1988 ITU reproduced with permission)

where f is the frequency in GHz. Values of γ_0 between 57 and 63 GHz may be obtained from Fig. 5 of Reference 23. Eqn. 3.35b is valid up to a frequency of 350 GHz. The corresponding specific attenuation for water vapour γ_w is [23]

$$\gamma_w = \left\{ 0 \cdot 050 + 0 \cdot 0021 \varrho + \frac{3 \cdot 6}{(f - 22 \cdot 2)^2 + 8 \cdot 5} + \frac{10 \cdot 6}{(f - 183 \cdot 3)^2 + 9 \cdot 0} \right.$$

$$\left. + \frac{8 \cdot 9}{(f - 325 \cdot 4)^2 + 26 \cdot 3} \right\} f^2 \varrho \; 10^{-4} \; \text{decibels/km} \tag{3.36}$$

where f is the frequency in gigahertz and ϱ is the water vapour density in g/m³. Eqn. 3.36 is valid, within an overall accuracy of $\pm 15\%$, over the range 0–50 g/m³, with a temperature dependence of -0.6% per deg C from 15°C over the temperature range $-20°$ to $+40°$C.

Note that the attenuation increases with decreasing temperature. A similar correction for the dry air formulae in eqns. 3.35a and b can be made with a correction factor of $-1 \cdot 0\%$ per deg C from 15°C. Again, the range is $-20°$ to $+40°$C and the attenuation increases with decreasing temperature. Except for supersaturation cases (i.e. clouds), the water vapour density may not exceed the saturation value for the given temperature [23].

To calculate the total dry air (oxygen) and water vapour attenuation through the atmosphere along a path to a satellite, the specific attenuation values should be integrated along the given path taking into account the variations in pressure and water vapour densities with height. For vertical paths and earth stations at sea level, an exponential decay with height is assumed with equivalent heights postulated for dry air and water vapour [23]. Outside the absorption bands, these equivalent heights (i.e. the heights at which the contributions due to dry air and water vapour are assumed to reach zero) are approximated by 6 km for dry air and around 2 km for water vapour, depending on the weather conditions. More accurate formulations are [23]

$$h_0 = 6 \, \text{km} \qquad \text{for } f < 57 \, \text{GHz} \tag{3.37a}$$

$$h_0 = 6 + \frac{40}{(f - 118 \cdot 7)^2 + 1} \, \text{kilometres} \qquad \text{for } 63 < f < 350 \, \text{GHz} \tag{3.37b}$$

$$h_w = h_{w0} \left\{ 1 + \frac{3 \cdot 0}{(f - 22 \cdot 2)^2 + 5} + \frac{5 \cdot 0}{(f - 183 \cdot 3)^2 + 6} \right.$$

$$\left. + \frac{2 \cdot 5}{(f - 325 \cdot 4)^2 + 4} \right\} \, \text{kilometres} \qquad \text{for } f < 350 \, \text{GHz} \tag{3.38}$$

where h_{w0} is the water vapour equivalent height in the window regions and takes the values

$$h_{w0} = 1 \cdot 6 \, \text{km in clear weather}$$

$$h_{w0} = 2 \cdot 1 \, \text{km in rain}$$

The zenith attenuation values of dry air A_0 and water vapour A_w absorption are given by

$$A_0 = h_0 \gamma_0 \tag{3.39a}$$

and

$$A_w = h_w \gamma_w \tag{3.39b}$$

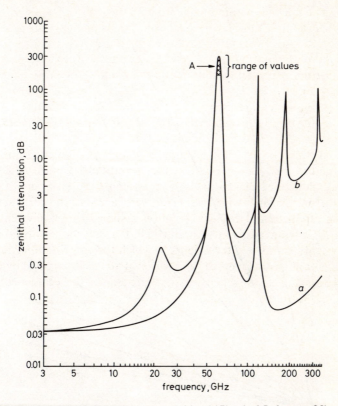

Fig. 3.15A *Total zenithal attenuation at ground level* (*Fig. 4 of Reference 23*)
  Pressure: 1013 mb
  Temperature: 15° C
  a For a dry atmosphere
  b With an exponential water-vapour atmosphere of 7·5 g/m³ at ground level and
    a scale height of 2 km
  (Copyright © 1988 ITU, reproduced with permission)

with A_z, the total zenith attenuation due to a moist atmosphere, being

$$A_z = A_0 + A_w \tag{3.40}$$

Fig. 3.15A gives a sample calculation of the total one-way zenith attenuation through the atmosphere versus frequency. It is instructive to compare Fig. 3.15A with Fig. 3.14.

If the elevation angle θ is 10° or greater, a cosecant law can be used to find the total one-way absorption A_a due to a moist atmosphere, namely

$$A_a = A_z/\sin\theta \text{ decibels} \tag{3.41}$$

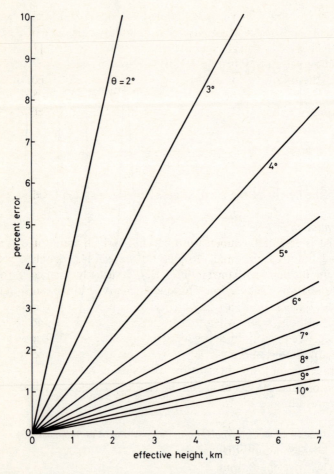

Fig. 3.15B *Percentage error produced by assuming no curvature of the Earth (Fig. 1 of Reference 73)*
(Copyright © 1986 IEEE, reproduced with permission)

For elevation angles between 0° and 10°,

$$A_a = \frac{\sqrt{R}}{\cos\theta}\left[\gamma_0\sqrt{h_0}\,F\left(\tan\theta\,\sqrt{\frac{R}{h_0}}\right) + \gamma_w\sqrt{h_w}\,F\left(\tan\theta\,\sqrt{\frac{R}{h_w}}\right)\right] \text{ decibels} \tag{3.42}$$

where R is the effective radius of the Earth in kilometres, including refraction (close to the Earth's surface $R = 8500 \, \text{km}$), and the function F is of the form

$$F(x) = \frac{1}{0 \cdot 661x + 0 \cdot 339(x^2 + 5 \cdot 51)^{1/2}} \tag{3.43}$$

A simpler correction which can be applied to eqn. 3.41 for elevation angles below $10°$ is to reduce the attenuation calculated by eqn. 3.41 by a percentage error that is incurred by neglecting the curvature of the Earth [73]. The correction percentages are given in Fig. 3.15B.

If the earth station is not at or close to sea level, corrected values of the various parameters should be used [23]. For frequencies up to 50 GHz and for elevation angles θ above $10°$, A_a can be approximated with acceptable accuracy by [25]

$$A_a = \frac{\gamma_0 h_0 \, e^{-h_s/h_0} + \gamma_w h_w}{\sin \theta} \text{ decibels} \tag{3.44}$$

where h_s is the height of the earth station above mean sea level in kilometres.

3.4.3 Attenuation in Fog
Fog or mist is essentially supersaturated air in which some of the water has precipitated out to form small droplets of water. The droplets are usually smaller than 0.1 mm in diameter [26]. Fog is usually formed through two processes, radiation or advection, although it can exist in frontal situations and on upslopes [7] where moist air, having been blown up a slope, cools to form the fog. Radiation fog occurs when the ground cools at night and the moist air above it is also cooled until it becomes supersaturated. Advection fog occurs when a warm, moist air mass is blown over a cool surface and becomes supersaturated upon losing heat to the cool surface [26].

A regression analysis has been conducted [26] on the theoretical attenuation due to fog, resulting in the following expression:

$$A = -1 \cdot 347 + 0 \cdot 0372\lambda + 18/\lambda - 0 \cdot 022T \tag{3.45}$$

where A = attenuation, dB/km/g/m^3
λ = wavelength, mm
and
T = temperature, deg C

The regression fit is only good between wavelengths of 3 mm and 3 cm and for temperatures between -8 and $+25°$C. Fig. 3.16 [26] shows the variation of A with temperature and wavelength.

To obtain the total attenuation, the fog density M in g/m^3 and the extent of the fog are needed. M can be obtained from [26, 27]

$$M = (0 \cdot 024/V)^{1 \cdot 54} \tag{3.46}$$

where V is the visibility in kilometres. A typical visibility in dense fog is 20 m. This will give a value of M of $1\cdot32$ g/m^3 and, at a frequency of 10 GHz, a value of A of $0\cdot63$ dB/km. Eqn. 3·46 is generally an underestimate for fogs in which

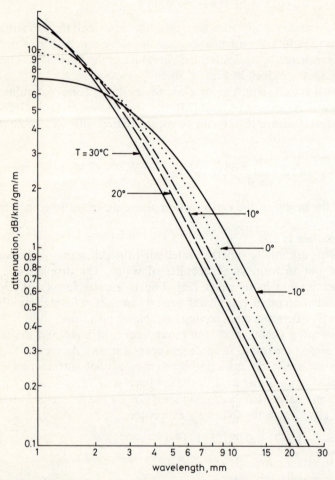

Fig. 3.16 *Attenuation due to fog at millimetre wavelengths (Fig. 1 of Reference 26)*

———	30° C
– – – –	20°
—·—	10°
· · · · ·	0°
———	−10°

the droplet size exceeds 10 microns. A fog layer is not usually very thick, a hundred metres being a typical height, and so fog attenuation is only a minor element in satellite-to-ground link design for frequencies below 100 GHz.

3.5 Tropospheric scintillation effects

When the atmosphere is still, the refractive index varies slowly with height and even more slowly in the horizontal plane. Ray bending and multipath at low elevation angles are likey to occur in these still-air situations as discussed in Sections 3.2.3 and 3.2.5. The presence of wind, however, causes the atmosphere to become mixed rather than stratified, and to cause relatively rapid variations in refractive index to occur over small intervals, referred to as scale sizes.

Small-scale fluctuations of the refractive index along a propagation path cause amplitude and phase scintillations which are detected as amplitude variations in the received signal level. The same conditions of Fresnel zone size (Section 2.3.1) and scintillation indices (Section 2.3.3) apply and, like ionospheric scintillation, the effect is generally non-absorptive and occurs on-axis. Unlike ionospheric scintillation, however, tropospheric scintillation effects increase as the frequency increases. As the use of radio telescopes extended to frequencies well above 1 GHz, the impairing effects of tropospheric scintillations were noted [29]. For radio telescopes, particularly for two or more operating as interferometers, it was the differential phase scintillations along the separate paths that caused the problem. Amplitude scintillations, however, were more of a problem when calibrating very large antennas.

Earth stations used in communications satellite systems are characterised in the receive mode by the ratio G/T (dBK), where G is the gain of the antenna on axis and T is the perceived excess noise temperature of the antenna, both values being logarithmic in the above ratio [30]. Celestial sources, some of which are quite powerful radiowave emitters in the lower microwave frequencies, are used to calculate the ratio G/T and a paper [61] identifying all the error sources, including scintillation, has been used to evolve a standard measurement technique [30].

Tropospheric scintillations were observed to be far worse at low elevation angles compared to high elevation angles and, in addition, at low elevation angles the celestial sources were neither small enough (in terms of angular extent) or quite powerful enough emitters to enable accurate measurements to be made. The solution to this problem had to await the availability of satellite beacon sources. Interest in tropospheric scintillation, as it affects satellite communications and Earth sensing, was not very great initially owing to the apparently smaller impairments induced by tropospheric scintillation when compared to other sources of errors.

In the case of satellite communications, the first antennas used were very large and the signal frequencies were around 4 and 6 GHz; both factors reducing the impact of tropospheric scintillations as will be seen later. The first low elevation angle measurements were therefore made on an opportunistic basis using satellites 'drifting' in quasi-geosynchronous orbits. Later, short-term measurements were conducted both at high latitudes and using mobile platforms utilising geostationary satellite sources. Only recently have long-term measurements been undertaken that were specifically aimed at quantifying tropospheric scintillation impairments. The evolution of these measurements and some of their results are reviewed below.

3.5.1. Drift measurements

The first reported tropospheric scintillation measurements at low elevation angles using a frequency above 1 GHz were made in Canada in 1970 using the 7·3 GHz beacon from the US satellite TACSATCOM–1 [31]. TACSATCOM–1 was drifting westerly and measurments were made over a period of 22 days. During the time, the satellite moved from an elevation angle of 6° to 0°. Two antennas, spaced 23 m apart, were used for the experiment. The diameters of the antennas were 9 m and 1·8 m. A typical data segment is shown in Fig. 3.17.

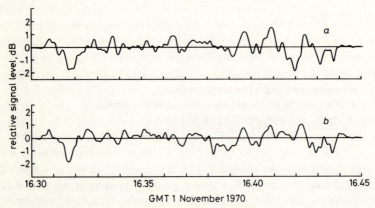

Fig. 3.17 *Signal strength versus time for a 15 min period on 1 November 1970 (from Fig. 4 of Reference 31)*
   (a) 9 m antenna
   (b) 18 m antenna
(Copyright © 1971 IEEE, reproduced with permission)

It is clear from Fig. 3.17 that there is quite good correlation between the scintillations occuring along the two paths. Cross-correlation studies undertaken between the two sets of data [31] revealed a scale size of up to 300 m for the turbulence structure, with a median scale size of the order of 30 m. Later analysis of the data [16] revealed an unexpected element in the fading statistics of the median signal level averaged over 15 min. Below an elevation angle of 2°, the median signal was well below the predicted value. Fig. 3.18 illustrates the range of the effect. This effect is now believed to be due to multipath effects skewing the data away from a symmetrical distribution.

During 1975 and 1976, the opportunity was taken by a number of experimenters to measure low-elevation-angle tropospheric scintillation effects at

2, 4, 20 and 30 GHz while the NASA satellite ATS–6 moved to and from 35° E longitude [33–37]. Unfortunately, the limited observation time available precluded any statistically meaningful results being obtained.

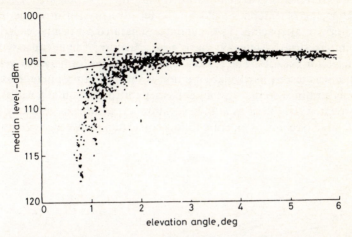

Fig. 3.18 *Median signal level as a function of elevation angle (from Fig. 1 of Reference 31)*
    Each point represents a 15 min average of received signal strength
    – – – Median clear-sky level in the absence of attenuation
    —— Includes effect of atmospheric attenuation
    (Copyright © 1971 IEEE, reproduced with permission)

3.5.2 High latitude measurements

At latitudes above 70°, the elevation angle of geostationary satellites is a maximum of about 11°. This decreases to zero as the latitude approaches 82°. Measurements at these latitudes is not only limited by the availability of beacon sources but by the local weather conditions which can be very severe in terms of high winds and low temperatures. High latitude measurements have been made at frequencies of 4 and 6 GHz at elevation angles down to 1° using signals from ANIK and LES satellites [32, 38] and INTELSAT and SYMPHONIE satellites [39], the former in Canada and the latter in Norway. Later single-site and dual-site diversity measurements were conducted in Norway using the OTS satellite at frequencies of 14 and 11 GHz [40]. The dual-site diversity measurements complemented earlier diversity measurments made at 6 and 4 GHz [41] and more recent site diversity measurements [74] at 38 GHz in Canada.

All the measurements showed that the cold, less humid climate at high latitudes greatly reduced the tropospheric scintillation effects. A comparison of high latitude and mid latitude data at approximately the same elevation angle [16] showed the high latitude fading distribution to be in between the winter and summer characteristics for the mid latitudes. Fig. 3.19 [16] illustrates this. The characteristics shown in Fig. 3.19 are not truly symmetrical, which could be explained by equipment effects, by rain along the path, or by multipath effects.

The fairly long-duration high latitude measurements confirmed the strong positive correlation between scintillation amplitude and both temperature and humidity that had been suspected in the earlier drift measurements. The site diversity measurements also demonstrated that, for elevation angles below 3°, a vertical spacing of antennas yielded a better diversity performance than an identical horizontal spacing. Two antennas separated vertically, called space diversity in terrestrial systems [9], have been a feature of high-capacity microwave systems for decades. The reason for the better performance of vertically spaced antennas in this particular situation is that the atmosphere tends to be horizontally stratified so that two antennas separated horizontally will still 'see' highly correlated scintillations at low elevation angles while a vertically spaced pair will tend to 'see' only moderate to low correlation between the scintillation

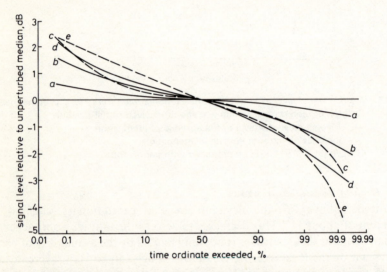

Fig. 3.19 *Distribution of received signal level for the elevation angle interval 5–6° showing the seasonal dependence (from Fig. 4 of Reference 16)*
a Ottawa, Feb. 1971
b Resolute, Aug. 1970
c Ottawa, May 1971
d Ottawa, Oct., Nov. 1970
e Ottawa, July 1971
Ottawa (latitude 45°N) and Resolute (latitude 75°N) are in Canada. The frequency of measurement was 7·3 GHz
(Copyright © 1972 IEE, reproduced with permission)

on the two paths. In warmer climates, where rain is the major propagation impairment, almost the opposite is the case. Site diversity, the spacing of two antennas horizontally to circumvent rain attenuation, is considered in more depth in Chapters 4 and 7. If control of the uplink power is required to keep a constant flux density at the satellite or if the signal must be switched between two receivers or transmitters operated in diversity, the rate of change of the scintillations is an important parameter. This can be determined through spectral analyses.

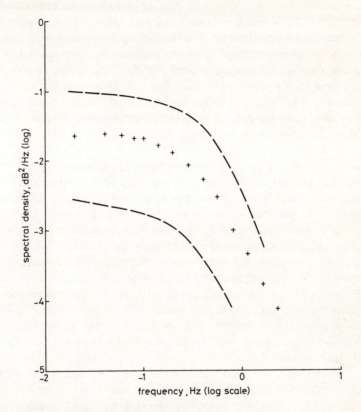

Fig. 3.20 *Amplitude scintillation power spectra from an 11·6 GHz tropospheric scintillation measurement (from Fig. 4 of Reference 42)*
   + Mean spectrum
   − Limits containing 90% of spectrum
   (Copyright © 1981, IEE, reproduced with permission)

3.5.3 Spectral analyses

Data from three summer months of scintillation observations in 1980 using the 11·8 and 11·6 GHz OTS beacons were analyzed to give the frequency components of the scintillations [42]. Fig. 3.20 shows the mean and 90% limits of the spectra.

The Fresnel frequency or critical frequency (Section 2.3.4) is around 0·3 Hz and the roll-off has a slope of about − 8/3, as expected [43]. Like ionospheric scintillation, spectra from individual events show a large variation in corner frequency and in roll-off slope. Other scintillation data at approximately the same 30° elevation angle from the same satellite [44, 47, 48] yielded the same corner frequency and roll-off slope but results at 6·2 GHz [45] on a 6·9° path and 11·2 GHz [46, 47] on 8·9° and 7·1° paths exhibited a much lower corner frequency. A critical frequency closer to 0·06 Hz [46] was found for these lower elevation angles. The probable explanation is that, while the longer path lengths at low elevation angles produce larger scintillations than on high elevation angle paths, the larger 'bulk' effect of the atmosphere at the lower elevation angles prevents the changes in scintillation amplitude from taking place quickly, somewhat akin to a smoothing action. The roll-off at low elevation angles, however, is still − 8/3 as at the higher elevation angles, but it can appear to be affected by rain along the path. It is therefore necessary to be able to separate 'wet' and 'dry' scintillations.

3.5.4 Separation of 'wet' and 'dry' tropospheric scintillations

The somewhat arbitrary distinction between wet and dry tropospheric scintillations is considered to be the presence or absence of rain along the path. Until quite recently, the presence of rain along the path was thought to reduce the amplitude of tropospheric scintillations significantly at all fade levels. Three independent experiments reported in 1985 using the OTS satellite [44, 48] and an INTELSAT V satellite [49] demonstrated that rain along the path produced no net reduction in the amplitude of the scintillations, at least until the mean path attenuations were in excess of about 5 dB, or of their characteristics; the critical frequency was unchanged and the roll-off was still − 8/3. Fig. 3.21 [49] shows the original and smoothed data, together with the net scintillations.

Fig. 3.22 [44] illustrates the − 8/3 slope after the attenuation effects have been removed. In Reference 49, a moving average of the data, with an averaging period of one minute, was used while in References 44 and 48, the data were smoothed over a fixed period of from many seconds to a minute.

Supplementary analyses indicated that the height of the turbulence causing the scintillations lay between 1·5 and 4 km [48] and around 2 km [44]. The amplitude distribution of the scintillations showed that, for the same percentage time, wet scintillations had a higher amplitude than dry scintillations. For small amplitude fluctuations (less than 0·5 dB peak-to-peak) the statistical data were well described by a gaussian distribution. At larger amplitudes, the distribution departed from a gaussian fit as had been observed consistently before [50] and became assymetric [47] with a definite negative bias; that is, there were more deep fades than corresponding enhancements. As with ionospheric scintillations of large amplitude, a Nakagami–Rice distribution was the best fit for large amplitude tropospheric scintillations. Another area where extremely large scintillations can occur is in maritime mobile communications.

3.5.5 Maritime mobile communications

Maritime mobile communications use a complex 'double-hop' arrangement [30] with 6/4 GHz being used between the satellite and the coast earth stations

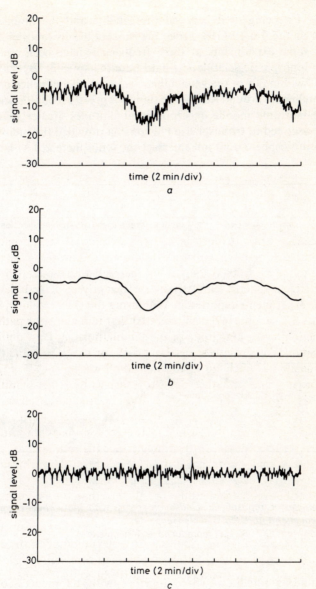

Fig. 3.21 *Separation of atmospheric scintillation and rain attenuation using a moving 1 min averaging technique, 29 July 1982 (from Fig. 3 of Reference 49)*
a original 11 GHz data at Yamaguchi
b Data smoothed by the moving average procedure at 1 min intervals (11 GHz attenuatior
c Difference between original and smoothed data (11 GHz scintillation)
(Copyright © 1988 IEEE, reproduced with permission)

and 1·6/1·5 GHz being used between the satellite and the mobile stations. Frequencies in the 1·6/1·5 GHz range are susceptible to ionospheric scintillations and some experiments at these frequencies have been conducted to establish an ionospheric scintillation data base (Chapter 2). By staying away from regions within 30° of the geomagnetic equator, the influence of the ionosphere on satellite-to-ground radiowave signals is greatly reduced. Scintillations observed in maritime mobile experiments at latitudes greater than 30° can therefore be ascribed to tropospheric phenomena provided the radiowave path through the atmosphere does not intersect the ionosphere at low latitudes.

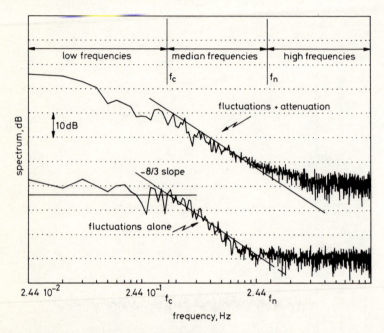

Fig. 3.22 *Examples of tropospheric scintillation spectra with and without the effect of rain attenuation (from Fig. 5 of Reference 44)*
(Copyright © 1985 IEE, reproduced with permission)

The ubiquitous satellite ATS–6, together with its predecessor ATS–5, provided the first data on maritime propagation effects at frequencies above 1 GHz. Longer-term measurements had to await the first satellite devoted exclusively to maritime communications: the Marisat spacecraft. Continuous data acquired near Japan [51] and the USA [52] over a period of many days established the first working model for maritime mobile communications. Fig. 3.23 [51] and Fig. 3.24 [52] show the effect of elevation angle on the amplitude of the fluctuations.

Note that the 50% curve in Fig. 3.23 corresponds to the 50% (or 0 dB) level in Fig. 3.24. The problem in defining the mean level, about which the scintillations

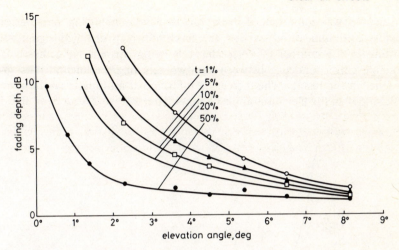

Fig. 3.23 *Estimated values of 1·5 GHz scintillation fading depth as a function of elevation angle (from Fig. 1 of Reference 51)*
(Copyright © 1986, ITU, reproduced with permission)

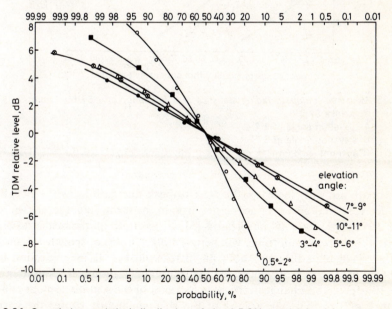

Fig. 3.24 *Cumulative statistical distribution of the 1·5 GHz carrier level in a tropospheric scintillation experiment (from Fig. 8 of Reference 52)*
The 50% probability level is referenced as the mean signal level of 0 dB
(Copyright © 1982 IEEE, reproduced with permission)

are measured, was compounded by the fact that the signal being measured was an active communications carrier, the level of which changed significantly according to the number of voice channels being carried. Nevertheless, the reasonable correspondence between the two sets of experimental data gave confidence in the results. The increasing influence of multipath can be seen in Fig. 3.25 [51] as the elevation angle changes from 5.7° to 2.4°.

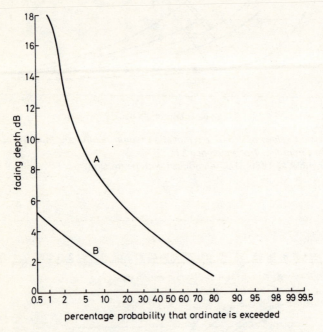

Fig. 3.25 *Measured multipath fading statistics at a frequency of 1·5 GHz (from Fig 6 of Reference 51)*
A: elevation angle of 2·4°
B: elevation angle of 5·7°
(Copyright © 1986 ITU, reproduced with permission)

Multipath not only increases the peak-to-peak fluctuations as the elevation angle decreases but also produces a net drop in the mean signal level over and above the simple gaseous absorption [51]. The effect of the sea is also an important factor. A smooth sea will permit a high degree of specular reflection while a rough sea will cause both an increase in the diffuse reflection and movement of the vessel on which the mobile terminal is mounted. The latter can cause an appreciable change in the power spectra of the scintillations, as can be seen in Fig. 3.26 [53]. The rolling of the ship apparently causes the corner frequency of the spectra to move up in frequency. The full effects of the rough sea in particular, and multipath measurements and modelling in general, are discussed in Chapter 6.

3.5.6 Tropospheric scintillation characteristics

Even though the world-wide data base from long-term measurements (i.e. those measurements that contain data from experiments that provided more than one

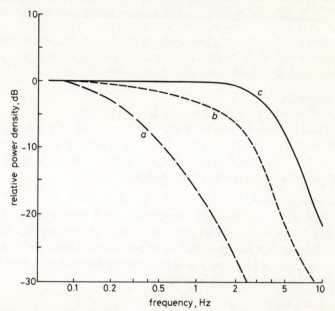

Fig. 3.26 *Frequency power spectra of multipath fading (Fig. 9 of Reference 53)*
   a Slow speed fading observed in the experiments
   b Rapid speed fading observed in the experiments
   c Extremely rapid fading estimated theoretically

Conditions	E_l	H	V_s	Rolling
a	5°	0·5 m	11 knots	1°
b	10°	3 m	11 knots	5°
c	10°	5 m	20 knots	30°

E_l = elevation angle,
H = significant wave height,
V_s = ship's speed,
Rolling is the angle the ship rolls from the vertical
(Copyright © 1986 IECE, reproduced with permission)

year of continuous observations) is inadequate at present, the general characteristics of tropospheric scintillation have been identified and can be summarised as follows:

Meteorological dependence

● Strong correlation with temperature and humidity. High temperature and high humidity give the greatest scintillation amplitudes for a given path.

- Evidence of correlation with wind; as wind velocity increases so does scintillation amplitude. Only weak correlation with wind direction as a large component of the wind movement is vertical owing to strong convective effects during peak scintillation events.
- Presence of rain does not significantly effect the amplitude of the scintillations until the path attenuations exceed about 5 dB. The low-frequency end of the power spectra is modfied by rain but the corner frequency and the frequency roll-off remain essentially unchanged.

Temporal dependence
- Strong correlation with the seasonal cycle and weak correlation with the diurnal cycle coinciding with the hottest and most humid parts of the period in question. Peaks of activity are in the early afternoon and in mid-summer for temperate latitudes; tropical and sub-tropical climates show peaks of activity corresponding to their wet seasons.

Geographic dependence
- Major correlation is with high temperatures and humidities, so there is a concomitant dependence on latitude; the higher the latitude, the colder is the average temperature of the atmosphere and so the lower the amplitude of scintillations over a given path will be. For a given latitude, there does not appear to be a longitude dependence.

Frequency dependence
- When the same antenna is used to measure the tropospheric scintillations at two, or more, frequencies that have been derived from a common source (i.e. the signals are coherently derived), high correlation exists over very large bandwidths [11, 49, 55, 56]. An even better correlation exists between the standard deviation of the scintillations measured at the two frequencies along the same path at the same time [49]. The frequency dependence for the amplitude of the scintillations is approximately the ratio of the frequencies raised to the power 7/12 [28].
- A lack of frequency dependence generally signifies that multipath effects are dominant on that particular path.
- The rate of change of amplitude is slower than that for ionospheric scintillations, the corner frequency is lower, and the roll-off is less steep. The power spectra rolls off as $f^{-8/3}$, generally independent of the elevation angle and frequency. The corner frequency and rate of change of amplitude vary with elevation angle.

Systematic dependence
- As the elevation angle goes down for a given location:

Scintillation amplitude increases on the average
Period of the scintillations increases
Corner frequency decreases
There is an increasing tendency for non-symmetrical scintillation distribution (i.e. an increasing impact of multipath).

● As the antenna diameter decreases for a given path

Scintillation amplitude increases
Probability of multipath increases

● Separation of diversity antennas horizontally by about 500 m effectively decorrelates the scintillation effects along the two paths; at very low elevation angles (below 3°), separating the two antennas vertically produces a greater decorrelation than separating them the same distance horizontally.

3.6 Theory and predictive modelling of clear air effects

All transmission systems designed to operate through a medium require an accurate assessment of the losses inherent in the medium. Transmission line systems have the advantage of a stable, well known loss mechanism that is usually invariable with time. Free space systems through the atmosphere, on the other hand, have to cope with the variabilities of the medium in both time and space that can cause wide changes in the received signal level. Part of the problem in calculating a link budget is deciding on the clear-sky level. At present, no universally agreed definition exists, but the tendency for most systems designers is to use the 50% level; i.e. the level of the received signal that pertains for 50% of the measured time (at least one year). As has been noted in previous Sections, the level of the signal can change appreciably under what are essentially clear-sky conditions. It is therefore necessary to be able to predict the magnitude of the individual effects that cause a variation in the clear-sky level and to assess if the effects are additive or mutually exclusive.

The four major classes of clear-air effects — refractive, reflective, absorptive, and turbulent (scintillation) — all depend on path length. The higher the elevation angle, the smaller the impairment in each case. Above an elevation angle of about 5°, the various effects are generally separable, with the exception of gain degradation, and the individual effects can be added on a root-sum-squared basis. Below 5°, the effects become inter-related and it is difficult to separate out the effects of the various phenomena; e.g. multipath and turbulent scintillation effects at, say, 1° elevation angle. It is also difficult to decide on how to sum the effects at such low elevation angles and usually empirical models have to be evolved based on whatever measured data exist. At elevation angles below 5°, however, the variability of the clear-sky impairment is so large that there is very little utility in attempting to calculate a mean value for a practical communications link. For this reason, most commercial systems impose a minimum elevation angle of 5°, and in some cases 10°, below which performance levels are not guaranteed.

The calculation of each of the clear-air effects has been given in preceding Sections with the exception of multipath and scintillation. Apart from resonance absorption, all of the effects are typically much smaller than multipath or scintillation. At an elevation angle of 5°, for instance, Table 3.2 gives an RMS ray bending error of 1·9 millideg, which is undetectable in terms of power level changes in most microwave communications system. Fig. 3.5 shows the defocusing loss to be about 0·1 dB for the same elevation angle and Fig. 3.6 gives the RMS angle of arrival error as less than 5 millideg, in agreement with recent results [57]. For an 11 GHz, 10 m diameter antenna, the 1 dB beamwidth is 94 millideg. A 5 millideg error will again be immeasurable in terms of a change in the received power level. The total ray bending, defocusing, and angle-of-arrival losses for the above example are therefore of the order of 0·1 dB. Gain degradation can apparently amount to several decibels depending on the frequency, and complex equations have been derived [15] to calculate it. There is some doubt, however, that gain degradation can be uniquely separated from scintillation effects at low elevation angles. For this reason, some models of clear-air effects assume that the effects of gain degradation are embedded in the calculation of scintillation effects, particularly if empirical models are used.

The severe effects of the resonance absorption lines are usually avoided by designing the system to operate well away from these lines. Some systems, such as a 60 GHz inter-satellite link or a 60 GHz battlefield communications manpack, rely on the high losses to avoid interference or, in the latter case, eavesdropping. Calculation of the losses close to, or within, the resonance absorption lines can be obtained by reference to the work of Liebe, summarised in Reference 24.

Multipath is a complex phenomenon that is only of major interest to mobile communications systems on Earth–space paths since these have to operate with wide-beam antennas at low elevation angles. Mobile systems have other unique problems and these, together with the effects of multipath, are dealt with in Chapter 6. The prediction of the remaining clear-air effect, that of scintillation, is addressed below.

3.6.1 Summary of current theories on tropospheric scintillation

The presentation of tropospheric scintillation data has not yet been standardised. Data have been presented in many ways, such as peak-to-peak amplitude distributions, e.g. Reference 54, RMS fluctuations [14], and as frequency of occurrence [45]. The effect that antenna diameter has on the results and the inadequacies of the data base obtained from long-term measurements inhibited the development of a general predictive model. Early theoretical work in CCIR Report 881 [28], based on the work of Crane [14], used RMS fluctuation level as the parameter modelled.

The RMS fluctuation, or standard deviation is perhaps the most relevant parameter for system design as it permits the designer to estimate the mean loss directly. Increasingly, experimental results are being reported in terms of the RMS fluctuations.

The early modelling of tropospheric scintillation was based on an empirical relationship developed between the parameters of the equipment and path used to obtain the reference measured data and those of the desired path and equipment [14]. The amplitude scaling with frequency was observed to follow an $f^{7/12}$ trend, the elevation angle a (sec θ) characteristic, and the antenna aperture averaging G a $(G)^{1/2}$ dependence. The RMS fluctuation level at a frequency f and an elevation angle θ was therefore related to the level at a frequency f_0 and an elevation angle θ_0, by [28]

$$\sigma(f, \theta) = [f/f_0]^{7/12}[\sin \theta_0/\sin \theta]^{11/12}[G(R)/G(R_0)]^{1/2}\sigma(f_0, \theta_0)$$

(3.47)

$G(R)$ is an approximate aperture averaging factor dependent on antenna diameter and efficiency, wavelength, and the distance to the turbulent portion of the atmosphere responsible for the scintillations. $G(R)$ is given by

$$G(R) = \begin{cases} 1\cdot0 - 1\cdot4(R/\sqrt{\lambda L}), & 0 \leqslant (R/\sqrt{\lambda L}) \leqslant 0\cdot5 \\ 0\cdot5 - 0\cdot4(R/\sqrt{\lambda L}), & 0\cdot5 < (R/\sqrt{\lambda L}) \leqslant 1\cdot0 \\ 0\cdot1, & 1\cdot0 < (R/\sqrt{\lambda L}) \end{cases}$$

(3.48)

where R is the effective radius of the antenna, expressed in terms of its physical diameter D (in metres) and the antenna efficiency η by

$$R = \eta^{1/2}(D/2)$$

(3.49)

and L is the slant path length in metres to the turbulent layer at height h (in metres), given by

$$L = \{h^2 + 2R_e h + [R_e \sin \theta]^2\}^{1/2} - R_e \sin \theta$$

(3.50) .

where R_e is the effective radius of the Earth in metres, accounting for refraction ($R_e \approx 8\,500\,000\,\text{m}$), and λ is the wavelength in metres. The height of the turbulent layer has been found to be between 1·5 and 4 km in two experiments [44, 48] and to be seasonally dependent [59], but its precise value does not alter the value of L to a significant degree, and a value of $h = 1\,\text{km}$ is usually used in eqn. 3.50. Fig. 3.27 indicates the strong dependence of $G(R)$ on R, and hence on the antenna diameter.

It is clear that tropospheric scintillation amplitudes are larger for smaller antenna diameters. The initial empirical model above was based on data obtained in Kansas, USA, and did not contain a latitude and/or a humidty variable. The early model, however, was found to provide generally good results up to frequencies of 20 GHz.

A later model, called the Moulsley–Vilar model by its inventors [50], was based on the premise that the scintillations (either their peak-to-peak amplitudes or the envelope of the scintillation peaks) can be described as a gaussian process with a time-variable standard deviation. In essence, as the percentage time for the prediction is reduced, the standard deviation increases. Although the model appears to perform quite well [50], it requires two complicated parameters to be determined accurately as an input: the mean of the scintillation intensity and the standard deviation of the log variance of the scintillation intensity. For the measurements described in Reference 50, these values were 0·09 dB and between 1 and 1·8 dB, respectively, for a 30° elevation angle path. Obtaining such values to the accuracy required is not usually possible and the model appeared to lack engineering utility.

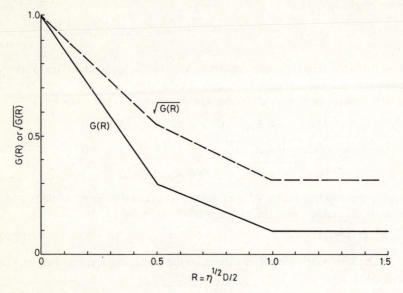

Fig. 3.27 *Antenna aperture factor as a function of effective antenna radius (from Fig. 1 of Reference 58)*
(Copyright © 1987 IEE, reproduced with permission)

There are basically two approaches to obtaining an 'engineering' model of tropospheric scintillation amplitudes. The first is aimed at describing the worst-case envelope of the scintillation peaks, the so-called three-sigma value, and the second at developing a long-term mean value of the standard deviation of the scintillation amplitude at various average annual percentage times. Both approaches are described below.

3.6.2 Two approaches to an engineering model for the prediction of tropospheric scintillation amplitude

The tropospheric scintillation amplitude is a function of the size of the receiving antenna. Large earth station antennas, such as those used initially in 6/4 GHz communications satellite operations, were typically 30 m in diameter. With these diameters, and the diameters of the large radio telescopes, angle-of-arrival (phase) scintillations become important because of the very small beamwidths involved, while amplitude scintillations tend to be suppressed by an aperture averaging effect [59] due to the phase incoherence across the antenna aperture [11]. However, as the frequency goes up and/or the antenna diameter is reduced, so the magnitude of the tropospheric amplitude scintillations increases. At the same time as the antenna diameter is reduced, so usually is the operating margin for the system being designed. For example, in the 14/11 GHz bands, typical margins for large antennas carrying trunk traffic are of the order of 7 to 10 dB; smaller antennas carrying small streams of traffic have typical margins of 1·5 to 2·5 dB. Clearly, 3 dB peak-to-peak scintillations should not adversely affect the trunk traffic, protected as it is with at least a 7 dB margin, while the small earth station might conceivably suffer short outages during the 3 dB peak-to-peak scintillation event. Modelling of tropospheric scintillation is therefore considered to be more important for small earth stations (typically less than 5 m in diameter). Both of the modelling approaches have been adopted experimentally at INTELSAT and are described in detail below.

(a) Three-sigma model
The tentative modelling approach [58] was to collect as much statistically valid data as possible and to scale the results to a particular frequency, elevation angle and antenna diameter in order to derive a reference curve. These scaled data could then be used to establish climatic and latitude dependence trends. Unfortunately, the statistically valid data available [58] did not cover enough latitudes to establish a latitudinal trend. The reference curve was therefore set at the upper bound of the essentially mid- to high-latitude data in anticipation that later data, from more humid, lower latitudes, would exhibit larger scintillation fluctuations. The worst-case bound used in the above model would therefore be closer to the anticipated median curve for more humid climates. The derived reference curve is shown in Fig. 3.28.

Using this curve and eqns. 3.47–3.50, inclusive, the reference distribution can be scaled to the required elevation angle, frequency and antenna diameter. Amplitude scintillations observed with small earth stations tend to have a component of antenna grain degradation included. The two effects are indistinguishable at the receiver and so the above engineering model is really used to predict the net effect of antenna gain degradation and tropospheric scintillations for typical Earth-space transmission systems under essentially worst-case situations.

(b) Long-term average model
This model is based on the orignal work of KDD in Japan [49,76] as amended [75]. As with the attenuation prediction method in CCIR Vol. V, a step-by-step procedure has been laid down [78], based on References 75 and 76. This is reproduced in its entirety below from Reference 78 with the exception that the numbers of the equations (eqns 3.51–3.59) have been changed so that they stay in sequence.

"Input parameters to the prediction procedure are the averaged ambient temperature t (°C), relative humidity H ($0 \leqslant H \leqslant 100\%$), antenna diameter D_{geom}, antenna efficiency η, frequency f, and elevation angle θ.

Step 1: Obtain average values of ambient temperature t (°C) and relative humidity, H, for the period of interest. (This period should be not less than about one month.)

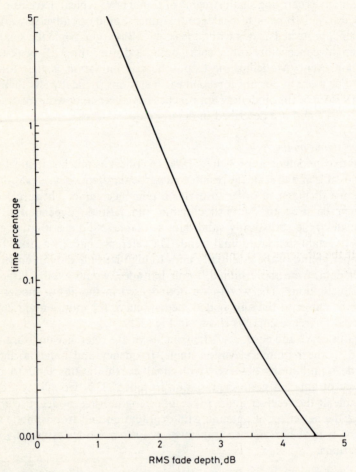

Fig. 3.28 *Reference scintillation fade distribution (from Fig. 3 of Reference 58)*
Frequency = 11 GHz
Elevation angle = 5°
Antenna diameter = 3 m
(Copyright © 1987 IEE, reproduced with permission)

Step 2: calculate the corresponding saturation water vapour pressure

$$e_s = \frac{5854 \times 10^{(20 - 2950/(273 + t))}}{(273 + t)^5} \text{ millibars} \tag{3.51}$$

Step 3: Estimate the 'wet-term' N_{wet} of the radio refractivity from e_s and H:

$$N_{wet} = \frac{3730 H e_s}{(273 + t)^2} \tag{3.52}$$

Step 4: Calculate the standard deviation of the signal amplitude σ_{ref} serving as reference (see Report 718) by:

$$\sigma_{ref} = (3\cdot6 \times 10^{-3}) + (1 \times 10^{-4}) \times N_{wet} \text{ decibels} \tag{3.53}$$

Step 5: Calculate the effective path length L according to

$$L = \frac{2000}{\sqrt{(\sin^2 \theta + 2\cdot35 \times 10^{-4})} + \sin \theta} \text{ m} \tag{3.54}$$

the elevation angle θ should be greater than about 4°.

Step 6: Estimate the effective antenna diameter D_{eff} from the geometrical diameter D_{geom}, and the antenna's efficiency η:

$$D_{eff} = \sqrt{\eta} \, D_{geom} \text{ m} \tag{3.55}$$

(If the efficiency η is unknown, $\eta = 0\cdot5$ is a conservative estimate.)

Step 7: Calculate the antenna averaging factor from:

$$g(x) = \sqrt{3\cdot86(x^2 + 1)^{11/12} . \sin\left[\left(\frac{11}{6}\right) \arctan\left(\frac{1}{x}\right)\right] - 7\cdot08 x^{5/6}}$$

with

$$x = 1\cdot22 D_{eff}^2 \times f/L \tag{3.56}$$

where f is the carrier frequency in gigahertz

Step 8: Calculate the standard deviation of the signal for the considered period and propagation path:

$$\sigma = \sigma_{ref} f^{7/12} [g(x)/(\sin \theta)^{1\cdot2}] \tag{3.57}$$

Step 9: Calculate the time percentage factor a for the time percentage p of concern from:

$$a = -0\cdot061 (\log_{10} p)^3 + 0\cdot072 (\log_{10} p)^2$$
$$- 1\cdot71 (\log_{10} p) + 3\cdot0$$
$$\text{for } 0\cdot01\% \leqslant p \leqslant 50\% \tag{3.58}$$

Step 10: Calculate the scintillation fade depth for the time percentage p by:

$$x_{(p)} = a\sigma \text{ decibels"} \tag{3.59}$$

3.7 System impact

3.7.1 Phase effects

The prediction of signal variations in essentially clear air is a complex mix of impairments [63]. Recognising the cause of the observed signal variations is equally difficult. Generally speaking, phase scintillations are not significant compared to amplitude scintillations except at very low elevation angles or for very large antennas [79]. For the large diameter radio telescopes, the effect is to broaden the apparent diameter of the source or, conversely, to limit the angular resolution of the observing system. In stable, clear air, calculations can be made to correct for such limitations [61]. Single frequency ranging systems will also be limited by phase effects but dual-frequency measurements, even at optical frequencies [62], effectively compensate for these variations. Interestingly, scintillations at optical frequencies can be used to predict the average rainfall rate along the transmission path [64, 65].

Many communications systems utilise large instantaneous bandwidths in order to transmit at high data rates. A communications satellite transponder bandwidth of 72 MHz is common and instantaneous bandwidths in excess of 100 MHz will probably be required for some high capacity systems. Differential effects over the bandwidth being used can cause errors in the received signal. Phase and amplitude effects are variable with frequency, the former essentially spreading the signal in time and the latter affecting the received signal amplitude. By applying pre-emphasis across the bandwidth at the transmitting end [30], however, the received signal will appear to have reasonably uniform characteristics in phase and amplitude across the whole bandwidth. Pre-emphasis assumes that the amplitude and phase effects are monotonic with frequency; i.e. they either increase or decrease with frequency. A departure from this assumption will cause dispersion effects to occur. An analysis of such effects in the 10–30 GHz frequency range [66] showed that dispersion effects could be ignored in clear-sky conditions. It is very likely that dispersion effects can be ignored in most communications systems utilising instantaneous bandwidths of less than 1% of the carrier frequency.

3.7.2 Amplitude effects

The system impact of clear-air amplitude effects can be broadly classified into two categories: those that vary very slowly (of the order of hours or longer) and those that vary relatively rapidly. The former can be called 'bulk' effects and the latter 'short-term' effects.

Bulk effects: The principal impact of these effects is to cause a variation in the clear-sky level. Changes in relative humidity over a few hours can cause variations on the order of 1 dB on some paths at frequenceis above 10 GHz. In itself, the magnitude of a change in the clear-sky level is not serious but if a system of uplink power control (see Chapter 7) is being used that calculates the additional power to be added to the uplink transmissions by measuring changes in the received signal power on the downlink, the impact can be quite serious. For example, a typical ratio between rain attenuation at 14 GHz to rain attenuation at 11 GHz is 1·45. If a rain event occurs causing 3 dB of attenuation on the downlink, the calculated uplink fade (at 14 GHz) will be 4·35 dB. Increasing the uplink power by 4·35 dB should, in theory, compensate for the rain attenuation on the uplink. However, if humidity variations in the atmosphere (common in periods of rain) have caused the mean level of the received signal to drop by 0·5 dB, instead of measuring a 3 dB fade, the equipment will record an apparent 3·5 dB fade. The computer will therefore command an increase in the uplink power of 5·075 dB, an error of almost 1 dB. The decrease in the mean clear-sky level that occurs on the uplink due to the same change in the humidity will not compensate for the erroneously high increase in uplink power. To overcome these errors, an absolute reference could be established from which the rain fades are measured, but this would require constant re-calibration. Chapter 7 discusses this situation at more length.

Short-term effects: Tropospheric scintillation is essentially a non-absorptive effect; i.e. the mean level of the signal does not change significantly when compared to the peak-to-peak amplitude excursions. This is unlike multipath effects where there is a significant net lowering of the apparent mean level. This is evident from the skewed distributions of the low elevation angle scintillation measurements, which show an increasingly large 'tail' in the negatively-going direction as the elevation angle goes down and/or the receiving antenna beam-width increases. Tropospheric scintillation is therefore an 'on-axis' effect. This is important for some antenna systems that exhibit poor cross-polarisation performance off the main beam axis. Even severe tropospheric scintillations will therefore cause no appreciable change in the measured XPD [40, 66]. How-ever, since XPD statistics at frequencies above 6 GHz are calculated from the measured attenuation statistics, errors in the calculation of XPD from attenu-ation values of less than 5 dB can occur owing to the combining of 'scintillation attenuation' into the overall rain attenuation statistics [66].

Tropospheric scintillation can adversely affect uplink power control systems if the response time of the system is too slow. The rate of change of amplitude caused by tropospheric scintillations actually goes down as elevation angle goes down, exactly the opposite of multipath effects. Statistics reported by Strickland [38] showed fade rates in excess of 10 dB/s for about 0·03% of the time. This is almost certainly a multipath effect. Tropospheric scintillations usually do not exceed effective fade rates above 1 dB/s except in very strong, rare events [47]. This level of rate of change is within the response time of most control loops associated with receiving systems. The magnitude of the tropospheric scintil-lations, however, can also cause a problem in the satellite receiver.

Most satellite transponders are carefully designed to operate at a precise received power flux density. This is particularly important if more than one carrier is accessing the same transponder. Multi-carrier operation is the rule rather than the exception in current communications satellite systems and it requires that all of the carriers remain within about 1 dB of their prescribed power settings. If this is not done, the carriers will tend to interfere with each other through intermodulation of their carrier frequencies at above the designed levels. Carrier signals subject to tropospheric scintillations that exceed peak-to-peak amplitudes of 2 dB for appreciable portions of the time are therefore of considerable concern to systems designers. Without some sort of signal restoration technique (e.g. uplink power control), the only recourse is to reduce the number of carriers in a given transponder. By permitting a larger frequency gap between carriers, the intermodulation products can be made to fall outside the signal bandwidths. A capacity reduction on the order of 20% can result with these greater separations of carrier, and this is a significant penalty to pay in terms of lost revenue.

The statistical impact of the amplitude of tropospheric scintillations can be assessed by calculating the RMS level of the fluctuations. The three-month experiment in the UK at a frequency of 11·198 GHz and an elevation angle of 8·9° [46] recorded on RMS value of 0·44dB. The measurement period was between July and September 1983. In Japan [49, 66], the RMS value for August 1983 on a path at an elevation angle of 6·5° and a frequency of 11·452 GHz was about 0·85 dB. The corresponding figure for February in the Japan experiment was about 0·3 dB. Since tropospheric scintillations are generally additive to rain attenuation at these levels, an additional margin must be allocated to allow for the net effective signal reduction due to tropospheric scintillation. The variability of tropospheric scintillation, both diurnally and seasonally, is also of importance since many sytems are designed to meet 'worst month' criteria (see Chapter 4). The limited data available seem to indicate that the ratio between worst month and the average annual statistics for tropospheric scintillations is about the same as that for rain attenuation [54]. Tropospheric scintillations, however, may have a different impact depending on the precise definition of what constitutes an attenuating event.

International communications systems are governed by the interfaces and recommendations set down, by mutual agreement, in the CCIR and CCITT volumes. Of particular importance are the hypothetical reference circuits and, for the fixed-satellite service using digital transmission techniques, these are embedded in Report 997 of CCIR Volume IV. Criteria are laid down in this report that deal with the permitted error rates for given percentage times. Tropospheric scintillations, because they exhibit a fairly rapid variation with time, are affected by the 'short-term' objective of Recommendation G.821 [67] in which a dividing line of 10 s is used to differentiate between different types of fades. The definition is not simple, however, and reference should be made to Fig. 3.29 in which three separate fades are depicted, each lasting 5 s below the given threshold.

Historically, attenuation measurements have simply summed the total time below a given threshold; in the case of Fig. 3.29 this is 15 s. The 'strict 10 second' rule brought in for digital systems said that fades of 9 s or less would not be counted as unavailable time (i.e. a propagation-induced outage). The revised version of G.821 stipulates that a fade of 9 s or less must recover and remain above the threshold for at least 10 s; i.e. there must have been an effective 'change of state' of the signal about the threshold. Fade A meets this criterion but Fade B does not. In the case of fade B, the length of the fade is the total time below the threshold until the signal has recovered above the threshold for at least 10 s. Fades B and C, plus the time in between them, would be counted as unavailable time. The two versions of calculating the fading time come up with the following available times: zero ('strict 10 s') and 17 s (revised G.821).

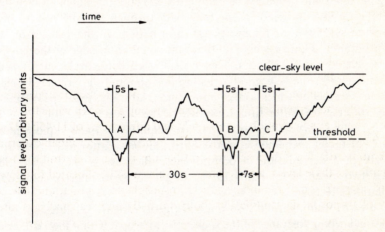

Fig. 3.29 *Schematic illustrating the difference between 'available' and 'unavailable' time in fading statistics calculated using Recommendation G.821* [67]

A, B and C are all 5 s fades. After fade A, the signal recovers (goes above the threshold) for more than 10 s. Fade A is therefore 'available' time and does not count in the cumulative fading statistics. The signal does not recover for 10 s after fade B and so fades B and C, plus the time in between (a total of 17 s) are called unavailable time

The first calculations using actual measured data [68] showed that the application of Recommendation G.821 reduced the fading time that had previously been allotted to available time under the strict 10 s rule. Later work [77] appeared to indicate that there would be little overall difference in the statistics of path attenuation between any of the methods of calculation. The full implications of G.821 to tropospheric scintillations have yet to be assessed.

3.7.3 System effects

These are caused mainly by the imperfections of the earth station equipment. Earth stations that employ large antennas utilise a monopulse tracking system [30] that samples the received singal from four 90° quadrants, taking the sum of two signals and the difference of the other two. Amplitude and phase variations between the sum and difference channels provide very sensitive tracking information. However, in the presence of severe tropospheric scintillation, the tracking accuracy can be severely degraded. In some cases, the tracking mechanism is switched off at a predetermined threshold to prevent complete loss of signal due to antenna mis-pointing.

Step-tracking systems, utilised by medium-sized earth stations, are even more prone to errors in the presence of scintillation [69]. Some of the problems can be overcome if the time constant of the tracking system can be made long enough. If an averaging period of about one minute is used, most of the major amplitude peaks are removed. For low elevation angles (around 5°), a longer averaging time will probably be required owing to the much slower 'bulk' scintillation effects incurred by the long path through the lower atmosphere.

Tracking systems that require phase information will have to resort to a quasi-program tracking system [70] that relies on a knowledge of the satellite's location over the last 24 hours. This system essentially checks each tracking command against the last 'good' tracking position and the predicted position based on the satellite's location 24 hours ago. Naturally, this type of tracking is only suitable for those situationss where the object to be tracked is in a precisely defined orbit. Antennas that are required to track rapidly moving satellites or targets, whether from the Earth, mobile platforms or from space, must accept a degradation in pointing acccuracy unless they can establish at least two independent paths to the target so as to remove some of the scintillation effects.

Small earth stations usually do not track geostationary satellites but rely on the station keeping tolerance of the spacecraft to keep it within the 3 dB, or even 1 dB, beamwidth of the earth station antenna. Most communications satellites are maintained to within $\pm 0.1°$ in both longitude and latitude, but even this small variation can lead to changes in the clear-sky level of several tenths of a decibel. The anticipated variation due to diurnal motion of the satellite must be factored into the link budget of small earth stations. In many cases, the changes in equipment parameters due to heating, cooling, tracking and ageing can amount to more than the clear-sky signal variations induced by the atmosphere [63] and some care must be taken in separating equipment effects from atmospheric effects when introducing signal restoration techniques (see Chapter 7).

3.8 References

1   'Electronics engineer's reference Book' MAZDA, F. F. (Ed.): (Butterworths, 1983, 5th Edn.) Chap. 11

2   JOHNSON, M. A.: 'A review of tropospheric scatter propagation theory and its application', *Proc. IEE*, 1958, Paper 2534R, pp. 165–176.

3   CRANE, R. K.: 'A review of transhorizon propagation phenomena', Review paper for the URSI Commision F Open Symposium, Lennoxville, Quebec, May 1980

4   PEKERIS, C. L.: 'Wave theoretical interpretation of propagation of 10 cm and 3 cm waves in low-level ocean ducts', *Proc. Inst. Rad. Engrs.*, 1947, **35**, pp. 453 *et seq.*

5   Recommendations and Reports of the CCIR, XVIth. Plenary Assembly, Dubrovnik, 1986, Volume V (Propagation in non-ionized media); Report 563–2: 'Radiometerological data'.

6   HALL, M. P. M., and COMER, C. M.: 'Statistics of tropospheric radio-refractive index soundings taken over a 3-year period in the UK', *Proc. IEE.*, 1969, **116**, pp. 685–690

7   FLOCK, W. L.: 'Propagation effects on satellite systems at frequencies below 10 GHz', NASA Reference Publication 1108, Dec. 1983

8   BEAN, B. R., and DUTTON, E. J.: 'Radio meterology' (Dover, NY, USA, 1966)

9   HALL, M. P. M.: 'Effects of the troposphere on radio communications' (Peter Perigrinus Ltd., 1979)

10   CRANE, R. K.: 'Refraction effects in the neutral atmosphere' in MEEKS, M. L. (Ed.): 'Methods of experimental physics: Vol. 12. Astrophysics part B, Radio telescopes' (Academic Press, NY, USA, 1976)

11   YOKOI. H., YAMADA, M., and SATOH, T.: 'Atmospheric attenuation and scintillation of microwaves from outer space', *Astron. Soc. (Japan)*, 1970, **22**, pp. 511–524

12   CRANE R. K.: 'Propagation phenomena affecting satellite communications systems operating in the centimeter and millimeter bands', *Proc. IEEE*, 1971, **59**, pp. 173–188

13   Recommendations and Reports of the CCIR, XVth. Plenary Assembly, Geneva, 1982, Volume V (Propagation in non-ionized media); Report 718–1: 'Effects of large-scale tropospheric refraction on radio wave propagation'.

14   CRANE, R. K.: 'Low elevation angle measurement limitations imposed by the troposphere: An analysis of scintillation observations made at Haystack and Millstone'. MIT Lincoln Lab. Rep. 518, 1976, Lexington Mass., USA

15   THEOBALD, D. M., and HODGE, D. B.: 'Gain degradation and amplitude scintillation due to tropospheric turbulence'. The Ohio State University Electro Science Lab., 1978, Tech. Rep. No. 784229-6, Rev. 1, Ohio State Univ., Dept. of Elec. Engg., Columbus, Ohio 43212, USA

16   McCORMICK, K. S., and MAYNARD, L. A.: 'Measurement of SHF tropospheric fading along Earth-space paths at low elevation angles', *Electron. Lett.*, 1972, **8**, pp. 274–276

17   IPPOLITO, L. J., KAUL, R. D., and WALLACE, R. G.: 'Propagation effects handbook for satellite system design: A summary of propagation impairments on 10 to 100 GHz satellite links with techniques for system design'. NASA Reference Publication 1082(03), June 1983

18   MATHEWS, P. A.: 'Radio wave propagation, VHF and above' (Chapman and Hall, 1965)

19   BECKMANN, P., and SPIZZICHINO, A.: 'The scattering of electromagnetic waves from rough surfaces' (McMillan Co., NY, USA, 1963); republished in 1987 by Artech House, Massachusetts, USA.

20   DEBYE, P.: 'Polar molecules' (Chemical Catalogue Company, NY, USA, 1929)

21   COLE, S. C., and COLE, R. H.: 'Dielectric relaxation in glycerol, propylene glycol, and *n*-propanol', *Chem. Phys.*, 1941, **9**, pp. 341–351

22   ZHEVAKIN, S. A., and NAUMOV, A. P.: 'Absorption of centimeter and millimeter radio waves by atmospheric water vapour', *Radio Eng. & Electron. Phys.*, 1964, **9**, pp. 1097–1105

23   Conclusions of the Interim meeting of Study Group 5 (Propagation in non-ionized media), Geneva, 11–26 April 1988, Documents 5/204; Report 719-2 (MOD I): 'Attenuation by atmospheric gases'.

24   LIEBE, H. J.: 'An updated model for millimeter wave propagation in moist air', *Radio Science*, 1985, **20**, pp. 1069–1089

25   ROGERS, D. V.: 'Propagation considerations for satellite broadcasting at frequencies above 10 GHz', *IEEE J.*, 1985, **SAC-3**, pp. 100–110

26   ALTSHULER, E. E.: 'A simple expression for estimating attenuation by fog at millimeter wavelengths', *IEEE Trans.* 1984, **AP-32**, pp. 757–758

27   ELRIDGE, R. G.: Haze and fog aerosol distributions', *J. Atmos. Sci.*, 1966, **23**, pp. 605–613

28   Recommendations and Reports of the CCIR, XVIth. Plenary Assembly, Dubrovnik, 1986, Volume V (Propagation in non-ionized media); Report 881: 'Effects of small-scale spatial or temporal variations of refraction on radiowave propagation'.

29   HINDER, R. A.: 'Observations of atmospheric turbulence with a radio telescope', *Nature*, 1970, **225**, pp. 614–617

30   MIYA, K. (Ed.): 'Satellite communications technology' (KDD Engg. and Consulting, Inc., Japan, 1985 English edition) 2nd. Edn.

31   McCORMICK, K. S., and MAYNARD, L. A.: 'Low angle tropospheric fading in relation to satellite communications and broadcasting'. International Conf. on Comm., ICC-71-CIC, 1971, pp. 12–18 to 12–23

32   LAM, W. I.: 'Low angle signal fading at 38 GHz in the high Artic', *IEEE Trans.*, 1988, **AP-35**, pp. 1495–1499

33   BROWNING, D. J., and PRATT, T.: 'Low angle propagation from ATS–6 at 30 GHz', Proc. ATS–6 Meeting, 1977, ESTEC, Noordwijk, The Netherlands, pp. 149–153

34   VOGEL, W. J., STRAITON, A. W., and FANNIN, B. M.: 'ATS–6 ascending: near horizon measurements over water at 30 GHz', *Radio Science*, 1977, **12**, 757–765

35   DEVASIRVATHAM, D. M. J., and HODGE, D. B.: 'Amplitude scintillations on Earth-space paths at 2 and 30 GHz', Tech. Rep. 4299-4, 1977, Electro Science Lab., Ohio State University, Columbus, Ohio USA

36   WEBBER, R. V., and McCORMICK, K. S.: 'Low elevation angle measurements of the ATS-6 beacons at 4 and 30 GHz', *Ann. Telecomm.*, 1980, **35**, pp. 1/7–7/7

37   STUTZMAN, W. L., BOSTIAN, C. W., MANUS, E. A., MARSHALL, R. E., and WILEY, P. H.: 'ATS–6 satellite 20 GHz propagation measurements at low elevation angles', *Electron. Lett.*, 1975, **11**, pp. 635–636

38   STRICKLAND, J. I., OLSEN, R. I., and WESTIUK, H. L.: 'Measurement of low angle fading in the Canadian Arctic', *Ann. Telecomm.*, 1977, **32**, pp. 530–535

39   OSEN, O.: 'Propagation effects in high latitudes', Proc. International Symposium on Symphonie, 1980, Berlin, pp. 415–423

40   GUTTERBERG, O.: 'Measurement of atmospheric effects on satellite links at very low elevation angles', AGARD EPP Symposium on Characteristics of the lower atmosphere influencing radio wave propagation, 1983, Spatind, Norway, pp. 5–1 to 5–19

41   MIMIS, V., and SMALLEY, A.: 'Low elevation angle site diversity satellite communications for the Canadian Arctic'. ICC 82, 1982, Vol. 1 pp. 4A.4.1 to 4A.4.5

42   HADDON, J., LO, P., MOULSLEY, T. J., and VILAR, E.: 'Measurement of microwave scintillations on a satellite down-link at X-band'. IEE Conf. Publ. 195, 1981, pp. 113–117

43   ISHIMARU, A.: 'Temporal frequency spectra of multi-frequency waves in turbulent atmosphere', *IEEE Trans.* 1972, **AP-20**, pp. 10–19

44   VANHOENACKER, D., and VANDER VORST, A.: 'Tropospheric fluctuation spectra and radio systems implications'. IEE Conf. Publ. 248, 1985, pp. 67–71

45   WANG, C. N., CHEN, F. S., LIU, C. H., and FANG, D. J.: 'Tropospheric amplitude scintillations at C-band along satellite up-link'. *Electron. Lett*, 1984, **20**, pp. 90–91

46   LO, P. S. L., BANJO, O. P., and VILAR, E.: 'Observations of amplitude scintillations on a low-elevation angle Earth-space path', *Electron. Lett.*, 1984, **20**, pp. 307–308

47   BANJO, O. P., and VILAR, E.: 'Measurement and modelling of amplitude scintillations on low-elevation angle Earth-space paths and impact on cummunications systems'. *IEEE Trans.*, 1986, **COM-34**, pp. 774–780

48   ORTGIES, G.: 'Amplitude scintillations occurring simultaneously with rain attenuation on satellite links in the 11 GHz band'. IEE Conf. Publ. 248, 1985, pp. 72–76

49   KARASAWA, Y., YASUKAWA, K., and YAMADA, M.: 'Tropospheric scintillation in the 14/11 GHz bands on Earth-space paths with low elevation angles', *IEEE Trans.*, 1988, **AP-36**, pp. 563–569

50   MOULSLEY, T. J., and VILAR, E.: 'Experimental and theoretical statistics of microwave amplitude scintillations on satellite downlinks', *IEEE Trans.*, 1982, **AP-30**, pp. 1099–1106

51   Recommendations and Reports of the CCIR, XVIth Plenary Assembly, Dubrovnik, 1986, Volume VIII (Mobile Services); Report 920: 'Maritime satellite system performance at low elevation angles'.

52   FANG, D. J., and CALVITT, T. O.: 'A low elevation angle propagation measurement of 1.5 GHz satellite signals in the Gulf of Mexico', *IEEE Trans.*, 1982, **AP-30**, pp. 10–15

53   KARASAWA, Y., YASUNAGA, M., NOMOTO, S., and SHIOKAWA, T.: 'On-board experiments on L-band multipath fading and its reduction by use of the polarization shaping method', *Trans. IECE Japan*, 1986, **E69**, pp. 124–131

54   ALLNUTT, J. E.: 'Low elevation angle propagation measurements in the 6/4 GHz and 14/11 GHz bands'. IEE Conf. Publ. 248, 1985, pp. 62–66

55   THOMPSON, M. C., LOCKETT, W. E., JAMES, H. B., and SMITH, D.: 'Phase and amplitude scintillations in the 10 to 40 GHz band', *IEEE Trans.*, 1975, **AP-23**, pp. 792–797

56   COX, D. C., ARNOLD, H. W., and RUSTAKO, A. J.: Attenuation and depolarization by rain and ice along inclined radio paths through the atmosphere at frequencies above 10 GHz'. EASCON 1979, IEEE Publ. 79 CH 1476–1, Arlington, VA, USA, pp. 56–61

57   VILAR, E., and SMITH, H.: 'A theoretical and experimental study of angular scintillations in Earth-space paths', *IEEE Trans.*, 1986, **AP-34**, p. 2–10

58   ROGERS, D. V., and ALLNUTT, J. E.: 'A practical tropospheric scintillation model for low elevation angle satellite systems'. International Conf, on Antennas and Propagation ICAP 87, IEE Conf. Publ. 274, 1987, Pt. 2, pp. 273–276

59   'Project COST 205: Scintillations in Earth-satellite links', *Alta Frequenza*, 1985, **LIV**, pp. 209–211

60   LIEBE, H. J.: 'Modeling attenuation and phase of radio waves in air at frequencies below 1000 GHz', *Radio Science*, 1981, **16**, pp. 1183–1199

61   SATOH, T., and OGAWA, A.: Exact gain measurements of large aperture antennas using celestial radio sources', *IEEE Trans.*, 1982, **AP-30**, pp. 157–161

62   ASHINE, J. B., and GARDNER, C. S.: 'Atmospheric refractivity corrections in satellite laser ranging', *IEEE Trans.*, 1985, **GE-23**, pp. 414–425

63   ALLNUTT, J. E., and ARBESSER-RASTBURG, B.: 'Low elevation angle propagation modelling considerations for the INTELSAT business service'. IEE Conf. Publ. 248, 1985, pp. 57–61

64   WANG, T-I, KUMAR, P. N., and FANG, D. J.: 'Laser rain gauge: near-field effect', *Applied Optics*, 1983, **22**, pp. 4516–4524

65   SCHIESOW, R. L., CUPP, R. E., and CLIFFORD, S. F.: 'Phase difference power spectra in atmospheric propagation through rain at 10·6 microns', *Applied Optics*, 1985, **24**, pp. 4516-4524

66   YAMADA, M., YASUKAWA, K., FURUTA, O., KARASAWA, Y., and BABA, N.: 'A propagation experiment on Earth-space paths at low elevation angles in the 14 and 11 GHz bands using the INTELSAT V Satellite'. Proc. ISAP, 1985, Tokyo, Paper 053-2, pp. 309–312

67   Recommendation G.821: 'Error performance on an international digital connection forming part of an integrated services digital network'. CCITT Yellow Book, 1980 (Revised May 1984), Vol. III, Geneva, pp. 193–195

68   LARSEN, J. R.: 'The influence of rain attenuation on the error performance of satellite circuits', *Teleteknik*, 1985, pp. 1–6

69   RICHHARIA, M.: 'Effects of fades and scintillations on the performance of an earth station step track system', *Space Commun. Broadcasting*, 1985, **3**, pp. 309–319

70   EDWARDS, D. J., and TERRELL, P. M.: 'The smooth step-track antenna controller' *Int. J. Satellite Communi.*, 1983, **1**, pp. 133–140

71   BARRETT, E. C., and MARTIN, D. W.: 'The use of satellite data in rainfall monitoring' (Academic Press, 1981); from chapter 5 by R. S. Fraser in 'The manual of remote sensing, 1st edition, edited by F. J. Janza, The American Society for Photogrammetry and Remote Sensing, 1975.

72   GIBBONS, C. J.: 'Improved algorithms for the determination of specific attenuation at sea level by dry air and water vapour, in the frequency range 1–350 GHz', *Radio Science*, 1986, **21**, pp. 945–954

73   ALTSHULER, E. E.: 'Slant path absorption correction for low elevation angles', *IEEE Trans.*, 1986, **AP-34**, pp. 717–718

74   LAM, W. I., and OLSEN, R. L.: 'Measurement of site diversity performance at EHF', International Symposium on Radio Propagation, 1988, Beijing, China, pp. 560–563

75   Conclusions of the Interim Meeting of Study Group 5 (Propagation in non-ionized media), Geneva, 11–26 April 1988, Document 5/204; Report 718-2 (MOD I): 'Effects of tropospheric refraction on radiowave propagation'.

76   KARASAWA, Y., YAMADA, M., and ALLNUTT, J. E.: 'A new prediction method for tropospheric scintillation in satellite communications', *IEEE Trans.*, 1988, **AP-36**, pp. 1608–1614

77   BTRL Final Report on INTELSAT contract INTEL-608: 'Assessment of the impact short-term fades have on the overall availability of digital satellite circuits'. INTELSAT, 3400 International Drive N.W., Washington, DC 20008-3098, USA

78   Conclusions of the Interim Meeting of Study Group 5 (Propagation in non-ionized media), Geneva, 11–26 April 1988, Document 5/204; Report 564 3 (MOD I). 'Propagation data and prediction methods required for Earth-space telecommunications systems'

79   VILAR, E., HADDON, J., LO, P., and MOUSLEY, T. J.: 'Measurements and modelling of amplitude and phase scintillations in an Earth-space path'. *Journal of the Institution of Electronic and Radio Engineers*, 1985, **55**, pp. 87–96

Chapter 4

Attenuation effects

4.1 Introduction

In Chapter 3, the potential impairments along an Earth–space path during apparently clear-sky conditions were discussed. At frequencies below 100 GHz and outside the absorption lines of oxygen and water, the changes in the received power level induced either by variations in atmospheric humidity or tropospheric scintillation were generally quite small although significant in a number of cases. In most system designs, it is not usually the average, steady-state (clear-sky) performance that sets the operating limit but the variations about that steady-state, or clear-sky, level. To a first approximation, as the frequency increases, so does the magnitude of the potential signal variations about the mean. While the variations can be both positive and negative, it is usually the negative variations that are cause for most concern. These negative variations, or excess attenuations below the mean clear-sky level, must be accounted for in system design.

The excess attenuation of a radiowave is made up of two components, absorption and scattering. Absorption takes place when the incident radiowave energy is transformed into mechanical energy, thereby heating up the absorbing material. If the material is raised to a temperature above its surroundings, it will isotropically re-radiate the energy absorbed according to Kirchhoff's laws.

A radiowave is said to be scattered when its energy is redirected without loss of energy to the scattering particle or particles. The scatter of energy can be in any direction. Back scatter occurs when the redirected energy retraces its path; this mechanism is used by radars. Side scatter occurs when the redirected energy moves out of the transmission path; this mechanism gives rise to interference to other systems. Forward scatter is generally taken as energy that has been redirected after more than one scatter back into the original propagation direction. This forward scatter can be both coherent and incoherent with the main energy that is being transmitted along the path. These principles are demonstrated in Fig. 4.1. Note that side scatter usually contains a component in either the forward direction or the back direction.

4.1.1 Scattering and absorption

Mathematically, the signal attenuation on a path, sometimes referred to as the extinction, is the algebraic sum of the components due to scattering and absorption, namely

$$A_{ex} = A_{ab} + A_{sc} \text{ decibels} \tag{4.1}$$

where the subscripts *ex*, *ab* and *sc* refer to extinction, absorption and scattering, respectively, and A_{ex} is the total attenuation or extinction along the path. The relative importance of scattering and absorption is a function of the complex index of refraction of the absorbing/scattering particle, which is itself a function of signal wavelength and temperature, and the size of the particles relative to the wavelength of the radiowave.

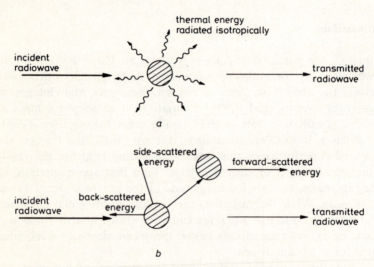

Fig. 4.1 *Schematic representation of the two attenuating mechanisms*
   a Absorption
   b Scattering
   In (*a*) the radiowave energy lost on transmission is radiated as thermal energy.
   In (*b*) the radiowave energy lost on transmission is re-directed in various directions

If the particle is very small compared to the wavelength of the radiowave, Rayleigh scattering theory can be applied. In these situations, the medium will scatter very little energy out of the path and any signal extinction will be mainly due to absorption. This condition generally holds for signals well below a frequency of 10 GHz propagating through an ensemble of hydrometeors. As the frequency goes up, not only does the size of the raindrop become an appreciable fraction of the wavelength but also the raindrop becomes absorbing. The imaginary part of the complex permittivity of water becomes significant at frequencies at or above 10 GHz, and the raindrop can no longer be considered a lossless dielectric. Rayleigh scattering theory is therefore not usually applied to an ensemble of raindrops at frequencies much above 1 GHz. For these cases, Mie scattering theory is used [1].

Mie defined an effective extinction cross-section σ_{ex} and an effective scattering cross section σ_{sc} and proposed an efficiency factor Q that would describe the relative efficiency with which a particle would extinguish and scatter a radiowave. These are

$$Q_{ex} = \sigma_{ex}/\pi r^2 \tag{4.2}$$
$$Q_{sc} = \sigma_{sc}/\pi r^2 \tag{4.3}$$

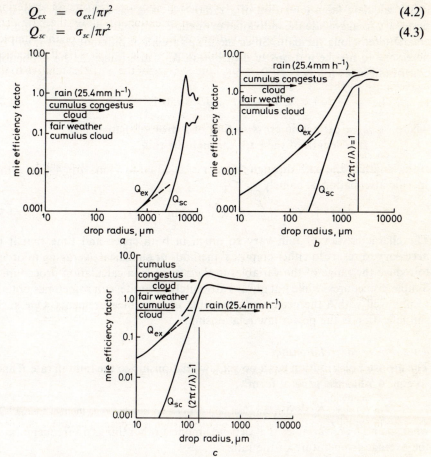

Fig. 4.2 *Mie efficiency factors for scattering and extinction as a function of drop radius (from Fig. 8.5 of Reference 2 after Figs. 5.33, 5.34, and 5.35 on pp. 207–208 in Reference 3, reproduced with permission from "The manual of remote sensing", 1st. edition © 1975 The American Society for Photogrammetry and Remote Sensing)*
(a) 3·0 GHz (b) 30 GHz (c) 300 GHz
– – – Rayleigh extinction

where the subscripts *ex* and *sc* refer to extinction and scattering, respectively, and r is the radius of the raindrop. The absorption efficiency Q_{ab} is simply the difference $Q_{ex} - Q_{sc}$. Values of the efficiency factor at three frequencies, 3, 30 and 300 GHz, are given in Fig. 4.2 [2, 3].

A number of interesting points emerge from Fig. 4.2: firstly, the increasing significance of scattering as the frequency and the size of the particle increase; secondly, the possible error in assuming Rayleigh scattering instead of Mie scattering for the various cases as the frequency increases; and thirdly, that heavy rain is more attenuating than other hydrometeors in the microwave and millimetre wave portion of the radio spectrum.

To calculate the attenuation of a radiowave as it passes through rain, it is necessary to aggregate the individual extinction contributions of each raindrop encountered along the path. Since the drops are all of different sizes, it will be necessary to invoke a dropsize distribution $N_{(D)}$, and integrate the extinction contributions as follows:

$$A_{ex} = 4\cdot343 \times L \int_0^\infty C_{t(D)} N_{(D)}\ dD \text{ decibels} \tag{4.4}$$

where $C_{t(D)}$ = extinction cross-section of a drop of diameter D
        L = length of the path through the rain.

If the length of the path through the rain is set equal to 1 km, eqn. 4.4 gives the specific attenuation α, namely

$$\alpha = 4\cdot343 \int_0^\infty C_{t(D)} N_{(D)}\ dD \text{ decibels/km} \tag{4.5}$$

The characteristics of rain vary so much in both space and time that it is necessary to resort to either empirical methods or statistical averaging in order to reduce the range of the variables in the attenuation calculation procedures. Simple procedures, while not always absolutely accurate, can sometimes obtain results well within the accuracy achievable in most measurements. One such simplification is the power law relationship.

4.1.2 Power law relationship
The dropsize distribution has a power law relationship to the rainfall rate R and so eqn. 4.5 has the general form

$$\alpha = a \times R^b \text{ decibels/km} \tag{4.6}$$

where a and b are variables that depend amongst other things on frequency and the average temperature of the rain.

Originally, the coefficient b was taken to be unity [4] but this was found to be too imprecise. A more recent investigation [5] calculated values for a and b for Marshall and Palmer, Laws and Parsons, and Joss dropsize distributions at temperatures of -10, 0, and 20° Celsius for frequencies between 1 and 1000 GHz. These data have been adopted by the CCIR and reproduced [6] for the frequency range 1–400 GHz. In the CCIR formulation, a is replaced by k and b by α, with the specific attenuation now given the notation γ, namely

$$\gamma = k \times R^\alpha \text{ decibels/km} \tag{4.7}$$

Table 4.1 depicts values for k and α at a temperature of 20° Celsius using the

Laws and Parsons dropsize distribution [7, but see Table 1.3], the Gunn and Kinzer terminal velocity of raindrops [8, but see Fig. 1.25], the refractive index of water due to Ray [9], and regression analyses due to Fedi [10] and Maggoiri [11]. Both horizontal (H) and vertical (V) polarisation coefficients are given.

To obtain k_H, k_V or α_H, α_V for a frequency f not in the table, logarithmic interpolation should be used for k and f, and linear interpolation for α. If values

Table 4.1 *Regression coefficients for estimating the attenu- ation coefficients for the specific attenuation γ, where $\gamma = kR^\alpha$ (from Table 1 of Reference 6)*

Frequency (GHz)	k_H	k_V	α_H	α_V
1	0·0000387	0·0000352	0·912	0·880
2	0·000154	0·000138	0·963	0·923
4	0·000650	0·000591	1·121	1·075
6	0·00175	0·00155	1·308	1·265
7	0·00301	0·00265	1·332	1·312
8	0·00454	0·00395	1·327	1·310
10	0·0101	0·00887	1·276	1·264
12	0·0188	0·0168	1·217	1·200
15	0·0367	0·0335	1·154	1·128
20	0·0751	0·0691	1·099	1·065
25	0·124	0·113	1·061	1·030
30	0·187	0·167	1·021	1·000
35	0·263	0·233	0·979	0·963
40	0·350	0·310	0·939	0·929
45	0·442	0·393	0·903	0·897
50	0·536	0·479	0·873	0·868
60	0·707	0·642	0·826	0·824
70	0·851	0·784	0·793	0·793
80	0·975	0·906	0·769	0·769
90	1·06	0·999	0·753	0·754
100	1·12	1·06	0·743	0·744
120	1·18	1·13	0·731	0·732
150	1·31	1·27	0·710	0·711
200	1·45	1·42	0·689	0·690
300	1·36	1·35	0·688	0·689
400	1·32	1·31	0·683	0·684

Laws and Parsons dropsize distribution [7]
Gunn and Ginzer terminal velocities [8]
Index of refraction of water at 20°C after Ray [9]
Values of k_H, k_V, α_H, and α_V for spheroidal drops [10, 11] in the range 1–150 mm/h

k_1, k_2, and α_1, α_2 (for either H or V polarisation) correspond to frequencies f_1, f_2, the interpolation procedure to the required frequency f can therefore be obtained with

$$k_{(f)} = \log^{-1} \{\log [k_2/k_1] \times [\log (f/f_1)/\log (f_2/f_1)] + \log k_1\}$$

(4.8)

and

$$\alpha_{(f)} = \{[\alpha_1 - \alpha_2] \times [\log (f/f_1)/\log (f_2/f_1)] + \alpha_1\}$$ (4.9)

For polarisations that are not horizontal or vertical but have a tilt angle τ with respect to the horizontal, resultant values of k and α can be computed via

$$k = \{k_H + k_V + [k_H - k_V] \cos^2 \theta \cos 2\tau\}/2$$ (4.10)

and

$$\alpha = \{k_H\alpha_H + k_V\alpha_V + [k_H\alpha_H - k_V\alpha_V] \times \cos^2 \theta \cos 2\tau\}/2k$$ (4.11)

where θ is the elevation angle of the slant path. Note that, for circular polarisation, τ can be set equal to 45°.

In most communications satellite systems that operate with a linear polarisation, the electric vector is selected to be either in the plane of the equator (horizontal polarisation) or perpendicular to the plane of the equator (vertical polarisation) when referenced to the sub-satellite point. The sub-satellite point of a geostationary satellite corresponds to the longitude on the equator at which the satellite's position is maintained. If the earth station is not on the meridian, the longitude of the satellite's position, the plane of the linearly polarised vector will differ from the local horizontal or vertical as perceived by an observer at the earth station. This rotation of the plane of the polarisation from the local vertical (or horizontal) is called the tilt angle. The tilt angle with respect to the horizontal, τ, can be found from [19]

$$\tau = \arctan (\tan \alpha/\sin \beta) \text{ degrees}$$ (4.12)

where α is the earth station latitude (positive for the northern hemisphere and negative for the southern hemisphere), and β is the satellite longitude minus the earth station longitude, with longitude expressed in degrees east. Eqn. 4.12 assumes that the polarisation vector of the satellite antenna is oriented west to east (i.e. parallel to the equator). For a given point rainfall rate, therefore, the specific attenuation can be calculated with any polarisation tilt angle. The results, however, can be influenced in a number of ways depending on the drop shapes, dropsize distribution, and temperatures assumed for the rain medium.

(a) Effect of drop shapes
The shape of a raindrop is influenced primarily by the aerodynamic forces acting upon it as it falls to earth. The larger the raindrop, the more it can be distorted from a spherical shape. The exact shape will not alter the volume of the water

present, but, for a linear polarisation, the attenuation will be dependent upon the relative orientation of the electric vector and the principal axis of the distorted raindrop. An oblate spheroidal raindrop, i.e. a raindrop with its principal axis horizontal, will cause more attenuation to be experienced by a horizontally polarised signal than a vertically polarised signal. This is illustrated for frequencies up to 50 GHz in Fig. 4.3 (from Fig. 8 of Reference 12).

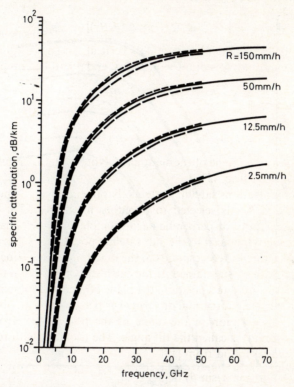

Fig. 4.3 *Specific attenuation for spherical raindrops compared to attenuations for vertically and horizontally polarised waves scattered by oblate spheroidal drops (after Oguchi and Hosoya 1974) (from Fig. 8 of Reference 12; after Rogers and Olsen, reproduced with permission of the Minister of Supply and Services, Canada)*
All curves are based on the Laws and Parson dropsize distribution (equivolume distribution for the distorted drops) and a rain temperature of 20°C.
——Spherical drops
Oblate spheroidal drops
— — Vertical polarisation
– – – Horizontal polarisation

(b) Effect of dropsize distribution
Raindrops cause more attenuation to a radiowave as the wavelength approaches the size of the raindrop. Below a frequency of 10 GHz, the effect of small raindrops is therefore not significant; only above 10 GHz will they make their

presence felt. The Marshall and Palmer dropsize distribution has many more small drops proportionately for a given rainfall rate than the Laws and Parsons dropsize distribution. Above 10 GHz, therefore, the Marshall and Palmer distribution will tend to give more attenuation for the same rainfall rate than the Laws and Parsons distribution. This is shown for some typical rainfall rates between frequencies of 1 and 1000 GHz in Fig. 4.4 (from Fig. 4 of Reference 12). Spherical raindrops are assumed.

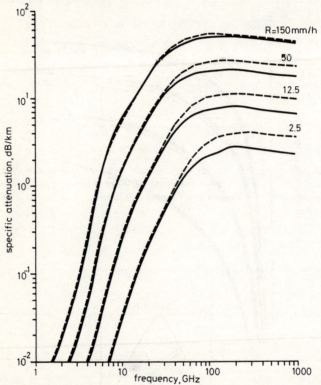

Fig. 4.4 *Comparison of specific attenuations for the Laws and Parsons and Marshall-Palmer dropsize distributions at several rainrates (from Fig. 4 of Reference 12; after Rogers and Olsen, reproduced with permission of the Minister of Supply and Services, Canada)*
Rain temperature of 20°C.
—— Laws and Parsons dropsize distribution
– – – Marshall-Palmer dropsize distribution

(c) *Effect of temperature*

The complex refractive index of water varies with temperature. While the real part generally decreases with frequency, the imaginary part peaks between 10 and 100 GHz.

For an assembly of raindrops, the bulk refractive index of the total volume of air and water is calculated first to enable the overall effect of a rain shower

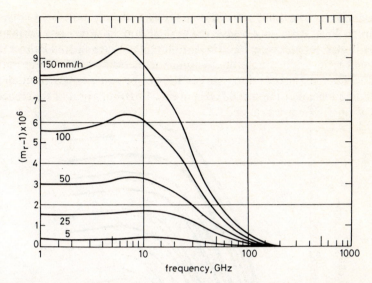

Fig. 4.5 *Real part* m_r *minus unity of the complex index of refraction multiplied by* 10^6 *[$(m_r - 1)10^6$] for a medium consisting of water drops in empty space for a temperature of 20° C and Laws and Parsons distribution. (Fig. 4.3a from Reference 13 after Reference 14)*

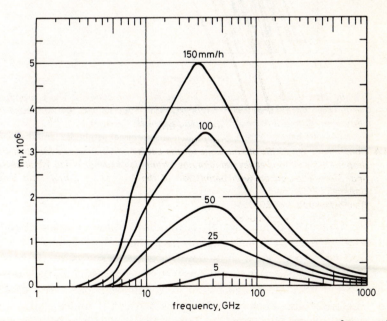

Fig. 4.6 *Imaginary part* m_i *of the complex index of refraction multiplied by* 10^6 *($m_i \times 10^6$) for the same medium as in Fig. 4.5 (Fig 4.3b of Reference 13 after Reference 14)*

to be estimated. The bulk refractive index is normally presented in refractivity units. Figs. 4.5 and 4.6 give the real and imaginary parts of the bulk refractive index, respectively, for different rainfall rates (from Reference 13 after Reference 14).

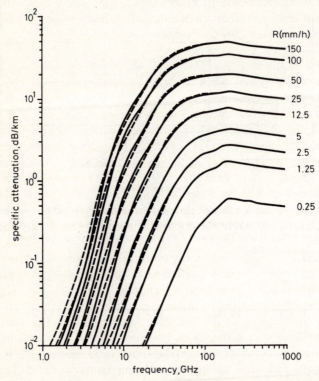

Fig. 4.7 *Specific attenuation as a function of frequency for coherent wave propagation through uniform rain (Fig. 3 of Reference 12; after Rogers and Olsen, reproduced with permission of the Minister of Supply and Services, Canada)*
The curves are based on the Laws and Parsons dropsize distribution and the terminal velocities of Gunn and Kinzer
More accurate values for the Laws and Parsons rainrates are 0·254, 1·27, 2·54, 5·08, 12·7, 25·4, 50·8, 101·6, and 152·4 mm/h.
—— Rain temperature of 20°C
– – – Rain temperature of 0°C

The bulk refractivity of an assembly of raindrops follows the same trends as that of water, particularly at high rainfall rates. The peak of the imaginary part of the refractive index of water shifts upwards in frequency as the temperature increases and this is reflected in the bulk refractivity of an assembly of raindrops. The attenuation of a radiowave passing through water, or water droplets, should therefore exhibit a corresponding peak between 10 and 100 GHz with a similar temperature dependence. Fig. 4.7 (from Fig. 3 of Reference 12) illustrates these effects between a frequency of 1 and 1000 GHz.

Note the reduction of specific attenuation in Fig. 4.7 as the frequency exceeds about 100 GHz and the cross-overs between the two temperature curves for a given rainfall rate in the relatively lower frequency ranges. Also evident is the lower excess attenuation levels experienced in the region approaching optical wavelengths when compared to millimetre wavelengths around 100 GHz. A laser beam will suffer less excess attenuation in a thunderstorm than a millimetre wave link and measurements confirm this [18].

4.1.3 Multiple scattering effects

Multiple scattering occurs when a single ray path involves more than one incidence of scattering before it emerges from the scattering medium. Most theories that calculate the scattering cross-sections of particles and then integrate to find the total scattered energy tend to neglect multiple scattering, assuming only single scattering. If the average distance separating the scattering particles is larger than the wavelength, any multiple scattering that occurs will give rise to an incoherent field due to the random phases of the scattered signals [15] irrespective of the direction [16]. In most rain systems, the average separation between adjacent raindrops is much larger than the wavelength for frequencies above 1 GHz. Multiple scattering in rain is therefore mainly incoherent.

Systems that rely on coherent energy transmission, such as satellite communications systems, will be largely unaffected by incoherent energy and, for this reason, multiple scattering tends to be ignored. An investigation [16] showed this was a correct assumption to make since the power law relation implicitly accounts for the majority of the multiple scattering effects in coherent transmissions. The incoherent scatter does, however, give rise to additional noise in the receiver which will reduce the signal-to-noise ratio. Energy that is absorbed by the attenuating particles, as opposed to being scattered, will also give rise to an increase in the received noise.

4.1.4 Sky noise temperature

An absorbing medium, if in equilibrium with its surroundings, will radiate as much energy as it absorbs. The emitted radiation will be isotropic [17]. In Fig. 4.8, an absorbing medium M has been raised to a temperature T_m by absorbing energy from around it, principally from the ground. The efficiency with which it is absorbing (and re-radiating) this energy can be described by its fractional transmissivity σ. The fractional transmissivity of a medium is the fraction of incident energy between zero and unity that passes through the medium and emerges on the other side. A fractional transmissivity of zero shows complete absorption takes place inside the medium and a fractional transmissivity of unity describes a completely transparent medium.

In Fig. 4.8, a signal of power S passes through the absorbing medium and emerges at a power level given by σS. By the same token, the amount of energy radiated as noise, and detectable as an increase in noise temperature, is given by

$(1 - \sigma)T_m$. The radiated noise temperature is isotropic and so a receiver that this detecting the signal will also detect an increase in noise temperature. The increase in noise temperature can be calculated equally from the fractional transmissivity or, more usually, from the attenuation experienced by the signal.

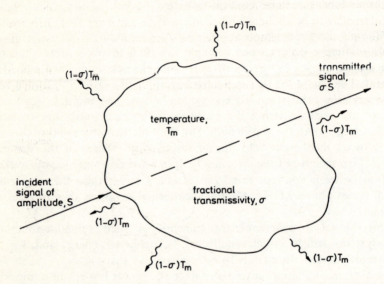

Fig. 4.8 *Schematic showing signal loss and temperature radiated due to an absorbing medium with a fractional transmissivity σ*

The increase in noise temperature T_r, also referred to as the radiated temperature, is given by both

$$T_r = T_m \times (1 - \sigma) \text{ degrees K} \tag{4.13}$$

and

$$T_r = T_m \times (1 - e^{-A/4.34}) \text{ degrees K} \tag{4.14}$$

where A is the attenuation in decibels of the signal. The attenuation is related to the fractional transmissivity by

$$A = 10 \log (1/\sigma) \text{ decibels} \tag{4.15}$$

To measure the attenuation on a satellite-to-ground path, both direct and indirect methods can be used. An indirect method usually measures a parameter that is related to attenuation, e.g. rainfall rate, and infers an attenuation along the slant path. A direct method measures the strength of a signal transmitted through the attenuating medium. There are advantages and disadvantages in both methods.

4.2 Measurement techniques

4.2.1 Raingauge measurements

Rainfall accumulation is usually one of the first meteorological parameters to be measured whenever such measurments are initiated. Records of rainfall accumulation measurements exist in many places for periods of over 100 years. The measurements therefore represent a large data bank from which reliable statistical trends can be predicted. Since attenuation can be related fairly readily to rainfall rate, the first attempts at inferring slant path attenuation were aimed at utilising raingauge data.

There are a number of errors in the measurement technique when related to slant paths. These are broadly subdivided into spatial errors, integration errors and inherent errors.

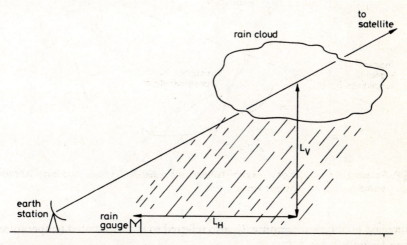

Fig. 4.9 *Schematic presentation of the possible horizontal (L_H) and vertical (L_V) spatial errors in measuring path rainfall rate*

(a) Spatial errors

The link to a satellite can intercept rain at many heights along the slant path. A raingauge on the ground, therefore, will not usually measure at any given instant the same rainfall encountered at a point along the path. The spatial error could not only be horizontal but also vertical. In Fig. 4.9 the possible horizontal and vertical separation between the raingauge measurement site on the ground and the rain on the path is indicated schematically. For very low elevation angle paths, it could be that no rain actually falls on the raingauge while a considerable volume of rain is present along the path some distance from the site. In general, therefore, no instantaneous correlation will be found between attenuation along a slant path and the rainfall rate measured at one point on the ground close to the receiving antenna.

Efforts to quantify the potential spatial errors have been made using data obtained from fields of raingauges operating simultaneously (e.g.[20–22]). The models have proved to be rather complicated when storm-to-storm correlation is required. Statistically, however, the annual variations in slant path attenuation have been shown to correlate very well with the annual variations in point rainfall rate [23]. The integration time of the raingauge was important, however.

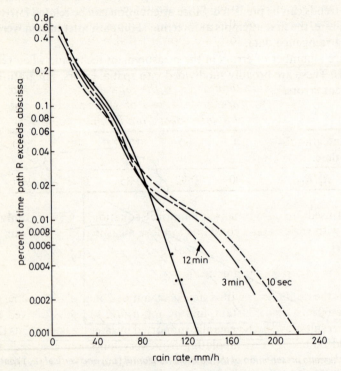

Fig. 4.10 *Percentage time that the rainfall rate exceeds the abscissa using different rain gauge integration times (from Fig. 8b of Reference 20)*
The data represent one summer (3144 h)
(Copyright © 1969 AT&T, reproduced with permission)

(b) Integration errors
The measured rate of a given rainfall will depend on the time constant, or integration time, of the measuring device. If the integration time is too slow, the high rainfall rates will be 'smoothed out'. This can be seen in Fig. 4.10 (from Fig. 8(b) of Reference 24), which was one of the first set of experimental data to quantify the required integration time.

The solid line in Fig. 4.10 is the average of a number of raingauges during a summer measurement campaign containing about 3000 hours of observations. It is evident from Fig. 4.10 that, with long integration times, rainfall rate measurement errors can be quite large. Typically, an integration time of one minute is a good compromise between experimental accuracy and equipment complexity.

It is interesting to note the cross-over in the curves in Fig. 4.10. This is a direct consequence of the principle of conservation of rain; i.e. the total accumulation of rain is not dependent on the measurement integration time. This fact has been supported by later measurements [27].

The variations in the measured rainfall rate due to the integration time have been calculated [28] at the 0.01% annual percentage time point in a temperate climate. The 'correction factors' are shown in Table 4.2.

Table 4.2 *Ratio of the point rainfall rate R_τ for an integration time of τ s to the rainfall rate measured with an integration time of 10 seconds, R_{10} at 0.01% of a year (data from Reference 27)*

Integration time, τ(s)	10	60	120	300	600
R_τ / R_{10}	1.00	0.92	0.85	0.77	0.71

Lin [29] used an un-normalised empirical equation to relate rainfall rates measured with various time constants to those measured over a 60 min period, namely [29]

$$R_T = a_T \times R_{60}^{n_T} \text{ millimetres/h} \tag{4.16}$$

where R_T is the rainfall rate with a time constant of T minutes, R_{60} is the rainfall rate with a 60 min time constant, and a_T and n_T are variables derived by Lin. Lin's early results have been generally supported by later measurements [27], but care should be taken in extrapolating from very long time constants to very short time constants. Fairly good accuracy should normally be possible over a limited range of extrapolation.

(c) Inherent errors

These are due to mechanical and electrical causes that are inherent in the design of the equipment. The most common form of raingauge is a tipping bucket rainguage. At very low rainfall rates the buckets do not fill sufficiently to cause a 'tip' and so very light rainfall is usually missed. At very high rainfall rates, considerable splashing and eventual swamping of the tipping buckets takes place. Nevertheless, these types of raingauges can exhibit acceptable accuracy between 5 and 100 mm/h if properly maintained and calibrated.

More recent developments of fast response raingauges have utilised optical techniques either to image the drops as they fall through a standard diameter pipe [25] or to provide scintillation spectra with lasers [26]. In the case of the former, an acceptable accuracy is obtained for point rainfall rates well in excess of 100 mm/h. In the latter, the use is still restricted to path average rainfall rates, albeit of paths of the order of 30 m.

4.2.2 Radiometer measurements

A radiometer is a device for measuring the variation in noise power or brightness temeprature of a source. The measurement technique has been used for decades in radio astronomy [30].

The brightest star in our own galaxy, in terms of radio noise at microwave frequencies, is Cassiopeia A, closely followed by Cygnus A and Taurus A. These are essentially point sources of radio energy and they have been used extensively in calibrating large earth stations which have narrow beamwidths [31]. The Moon and the Sun are also emitters of radio energy, particularly the Sun, but they are not point sources and so are not generally used to calibrate earth station equipment. The angular diameters of the Moon and the Sun are both of the order of half a degree as viewed from the Earth.

The frequency at which the celestial radio sources are observed will, in most cases, determine the brightness temperature available. There is a considerable variation in brightness temperature with frequency for the radio stars and for the Sun. This is shown in Fig. 4.11 [32]. In addition, the Sun exhibits a periodic variation in brightness temperature that follows approximately an 11-year cycle.

The availability of radio sources beyond the atmosphere permits two types of radiometric measurements to be undertaken from which slant path attenuation can be derived: active measurements and passive measurements.

(a) Active radiometer measurements

An active radiometer measurement uses a natural radio source beyond the atmosphere of the Earth to measure the attenuation through the atmosphere. At frequencies above 1 GHz, Fig. 4.11 indicates that the only extra-terrestrial radio noise source that will provide sufficient energy for an active radiometric measurement having a useful dynamic range is the Sun. Radiometers that are designed specifically to follow the motion of the Sun are called sun-tracking radiometers.

A radiometer with a beamwidth smaller than the angular dimension of the Sun will detect an emission temperature T_e, given by

$$T_e = T_s \times e^{-A/4.34} \tag{4.17}$$

where T_s is the brightness temperature of the Sun and A is the attenuation along the path, in decibels. Variations in T_e will give changes in A directly. Unfortunately, the intervening medium that is causing the attenuation will, itself, emit radio energy due to Kirchhoff's law. If the intervening medium only absorbs energy, as opposed to scattering it, and is at a temperature of T_m, eq. 4.17 is modified to

$$T_e = T_s \times e^{-A/4.34} + T_m(1 - e^{-A/4.34}) \text{ degrees K} \tag{4.18}$$

Note the similarity of the last term in eqn. 4.18 with eqn. 4.14, both being the radiated temperature from the intervening medium. Two techniques have been used to eliminate the term due to the radiated temperature [33]. The first employs two feeds on the same antenna reflector: one that receives the energy from the Sun and the other directed slightly away from the Sun so that it only receives the energy radiated from the sky close in angular degrees to the position of the Sun. The second method employs a single feed on one antenna with a 'nodding mechanism' that directs the beam of the radiometer first at the Sun and

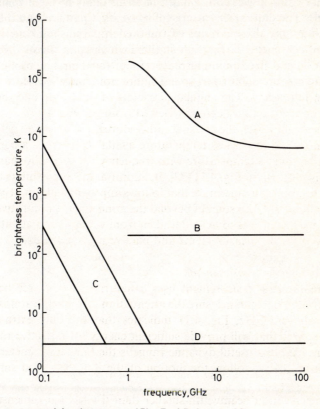

Fig. 4.11 *Extra-terrestrial noise sources (Fig. 7 of Reference 32)*
A: Quite sun ⎫
B: Moon ⎬ Diameter ~0·5°
C: Range of galactic noise
D: Cosmic background
(Copyright © 1986 ITU, reproduced with permission)

then away from the Sun, so that the radiometer measures alternately the brightness temperature of the Sun and then the brightness temperature of the sky close in angular degrees to the position of the Sun. The difference between the two measurements, one on the Sun and one on the sky, will enable the last term of eqn. 4.18 to be eliminated. Fig. 4.12 illustrates the technique schematically.

The dynamic range of a sun-tracking radiometer approaches 15 dB with a sensitive receiver [33] and so it can provide useful results up to frequencies of about 30 GHz. Above 30 GHz or so, the attenuation experienced on satellite-to-ground paths exceeds 20 dB for appreciable periods even in temperate climates. Another more serious problem is the applicability of the attenuation statistics generated by sun-tracking radiometers to satellite communications. Since the Sun is not a stationary source, the attenuation statistics will be for a multitude of different elevation angles and azimuths, and naturally no night-time data will be obtainable. The diurnal characteristics of weather patterns and their preferred orientations may also be completely masked in the sun-tracking radiometer data by the movement of the Sun itself. To overcome these defects, it is necessary to have a fixed-pointing radiometer. No natural 'stationary' extra-terrestrial radio sources exist and so the radiometers must therefore make use of passive techniques.

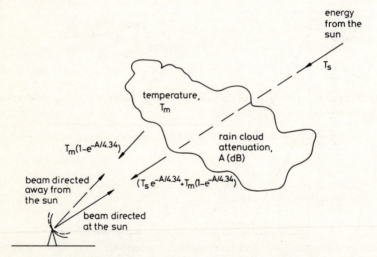

Fig. 4.12 *Schematic representation of a 'nodding' sun-tracking radiometer*
When directed away from the Sun the radiometer will receive only the thermal energy radiated from the cloud, $T_m(1 - e^{-A/4 \cdot 34})$. When directed at the Sun, the radiometer will detect the attenuated signal from the Sun, $T_s e^{-A/4 \cdot 34}$, plus the thermal energy radiated from the cloud.

(b) Passive radiometer measurements
These radiometers make use of only the radiated temperature T_r of the intervening medium to calculate the attenuation induced by the medium. Before examining the technique and the possible errors, the two basic types of passive radiometers will be discussed.

The two types were once called DC and Switched but are more usually called Total Power and Dicke radiometers now [34]. In the former case, the signal path from the antenna through the receiver to the detector remains open all the time and a continuous measurement is made of the total power incident upon the antenna. In the latter case, the same signal path is interrupted at a fairly rapid rate (about 1 kHz), the detector alternately measuring the power received from a known reference source and that received from the sky via the antenna. The two techniques are illustrated schematically in Fig. 4.13.

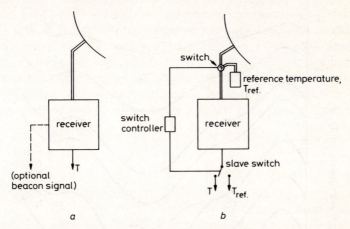

Fig. 4.13 *Schematic presentation of the basic differences between a total power radiometer (a) and a Dicke radiometer (b)*

Output of a total power radiometer will be a smooth signal

Output of a Dicke radiometer will be a square wave, alternating between a value of T and T_{ref}

A total power radiometer measurement is susceptible to gain changes in the receiving equipment but, since the signal path through the receiver from the antenna is not interrupted, other signals can be received at the same time. This is important where satellite beacon and radiometer data need to be correlated along the same path using the same antenna, e.g. Reference 35. A Dicke radiometer elliminates all errors due to gain changes in the receiver but will not allow any satellite beacon signal to be measured accurately using the same antenna and receiver. This is because the satellite signal needs to be detected coherently in a phase-locked receiving system. Interrupting the signal path will cause the phase-locked loop to drop lock unless great care is taken with the design of the equipment.

Dicke radiometers have been flown on meteorological satellites, the first being on COSMOS–243 in 1968 [2]. The later satellites, particularly the NOAA spacecraft of the USA, carry a variety of multi-frequency radiometers that can scan on either side of the track on the Earth that the satellite is following. The frequencies selected are usually close to the water vapour line at 22 GHz since the satellites are designed to detect moisture at various heights in the atmosphere. The meteorological satellites have narrow antenna beamwidths that can identify changes in brightness temperature over areas of the order of 1000 km²

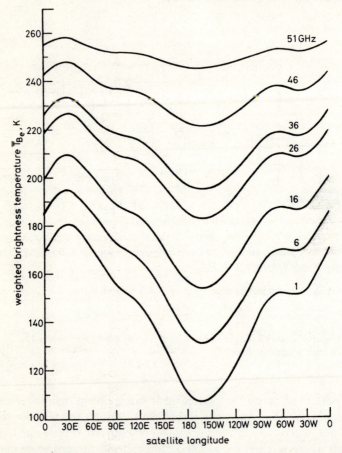

Fig. 4.14 *Weighted brightness temperature of the Earth as a function of longitude viewed from geostationary orbit at frequencies between 1 and 51 GHz (from Fig. 3 of Reference 36)*

Curves are for US Standard Atmosphere with 2·5 g/cm³ water vapor and 50% cloud cover. The Earth-coverage antenna pattern is given by $G(\varphi) = -3 [\varphi/8·715]^2$ decibels for $0 \leqslant \varphi \leqslant 8·715$ where φ is the angle off boresight

(Copyright © 1985 American Geophysical Union, reproduced with permission)

on, or above, the surface of the Earth. The Earth is itself, however, also emitting at these frequencies. If the antenna beam on the spacecraft is large enough to cover the whole of the Earth as viewed from space, the temperature measured would be as shown in Fig. 4.14 (Fig. 3 of Reference 36).

At a frequency of 51 GHz, a radiometer looking down at the Earth will tend to 'see' mostly the mid-to-upper atmosphere and so a fairly constant brightness temperature will be detected over most longitudes, as is evident from Fig. 4.14. As the frequency goes down, so the energy detected by the spaceborne radiometer corresponds to that emitted from closer to the surface of the Earth.

Dry land has a fairly high emissivity compared to water and so the variations in brightness temperature noted in Fig. 4.14 correspond to the changing ratios of land to sea surface in the radiometer's field of view. Around a longitude of 180°, the spacecraft is over the Pacific Ocean and a relatively low brightness temperature results. At 30° East longitude, the spacecraft is roughly over the middle of Africa with Europe to the north, and a relatively high brightness temperature results.

Passive radiometers, whether on the ground or in space, are conceptually very simple devices. They are also independent of artificial sources. A passive radiometer, for instance, can be located anywhere on the surface of the Earth and directed towards any point in the sky. Coherent, narrow-band radiowave signals (beacons) on geostationary satellites are not widespread and it is therefore not surprising that the bulk of the early slant path measurements used radiometers. There are a large number of potential error sources possible in passive radiometric measurements and, while the technique and the equipment are relatively simple, great care must be taken to eliminate as many errors as possible in the analysis. The possible error sources are considered below.

(c) Potential errors in passive radiometer measurements
The total flux density incident upon the radiometer antenna can be calculated by solving the radiative transfer equation [37] for each source and summing the results. This is a complex task and excellently set out in relation to radiometry in Reference 17. Taking a simpler approach to identify the error sources, it is best to begin with the idealised radiometer equation. The slant path attenuation A given in eqn. 4.14 can be inverted to make A the subject of the equation, namely

$$A = 10 \log \frac{T_m}{(T_m - T_r)} \text{ decibels} \qquad (4.19)$$

where T_m is the physical temperature of the absorbing medium and T_r is the radiated temperature.

Eqn. 4.19 is the idealised radiometer equation and assumes that the radiometer antenna is perfect with a lossless feed, that the radiating medium just fills the antenna beam and is perfectly absorbing (i.e. there is no scattering contribution), and that there is no other radiating source being detected within the antenna beam. The Sun and the Moon will occasionally enter the beam of a radiometer antenna directed towards a point in the geostationary orbit but these occasions will be predictable. There will, however, be a permanent background radiation contribution of about 2–4 K, thought to be due to the remnants of the

'Big Bang'. This cosmic temperature T_c will modify eqn. 4.19 to

$$A = 10 \log \frac{(T_m - T_c)}{(T_m - T_r)} \text{ decibels} \qquad (4.20)$$

The antenna and the medium effects are of a more subtle nature.

(i) *Antenna effects*

The antenna of a radiometer will have a main beam, defined by the half-power beamwidth, and a multitude of sidelobes. Since the energy being detected is incoherent, all contributions of energy entering the antenna from any direction will simply be summed by the detector to give the total power. The sidelobes of the antenna will intercept the ground at some point and, since the brightness temperature of the ground, T_g, is quite high (about 270 K), a significant increase in background temperature will be observed compared to the background temperatures of the idealised radiometer.

The sidelobe contribution is calculated by using the antenna integration factor H, which is the proportion of the antenna pattern that illuminates the sky. Conversely, $(1 - H)$ is the portion intercepting the ground. If the brightness temperature detected by the antenna is T_a, then

$$T_a = H T_s + (1 - H) T_g \text{ degrees K} \qquad (4.21)$$

where T_s is the brightness temperature of the sky. Inverting eqn. 4.21 to make T_s the subject of the equation yields

$$T_s = \frac{T_a - (1 - H) T_g}{H} \text{ degrees K} \qquad (4.22)$$

The measured antenna pattern will readily yield the factor H which, for most well designed antennas, will be of the order of 0·9. The detection and measurement of T_a will therefore give T_s directly from eqn. 4.22. The relationship between T_s in eqn. 4.22 and T_r in eqn. 4.20 now needs to be established.

In most cases, T_s is assumed to be the same as T_r; i.e. the perceived sky temperature T_s, calculated from a measurement of T_a, is taken to be the same as the radiated temperature T_r of the raincell or attenuating medium. Substituting the calculated value of T_s for T_r in eqn. 4.20 will give the attenuation along the path. If the radiating medium fills all parts of the antenna beam, then T_s will be the same as T_r for all practical purposes. Usually, the radiometer beam will detect parts of the sky that are less attenuating than others and an averaging effect will take place. This error, which amounts to an under-estimation of the attenuation through a lower than expected brightness temperature being detected, cannot be estimated with confidence, but, where simultaneous radiometer and satellite beacon measurements have taken place, the error has been observed to become more pronounced when intense raincells intercept the antenna beam at some distance from the radiometer site [38]. The differences between T_r and T_s arise out of the inhomogeneity of the rain medium that is emitting the noise temperature.

(ii) *Inhomogeneity effects*

These effects arise basically from the location of the raincell with respect to the radiometer, and hence the proportion of the raincell within the main beam and the sidelobe.

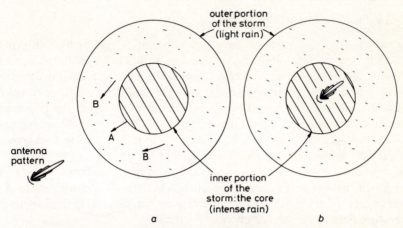

Fig. 4.15 *Schematic plan-view presentation of the inhomogeneity effect of rain on a radiometer*
a Radiometer is completely outside the rainstorm
b Radiometer is completely inside the intense core of the rainstorm
In case *a*, the radiometer will detect lower brightness temperatures from the light rain B through the side lobes than from the intense rain A. In case *b*, only the brightness temperature from the intense rain will be detected

In Fig. 4.15, two basic cases are illustrated. The first is with a radiometer antenna embedded in a raincell and the second with the radiometer located some distance from the raincell. In both cases, there is an outer shell of lighter rainfall surrounding a more intense core. For the radiometer located completely inside the intense core section of the raincell, the rain medium is to all intents and purposes homogeneous. A constant brightness temperature is seen by all parts of the antenna pattern. In addition, if the attenuation in all directions is large, the brightness temperature perceived will approach the physical temperature of the medium, T_m , even when partial scattering is occurring [38]. For the radiometer located well away from the raincell, not only will the antenna main beam and sidelobes 'see' different brightness temperatures, but also any energy that is scattered, as opposed to being absorbed and re-radiated within the raincell, will not generally enter the antenna beam. The brightness temperature observed will not therefore approach the radiated temperature of the medium.

Scattering will also reduce the perceived radiated temperature.

(iii) *Scattering effects*

If the attenuating medium is not a perfect absorber, but scatters some of the energy [39], the brightness temperature emitted from the medium will be reduced by the ratio of the absorbing efficiency Q_{ab} to the extinction efficiency Q_{ex}. That is

$$T_r' = \frac{Q_{ab}}{Q_{ex}} \times T_r, \text{ degrees K} \qquad (4.23)$$

Fig. 4.16 illustrates the whole process.

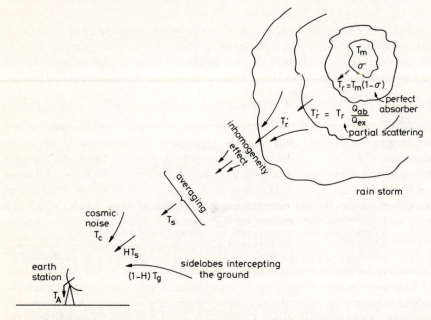

Fig. 4.16 *Graphical presentation of changes in perceived brightness temperature due to different causes*

Ideally, the radiated temperature, T_r, is required which, from a knowledge of the physical temperature of the medium T_m will give the fractional transmissivity σ, and hence the slant path attenuation A. (Note that $\sigma = e^{-A/4.34} = 10^{-A/10}$, where A is in decibels). T_r is reduced by the ratio of the absorption efficiency Q_{ab} to the total extinction, Q_{ex}; the inhomogeneity effect coupled with the wide beam of the radiometer will further reduce T_r to the sky temperature T_s; the cosmic background temperature T_c and the H factor of the antenna will yield T_A, the measured antenna temperature. The skill is going from T_A to T_r.

Despite all of the inherent errors that are possible, including reflections from the ground [40], the results provided by passive radiometers to date have been rather good. The dynamic range of passive radiometers is probably about 10 dB, somewhat less than a sun-tracking radiometer, and this may have reduced the obvious errors. As most of the measurements were at frequencies below 20 GHz, the scattering effects had not started to dominate and the major storms that give rise to inhomogeneous effects generally cause induced attenuations well in excess of the reliable dynamic range. Attenuation data in excess of 10 dB that have been measured by passive radiometers are usually discarded. Perhaps the major source of error in passive radiometers is an incorrect assumption of the physical temperature of the absorbing medium.

(iv) *Physical temperature*
Perhaps the most fundamental parameter in a passive radiometer experiment is the choice of the physical temperature of the medium. A variety of temperatures have been assumed by experimenters. Some were in the range of the actual temperature of the medium (273–293 K) while others, attempting to account for scattering and other effects, were artificially low. An early attempt to account for seasonal and climatic changes introduced a dependence of the physical medium temperature T_m on the ground temperature T_g, such that [41]

$$T_m = 1.12 \, T_g - 50 \text{ degrees K} \tag{4.24}$$

Eqn. 4.24 is an approximation that is actually accounting for the depth of the precipitation that is in liquid form as opposed to frozen. If the ground temperature is 0° Celsius, i.e. the freezing level that demarcates the liquid from the frozen precipitation is at ground level, then $T_m = 256$ K from eq. 4.24, clearly a non-physical temperature that will lead to an overestimate of attenuation. Conversely, if the ground temperature is 30° Celsius, $T_m = 289$ K, which is a more realistic physical temperature for rain in a summer thunderstorm. A careful evaluation of ESA data [17], however, has shown that a constant value of $T_m = 260$ K gives good results at 11 GHz. The low value of T_m accounted statistically for scattering and non-homogeneous effects in the 11 GHz data.

Other approaches to choosing T_m are to select a value for T_m that equals the highest value measured by the radiometer during the course of the experiment, or to select a range of values for T_m that change with the percentage time of the data base [42]. The former will cause the calculated path attenuations to be underestimated during medium and light rainfall events, and the latter method, while partially coping on a statistical basis with scattering and homogeneity effects, is not sufficiently tested to be acceptable. The only acceptable method at present to account for all the effects is to calibrate the passive radiometer with a satellite beacon measurement made along the same path at the same time.

4.2.3 Satellite beacon measurements
A satellite beacon signal is usually derived from an extremely stable crystal source that has very little phase and intermodulation noise [43]. This means that most of the energy is contained within a very narrow bandwidth centred on the desired frequency provided the carrier is not modulated. Since nearly all the energy is contained close to the carrier frequency, a satellite beacon receiver on the ground can be made with a very narrow bandwidth. The noise in the receiver, which is proportional to the bandwidth, will therefore be quite low, permitting either a large dynamic range (fade margin) to be achieved or a very small receiving antenna to be employed [44]. Given a satellite beacon in

geostationary orbit, in principle the direct detection of the received signal level will permit the excess attenuation along the path to be readily observed.

In Fig. 4.17 the satellite beacon signal level measured during a rain event is displayed. The difference between the signal level in clear-sky and that during the event will give the excess attenuation. By accumulating such data for a year, the annual statistics for that path can be calculated and presented as shown in Fig. 4.18.

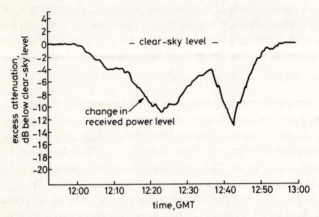

Fig. 4.17 *Example of the variation in received power from a satellite beacon during a thunderstorm*

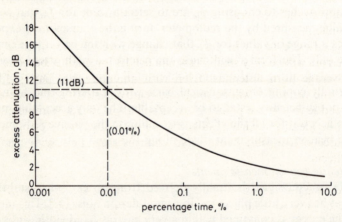

Fig. 4.18 *Cumulative statistics of excess rain attenuation on a slant path presented as annual exceedance percentages*

In Fig. 4.18, the excess attenuation exceeded for 0·01% of a year is 11 dB. That is, if the earth station has a margin of 11 dB, that particular path will only be below threshold for 0·01% of a year. Matters are not that simple, however.

(*a*) *Potential errors in satellite beacon measurements*

There are a number of possible errors that can occur in the measurement technique that can severely limit the accuracy of the data. Some relate only to the satellite and some only to the earth station, with yet others being a combination of the two.

(i) *Satellite-induced errors*

Pointing errors: There are two basic types of spacecraft: those that are spin-stabilised and those that are 3-axis, or body-, stabilised [45]. In the former case, the communications antennas and sensing devices are permanently directed towards the Earth by means of a de-spun platform that removes the spin of the main body of the spacecraft. There will always be a slight spin wobble and this translates into a slight, cyclic movement of the antenna coverages. Spin-stablised satellites rotate at rates on the order of 90 rev/min, and so any changes in received signal level will be at this rate. Unless very rapid sampling is being employed by the earth station, the rapid cyclic variations in the received signal will go undetected.

Body-stabilised spacecraft use reaction control jets to maintain correct pointing. There is a 'dead zone' in the pointing control mechanism. This means that the spacecraft antenna's direction will drift between the two extremes of the control limits before corrections are applied. The slow change in pointing is referred to as nutation and the period is of the order of 100 s. With more advanced spacecraft the antennas themselves have pointing mechanisms that are decoupled from the spacecraft body and so nutation effects could appear to be more complex.

In most cases, changes in perceived signal level due to satellite antenna pointing errors will be very small, less than 0·1 dB for satellite beacons that use global coverage horn antennas. Even with antennas that have quite small coverages, the variations will be small and relatively slow. The slow changes in signal level due to satellite pointing errors will allow the more rapid effects due to rain to be identified fairly readily and so normal pointing errors are not a major source of concern. More rapid changes can occur with transponder loading effects.

Transponder loading errors: Where a narrow-band satellite beacon source is not available, or if a wideband experiment is desired, an uplink signal can be transmitted by the earth station and returned along the same path via a satellite transponder. A transponder is usually a simple repeater, generally with linear characteristics.

Transponders have bandwidths on the order of tens of megahertz and are designed to carry more than one signal simultaneously in a frequency-multiplex mode [46]. As the number of signals being supported by a transponder changes, so does the distribution of power amongst the signals. In a like manner, if the satellite beacon can be modulated, the application of modulation will reduce the power in the carrier. If narrow-band detection of the beacon or transponded carrier is being undertaken without a knowledge of the modulation state of the beacon or the loading of the transponder, serious errors in estimating the true level of the signal can occur. If the changes in beacon modulation or transponder loading occur in clear-sky conditions they are easily detectable as abrupt changes in the level of the signal, but if they occur in the middle of a severe precipitation event, the changes may go unnoticed.

(ii) *Earth station induced errors*

Near-field effects: An electromagnetic wave emanating from a point source will not be established as a plane wave until the Rayleigh distance has been reached. The Rayleigh distance is given by $2D^2/\lambda$, where D is the diameter of the antenna and λ is the wavelength. As an example, the Rayleigh distance for a 6 m diameter antenna operating at a frequency of 30 GHz is 7·2 km. For slant paths at elevation angles above about 20°, the path through liquid precipitation is usually a lot less than 7·2 km. As a consequence, most of the rain effects can occur well within the Rayleigh distance, or near-field, of the earth station antenna.

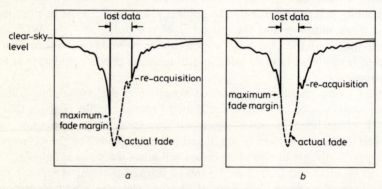

Fig. 4.19 *Illustration of the different aspects of data lost with differing phase-lock loop (PLL) bandwidths*
a Narrow-bandwidth PLL
b Medium-bandwidth PLL
In *a*, the narrow bandwidth (e.g. 25 Hz) will allow the beacon signal level to be tracked to a lower level than in (*b*), but, once the signal is lost, the search time to re-acquire it is generally much longer.

The Rayleigh distance, or far-field, is well known to antenna measurement engineers. For propagation scientists, the question was raised that signal level measurements on rain in the near-field of earth station antennas may not be the same as those taken in the far-field. The debate was settled [47] with the conclusion that near-field effects do not contribute to measurement errors of rain attenuation. A similar negative contribution to measurement errors was found for variations in beamwidth [48], at least for antennas having beamwidths larger than 0·1°.

Loss of lock errors: Most satellite beacon receivers employ phase-locked loop techniques [43, 44]. The bandwidth of the loop is a compromise between fade margin on the one hand and recovery of lock on the other. Essentially, the phase-locked loop acts as an automatic frequency control that tracks variations in the satellite beacon frequency while effectively maintaining a constant bandwidth. The smaller the bandwidth of the loop, the lower the signal level that can be tracked but, once lock is lost on the beacon signal, the longer it will take to re-acquire the signal. Fig. 4.19 illustrates the trade-off to be made.

In actual practice, the phase-lock loop bandwidth will be automatically switchable. A relatively narrow bandwidth (e.g. 100 Hz) will be maintained while the satellite beacon signal is locked. Once lock is lost, however, the bandwidth will be opened up to between 4 and 10 times the normal bandwidth to permit a relatively fast search rate and re-acquisition of signal.

(iii) *Dual-effects*
Errors due to dual-effects arise when a combination of earth station and satellite effects together produce errors. The major dual-effect is earth station tracking accuracy and satellite station keeping tolerances.

Most earth stations engaged in propagation measurements do not use active tracking techniques; i.e. they do not rely on a measurement of the satellite beacon to track the satellite's apparent motion. Instead, passive tracking techniques are employed by means of computer generated antenna pointing commands. In many cases, no antenna tracking is employed at all.

With program tracking, the accuracy essentially depends on the satellite ephemeris data [45] from which the required azimuth and elevation angles of the earth station antenna are predicted. The ephemeris data are usually very accurate but there will still be periods around satellite orbital manoeuvres when tracking accuracy can deteriorate. When no tracking is employed, the principal source of error is in defining the mean, or clear-sky, level. This can be compounded by slight changes in clear-sky path attenuation due to humidity effects.

The effect shown in Fig. 4.20 is typical of the diurnal variations that can be seen on a slant path link in the 10–30 GHz frequency range on a warm summer's day in temperate regions. Even accurate tracking will not eliminate the variations completely and it is now normal to use a radiometer to establish an independent measure of the mean, clear-sky level that is unaffected by satellite beacon variations.

4.2.4 *Radar measurements*
Unlike any of the other techniques for measuring or inferring slant path attenuation, the radar method actively probes a specific region along the path.

It is this attribute, which also lends itself to scanning a large volume around the radar site and not just one particular path, that makes a radar so attractive for many types of investigations. Variations in the received energy and the precise location of the source of the reflected energy is a complex mixture of scattering and attenuation effects.

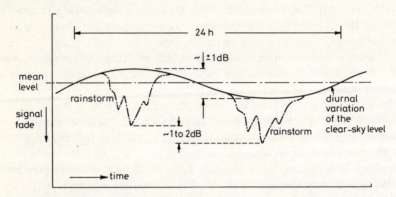

Fig. 4.20 *Potential errors in assessing the true rain fade level due to variations in the clear-sky level (from a presentation by J. Thirlwell)*
The diurnal variations detected in the satellite beacon level can be caused by spacecraft effects, either intrinsically due to variations in heating in various parts of the orbit or due to the non-zero inclination of the orbit moving the satellite from within the centre of the non-tracking earth-station antenna beam, or they can be caused by changes in atmospheric conditions, principally humidity and cloud cover variations.

(a) The radar equation
The power flux density (PFD) at a distance r from an antenna having a gain G_T is given by

$$\text{PFD} = \frac{P_T G_T}{4\pi r^2} \text{ watts/m}^2 \tag{4.25}$$

where P_T is the transmitted power. Suppose that, at this point, there is an object that intercepts the power over an area S, and scatters the energy intercepted isotropically. The energy scattered back and incident upon the radiating antenna, E_{SC}, is therefore

$$E_{SC} = \frac{P_T G_T}{4\pi r^2} \times \frac{S}{4\pi r^2} \text{ watts/m}^2 \tag{4.26}$$

If the aperture area of the receiving antenna is A_R, the received power P_R is

$$P_R = \frac{P_T G_T A_R S}{(4\pi r^2)^2} \text{ watts} \tag{4.27}$$

If the same antenna is used for transmission and reception, and given that the gain of an antenna is $(4\pi A)/\lambda^2$, eqn 4.27 will reduce to

$$P_R = \frac{P_T G^2 \lambda^2 S}{(4\pi)^3 \times r^4} \text{ watts} \tag{4.28}$$

Eqn. 4.28 is called the 'radar equation'. Many of the parameters in the radar equation are constant for a given radar system. If the radar frequency is constant, eqn. 4.28 can be expressed as

$$P_R = CP_T S/r^4 \text{ watts} \tag{4.29}$$

where

$$C = G^2 \lambda^2/(4\pi)^3 \tag{4.30}$$

The term C is called the radar constant. In general, the radar constant will contain all the factors that are constant for a particular radar, path geometry, and operational configuration.

The above radar equation assumes that all the energy incident on the area S is scattered. In most cases, the object or medium causing the energy to be reflected back to the radar neither completely fills the beam uniformly nor does it only scatter the energy. In many cases, absorption also takes place and, especially with precipitation particulates, there is a range of dropsizes that leads to a certain reflection factor.

(b) Reflectivity factor

The amount of energy returned back along the path to the transmitting antenna is a function of both the size of the particles intercepting the radar beam at that point and the scattering coefficients of the particles. For a single spherical drop of water with a diameter D (mm), that is small compared to the wavelength λ(m) of the radar signal, the Rayleigh scattering cross section σ is given by [50]

$$\sigma = \frac{\pi^5 |K|^2}{\lambda^4} \times 10^{-18} D^6 \text{ metre}^2 \tag{4.31}$$

where $K = (n^2 - 1)(n^2 + 1)$ and the term n is the complex refractive index; squaring n, giving n^2, yields the complex relative permittivity (see Section 3.3.1). The value of $|K|^2$ is 0·93 for liquid water drops and 0·20 for ice particles. This large difference in $|K|^2$ between the liquid and solid phases of water is significant, as will be seen later.

For an ensemble of drops with a size distribution given by $N_{(D)}$, the number of drops between diameter D and $D + dD$ in a unit volume, the scattering cross-section per unit volume, η, is given by [50]

$$\eta = \frac{\pi^5 |K|^2}{\lambda^4} \times 10^{-18} \int_0^{D_{max}} N_{(D)} D^6 \, dD \text{ metre}^2/\text{m}^3 \tag{4.32}$$

The term in eqn. 4.32 given by $\int N_{(D)} D^6$ depends only upon the dropsize distribution of the particles in the volume being illuminated by the radar and is usually called the reflectivity factor Z. That is

$$Z = \int_0^{D_{max}} N_{(D)}\, D^6\, dD \text{ millimetre}^6/\text{m}^3 \tag{4.33}$$

In order to derive the specific attenuation α for a path from a measure of the back-scattered energy, either the dropsize distribution must be measured directly or an assumption based on a statistical parameter that relates Z to rainfall rate R must be invoked. Usually the latter is employed and an expression is used of the form

$$Z = aR^b \text{ millimetre}^6/\text{m}^3 \tag{4.34}$$

where a and b are empirical constants. Table 4.3 gives some of the commonly used values for a and b.

Table 4.3 *Some typical values of the empirical factors a and b used to relate reflectivity factor Z to rainfall rate R, in the expression Z = aR^b*

a	b	Rain type	Dropsize distribution	Reference
140	1·5	Drizzle	Joss	[51]
220	1·6	All	Marshall and Palmer*	[52]
250	1·5	Widespread Rain	Joss	[51]
380	1·32	All	Marshall and Palmer	[54]
396	1·35	All	Marshall and Palmer	[55]
400	1·4	Recommended by the CCIR		[53]
500	1·5	Thunderstorms	Joss	[51]

* This relationship is also valid for the Laws and Parsons dropsize distribution

Eqn. 4.34 assumes spherical raindrops. Large raindrops will distort and cause different values of Z to be observed with linearly polarised radars as the orientation of the incident polarisation vector is varied. This will give rise to a differential reflectivity.

(c) Differential reflectivity
If the reflectivity factor can be measured in both the vertical (suffix V) and horizontal (suffix H) orientation, the differential reflectivity Z_{DR} is given by

$$Z_{DR} = 10 \log_{10} (Z_H/Z_V) \tag{4.35}$$

By measuring the differential reflectivity, the oblateness of the raindrop ensemble can be estimated. Since raindrop sizes, and hence oblateness, increase as the rainfall rate increases, an increasing value of Z_{DR} observed will denote an increasing rainfall rate in the volume radiated. This can be seen in Fig. 4.21 (from Fig. 3.16 in Reference 50).

Suddenly changing values of Z_{DR} along the path of the radar beam can denote the boundary between air and rain or rain and ice crystals. The large difference in both refractive index and oblateness between rain and ice particles will yield such large changes in Z_{DR}. There are other ways of detecting these changes in the physical phase of the particles and these will depend on the type of radar used.

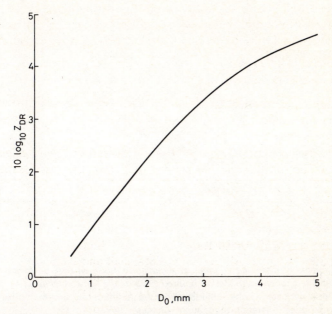

Fig. 4.21 *Variation of differential reflectivity factor Z_{DR} with median volume drop diameter D_0 for equivalent sphere (from Fig. 3.16 of Reference 50)*
Assuming a maximum drop diameter of 9 mm and a wavelength of 10 cm
(Copyright © 1979 IEE, reproduced with permission)

(d) Types of radar
(i) Single-frequency radars
This is the simplest form of radar. Even when these types of radar are accurately calibrated in gain, however, the accuracy of the estimates of slant path attenuation is poor [56, 57]. The errors arise from a number of causes with the principle contributors being statistical errors, dissimilar volumes, attenuating frequency and incorrect identification of the freezing level.

To obtain an estimate of the received radar power that is statistically significant, a large number of independent samples must be obtained for each volume being investigated. These samples are then integrated to average out the random fluctuations. If a radar-derived path attenuation is to be compared with a satellite beacon measurement made along the same path, care must be taken to ensure that the same overall integration time is used. Even then the volumes observed may be different.

In Fig. 4.22, the volume interrogated by the radar will be different from that illuminated by the beacon receiver antenna and the radar signal will also include sidelobe returns unless the radar antenna is well designed. The beacon receiver will essentially only respond to rain activity within the first Fresnel zone which will include a much smaller volume generally than that observed by the radar.

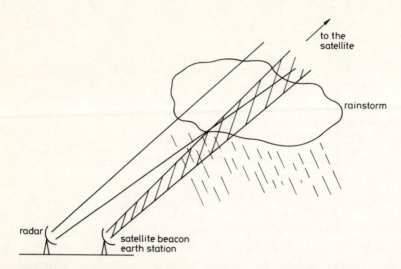

Fig. 4.22 *Example of dissimilar volumes of a rainstorm observed by a co-located radar and satellite beacon earth station pointed towards the same satellite*
Dissimilar beamwidths, including sidelobe contributions, and path offsets combine to produce different interrogation volumes.

The frequency used by the radar is also important from two considerations. Most meterological radars use the Rayleigh scattering equations that assume particle sizes are much smaller than the wavelength. This is usually true for radars that employ frequencies below about 10 Ghz. Above 10 GHz, Mie scattering theory should be used and an effective reflectivity factor Z_e calculated. Even at a frequency of 5 GHz, however, rain particles are becoming significantly attenuating and accurate knowledge of this attenuation is required in order to normalise the scattered energy received. Since the level of the detected scattered energy is being used to invoke a rainfall rate in the volume observed, an error in accounting for the attenuation of the radar signal in both directions through the other intervening volumes of rain can give rise to large errors in predicting the overall path attenuation. The largest errors, however, are due to an incorrect assumption of the physical phase of the precipitation.

In Fig. 4.23, a radar signal is sent through a large, precipitating rain cloud. At a height given approximately by the 0° isotherm, the ice particles that are falling start to melt. This region is known as the freezing level. The scattering albedo of ice particles is almost unity [15], leading to a large radar return. The attenuation of ice is insignificant, and so, if the presence of ice is not suspected, a large over-estimate of the path attenuation will result. The generally enhanced radar return at the freezing level has lead to the term 'Bright Band' being used for this portion of the data.

If the radar can be moved up and down fairly readily in elevation, referred to as nodding the radar, the variation in the position of any bright band can

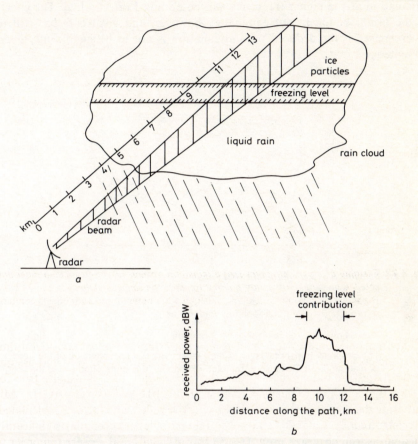

Fig. 4.23 *Illustration of the effect of the freezing level (sometimes called the melting level) on the received radar signal*
a Path geometry
b Received signal on the radar
The presentation of the received radar signal level versus distance shown in (*b*), after correcting for distance, is referred to as an 'A-scan'

sometimes be detected and its effects allowed for. Nodding the radar gives a range–height (*R–H*) display as opposed to a simple *A*-scan of a fixed radar shown in Fig. 4.23*b*. The *R–H* scan will give a 'snap-shot' of the propagation medium in a vertical plane. If there is a section of the *R–H* scan that shows a large increase in signal return at a fairly constant height, this is probably the freezing level. A typical *R–H* scan is shown in Fig. 4.24.

(ii) *Dual-frequency radars*

The effective reflectivity factor Z_e is essentially frequency independent up to about a frequency of 20 GHz [58]. If a non-attenuating frequency is transmitted by the radar together with another frequency in the range 10–20 GHz, the path attenuation at the higher frequency can be obtained directly [58]. The effects of the bright band are not eliminated, however, and great care has still to be excercised if the deduced path attenuation is not to be significantly over-estimated.

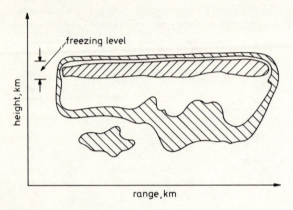

Fig. 4.24 *Schematic of a range/height (RH) scan*
Reflectivity levels

Low   Medium   High

The generally horizontal layer of high reflectivity is a characteristic of the freezing, or melting, level. Because of the high reflectivity, it is quite often called the 'Bright Band'.

(iii) *FM radars*

These are special cases of multi-frequency radars where the radar frequency is continuously swept over a bandwidth of a few megahertz. The carrier frequency of an FM radar is usually in an unattenuating band. The carrier is modulated with a linear frequency modulator. The received signal and a portion of the transmitted signal that has been coupled out of the feed and suitably attenuated are compared and the difference in frequency yields the range of the scattering volume in the path. If the scattering volume is moving with respect to the radar beam, a change in phase of the return signal will be noted. From this change in phase, the Doppler shift can be calculated, and hence the velocity component of the scattering volume along the radar beam. For this reason, these radars are sometimes referred to as Doppler radars and they are used increasingly to measure wind shear in weather systems [59, 60]. For propagation studies, the FM radars are limited to low rainfall rate situations (less than 10 mm/h) for two reasons. Firstly, the dynamic range is not large and, secondly, high rainfall rates are usually associated with heavy thunderstorms, the severe convective activity of which will exceed the dynamic range of the radar.

(iv) *Dual-polarised radars*

There are basically three types of dual-polarised radar that have been developed to date: a circular depolarisation ratio (CDR) radar, a linear depolarisation ratio (LDR) radar, and a differential reflectivity (ZDR) radar.

(a) *CDR dual-polarised radar*

First developed by McCormick and Hendry [61], the CDR radar transmits one sense of circular polarisation and receives the same sense (cross-polarised reflection) and the opposite sense (co-polarised reflection) (see Fig. 4.25a).

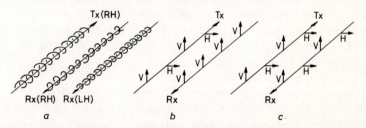

Fig. 4.25 *Schematic representation of the three different classes of dual-polarised radars*
  (a) C_{DR}: Transmit in one sense of polarisation continuously; receive in both the co- and cross-polarised senses continuously
  (b) L_{DR}: Transmit alternately in orthogonal linear polarisation senses; receive in only one linear polarisation sense. Owing to the switched transmit stream, reception is alternately co- and cross-polarised with respect to the transmit stream
  (c) Z_{DR}: Transmit alternately in orthogonal linear polarisation senses; receive alternately in orthogonal linear polarisation senses, synchronised to the transmit stream so that only co-polarised reception takes place

Four derived quantities are available from the measurements [62]. The first is the reflectivity for circular polarisation; the second is the circular depolarisation ratio which is a measure of the difference in received power between the two reflected polarisations; the third is the complex correlation between the two reflected polarisations; and the fourth is the apparent mean orientation angle of the precipitation causing the reflections.

The correlation between the two received signals has been inferred as a measure of the degree of alignment of the precipitation particles; good correlation implies a high degree of alignment. Since the received signals have both their amplitudes and their relative phases measured, the 'phase' giving the highest correlation will yield the apparent mean canting angle of the precipitation volume being inspected.

The difficulty with the CDR measurements is that it relies on a very sensitive, and well calibrated, radar in order to detect both the amplitude and the phase of the cross-polarised component. For this reason, light rain events or rain that occurs over very long path lengths are not usually successfully measured and so the usefullness of a CDR radar in gathering cumulative statistics is in doubt. On the other hand, the CDR radar has given some important insights into the physical processes occurring within rainstorms.

(b) LDR dual-polarised radar

An LDR radar can be of two types. One is similar to the CDR radar in that one polarisation sense is transmitted and the two, orthogonal, polarisation senses are received, although in this case the polarisations are linear. More recently [63], a switched polarisation transmission has been employed with a fixed, linearly polarised reception (see Fig. 4.25b).

The operation of the LDR radar is inhibited, like the CDR radar, by the very weak cross-polarised return signals, particularly if the polarisation orientations used are vertical and horizontal. Precipitation particles tend to align their principal axes close to the vertical and horizontal, thus yielding low cross-polarised reflections. Orientating the polarisations to be in the 45° planes is equivalent to a CDR radar [64].

(c) ZDR dual-polarised radar

A ZDR radar [65] only receives in the co-polarised sense, switching both the transmitted and the received linear polarisation senses synchronously (see Fig. 4.25c). Because no cross-polarised reception is attempted, the ZDR radar has a much larger dynamic range than any other dual-polarised radar.

The disadvantage of a ZDR is that a fast polarisation switch is required that is capable of handling high power levels. The need for rapid switching is dictated by the random variations in the medium being interrogated. If the sampling of the medium can be carried out in both polarisation senses before the precipitation particles have moved more than one quarter of a wavelength, the reflections in the two polarisation senses will be well correlated. The slower the switching rate, the more integration is required to achieve an adequate correlation between the polarisation senses.

By taking the ratio of the two reflectivity factors, a number of common terms in prior equations may be cancelled, yielding (arithmetically)

$$Z_H/Z_R = \frac{\int e^{-3\cdot67 D_H/D_0} D_H^6 \, dD}{\int e^{-3\cdot67 D_V/D_0} D_V^6 \, dD} \tag{4.36}$$

Remember that $N_{(D)} = N_0 e^{-3\cdot67 D/D_0}$ was the raindrop size distribution from eqn. 1.20. Fig. 4.21 shows values for D_0 that can be derived from values of ZDR. If a relationship between the equivolumetric diameter of a drop and its oblateness can be assumed, D_H and D_V can be calculated. From these, the raindrop size distribution $N_{(D)}$ can be determined by solving for N_0.

Small values of ZDR, however, will lead to large values of N_0 and to an over prediction of rainfall rate, and hence attenuation [66]. A ZDR radar is therefore not very effective in light rain conditions. To overcome this, a gamma raindrop size distribution was proposed of the form

$$N_{(D)} = N_0 D^\mu e^{-\lambda D} \tag{4.37}$$

with $\mu = 2$ [67]. Smaller fluctuations in attenuation predictions in light rain will occur since this gamma distribution minimises the influence of the smallest raindrops with the lowest values of ZDR [66].

A factor running through all the radar measurements, whether they be single parameter or multi-parameter [68], is the need to establish an independent check on the radar derived measurements by means of so-called 'ground truth' data. Satellite beacon or radiometer measurements along a path common to one of the radar scans will give a check on the radar-derived attenuation. In a like manner, rain gauges and distrometers will give a comparative value for the rainfall rate in the closest volume over the measuring devices. Once a consistent performance has been established for the radar by a comparison over many events between the radar data and the ground truth data, that radar can be used with confidence over wide areas and the results extrapolated to multiple paths and frequencies.

4.3 Experimental results

4.3.1 Radiometer experiments
That precipitation attenuated radiowave signals at microwave frequencies and above was appreciated decades ago [69]. The microwave transmissions were usually for terrestrial applications where multipath was the principal impairment to high availability. The system margin was therefore quite large in order to surmount the deep multipath fades, and the attenuation caused by rain was well within this margin. The increased noise temperature that accompanied the rain attenuation was of even less importance. It was only when the antennas were directed skywards to receive signals from artificial earth satellites that the added noise temperature became significant [70].

At frequencies in the 6 and 4 GHz communications satellite bands, where rain attenuation is not very high, passive sky-noise radiometers proved to be quite effective [71]. For frequencies well above 10 GHz, sun-tracking radiometric measurements were initiated in the mid-1960s, giving results up to frequencies of 90 GHz [72–74]. The basic problem with sun-tracking radiometers, as has been observed earlier, is the variation in the path elevation angle and the lack of data for those periods when the Sun is not above the horizon. For this reason, many experimenters turned to passive radiometers, particularly since the most pressing need was for information on the 14/11 GHz communications satellite bands where the limitations of this type of radiometer are not too serious. The bulk of these experimental data are now enshrined in the CCIR Study Group

5 data bank which is updated at every meeting of the Study Group. The present one is Reference 75. Of particular importance for the development of European satellite systems was the COST Project 205 that synthesised the available European Earth–space propagation data [76].

4.3.2 Radar experiments

The rapid development of radar during the Second World War led increasingly to the use of higher powers and more sensitive receivers. With these came the discovery that the usefulness of the radars deteriorated in bad weather conditions. After the war, what had been merely a nuisance, the presence of 'clutter' in the returns due to the presence of heavy rain in the beam of the radar, led directly to the development of meterological radars. The ready detection of rainstorms at a great distance, coupled with the ability to map their movement, produced much more accurate near-term weather forecasts.

It was clear that the intensity of the rain echoes was well correlated with the intensity of the rain causing the echoes but it was difficult to establish a reliable method of deducing the rainfall rate from the amplitude of the echoes. Work in the early 1950s [77], however, established a mathematical description of the raindrop size distribution that is used to this day (see eqn. 1.20).

The initial experiments using mono-polarised radars proved to be reasonably accurate provided melting snow (the bright band) and hail were not present in the beam [56, 78]. The circumvention of these potential errors through the use of dual-parameter radars [65, 68] is now well understood and a collection of measurement results has been published as a special issue of *Radio Science* [79].

4.3.3 Satellite beacon experiments

The first geostationary satellites operated in the 6/4 GHz bands [80]. In these bands, the attenuating effects of the atmosphere are very small except at very low elevation angles. One of the first series of low elevation angle experiments aimed at acquiring the annual statistics of path attenuation took place in Japan using INTELSAT IV-A spacecraft [81]. The variation of the effective pathlength through the precipitation medium was clearly established through this experiment.

The rapid increase in the bandwidth required for satellite telecommunications led to the forecast need for bands other than those at 6 and 4 GHz. The next pair of bands allocated for Fixed Services using geostationary satellites was at 14/11 GHz and 30/20 GHz. In view of the need to predict more accurately the likely effect of precipitation on these bands, a series of satellite beacon experiments was undertaken. Table 4.4 sets out the past and future geostationary satellites that incorporated beacons for experimental or operational requirements.

Despite the unstabilised attitude of ATS–5 when eventually in orbit, some usable results were obtained [82]. ATS–6 proved to be an outstanding experimental satellite. Following some initial measurements over North America [83], a well co-ordinated set of experiments in Europe [84] provided useful information about both the propagation medium itself at frequencies of 20 and 30 GHz and the pitfalls of satellite beacon experiments. With this background the scene was set for some very thorough experiments with CTS (Hermes), COMSTAR, ETS–II, SIRIO, CS, BS, OTS–2, and INTELSAT V satellites over the next decade. Given reasonable good fortune, a new series of experiments in the 30/20 GHz bands will begin with Europe's OLYMPUS and the US ACTS satellites in the early 1990s. The more data that are accumulated, the more reliable the attenuation modelling will be, as the variability in space and time of rain attenuation does not lend itself easily to the accurate prediction of that phenomenon.

Table 4.4 *Beacons available at frequencies above 10 GHz on geostationary satellites*

Satellite	Launch date	Beacon frequencies	Location
ATS-5	July 1969	15·3, 31·65 GHz	N. America (N.Am.)
ATS-6	May 1974	20, 30 GHz*	N.Am./Europe
CTS (Hermes)	Jan. 1976	11·6 GHz	N.Am.
COMSTAR	May 1976[†]	19·04, 28·56 GHz	N.Am.
ETS-II	Feb. 1977	11·5, 34·5 GHz	Japan
SIRIO	Sept. 1977	11·6 GHz**	N.Am/Europe/China
CS	Dec. 1977	***	Japan
BS	April 1978	12 GHz	Japan
OTS-II	May 1978	11·6 GHz****	Europe
INTELSAT V	Dec. 1980[†]	11·2, 11·4 GHz	World wide
INTELSAT VI	(1989)[‡†]	11·2, 11·4, 11·7, 12·5 GHz	AOR and IOR[††]
OLYMPUS	(1989)[‡]	30, 20, 12 GHz	Europe
ACTS	(1990)[‡]	30, 20 GHz	N.Am.
ITALSAT	(1990)[‡]	45 GHz	Italy

 [†] First launch of a series of spacecraft
 [††] Atlantic Ocean Region and Indian Ocean Region
 * An 18/13 GHz carrier experiment was also carried
 ** An 18/11·6 GHz carrier experiment was also carried
 *** No beacons carried; 30/20 GHz carrier experiment only
 **** A 14/11·6 GHz carrier experiment was also carried
 [‡] Estimate only

4.4 Variability of path attenuation in space and time

By its very nature, precipitation intensity varies both in space and time. The normal method of treating such variabilities is to express the results, and any consequent predictive models, in a statistical manner. Traditionally, measured results are usually presented as cumulative statistics with the variable (excess attenuation along the path, rainfall rate on the ground, etc.) plotted against the percentage time the variable exceeds the given value.

4.4.1 Cumulative statistics

The CCIR has currently divided up the world into 15 distinct rain climates and assigned average annual cumulative statistics of the rainfall for each of the climates. Figs. 4.26*a* and 4.62*b* show the cumulative statistics for the 15 rain climates. Table 1.4 sets out these data in tabular form and Fig. 1.29 delineates the 15 rain climates of the Earth. For interference calculations it is also useful to have cumulative rainfall rate statistics for percentage times greater than 1%. This is done [85] using the formula

$$R_{(p)} = R_{(0\cdot3)} \left(\frac{\log (p_c/p)}{\log (p_c/0\cdot3)} \right)^2 \text{ millimeter/h} \tag{4.38}$$

where $R_{(p)}$ = rainfall rate at the desired percentage time p
$R_{(0\cdot3)}$ = rainfall rate at 0·3% of the time
p_c = percentage time at which the rainfall rate decreases to zero

The term p_c can be obtained [85] from Table 4.5

Table 4.5 *Value of P_c at which the rainfall rate decreases to zero for the 15 CCIR rain climatic zones*

Rain climatic zone	$P_c(\%)$
A, B	2
C, D, E	3
F, G, H, J, K	5
L, M	7·5
N, P, Q*	10

* It is assumed that the new climate Q will fall into the same category as climates N and P in the percentage of raining time

In Figs. 4.26*a* and 4.26*b* it can be seen that the divergence in the curves becomes greater as the time percentage reduces. This is because the occurrence of the severe, convective storms that give rise to very high rainfall rates increases as the climates go from a tundra/desert classification (climate A) to an equatorial rain forest classification (climate P). In all of the climates, however, the time that these severe storms occur represents only a small fraction of the total time. A few such events, therefore, influence the overall statistics markedly. The cumulative rainfall statistics measured over a one year period can therefore show substantial differences from similar statistics measured at the same site in

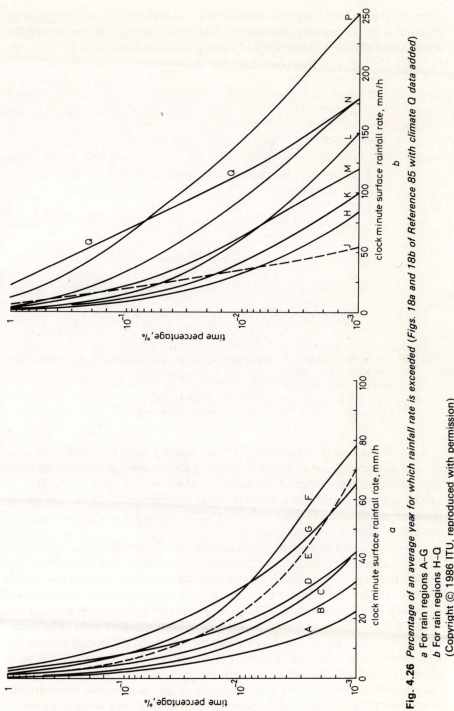

Fig. 4.26 *Percentage of an average year for which rainfall rate is exceeded (Figs. 18a and 18b of Reference 85 with climate Q data added)*
a For rain regions A–G
b For rain regions H–Q
(Copyright © 1986 ITU, reproduced with permission)

different years. This temporal variability of annual rainfall rate statistics is reflected in the temporal variability of annual path attenuation measurements. An example of such variations in a sky-noise radiometer measurement is shown in Fig. 4.27 (Fig. 1 of Reference 23)

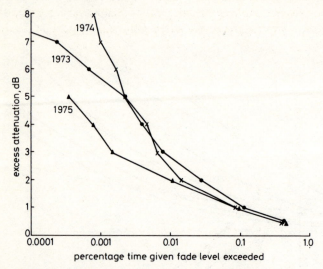

Fig. 4.27 *Cumulative statistics of excess attenuation at 11.6 GHz (Fig. 1 of Reference 23)*
Elevation angle: 29·5°
Azimuth angle: 198·25°
The 1973 curve excludes data for May and August
(Copyright © 1977 IEE, reproduced with permission)

At fairly high percentage times (0·1% or greater), there is little difference in the cumulative statistics in Fig. 4.27. At very low percentage times (0·001% or less), however, the variation between the three sets of data is marked. A percentage time of 0·001% represents about five minutes in a year, and so the presence, or absence, of a core of intense rain in the path can significantly affect the annual cumulative statistics.

One test that can be applied to a multi-year set of cumulative statistics to see if a 'true' long-term average curve has been obtained is to remove one of the sets of data that represent a year of measurements and observe the change that this makes to the average curve. If there is no noticeable change in the average curve when any set of annual data is removed, a satisfactory long-term average has been obtained. Measurements in Sweden have indicated seven years as being the minimum measurement period before such a condition is reached [86]. There are some indications, however, that there is an underlying trend in rainfall rate/climate phenomena that track the sunspot cycle variations [87] in a similar fashion to ionospheric effects. If this is so, an eleven year period is the minimum experimental period required to establish stable, long-term average statistics.

Within the annual variations there are also cyclic variations on a seasonal and diurnal basis.

(a) Seasonal variations

Most regions experience a variation in the rainfall accumulation with season. This variation in accumulation usually tracks similar variations in the rainfall rates that occur. An example of the latter is shown in Fig. 4.28 (from Fig. 4 of Reference 88).

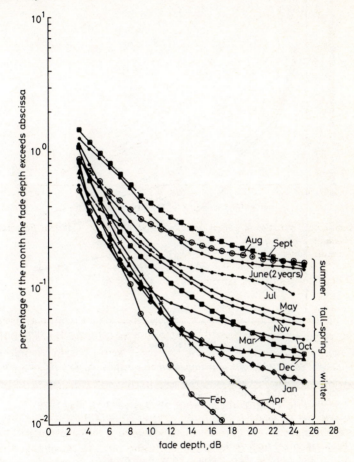

Fig. 4.28 *Comparison of average monthly cumulative fade distributions over three-year period at Wallops Island associated with COMSTAR beacon at 28·56 GHz (Fig. 4 of Reference 88)*
(Copyright © 1982 IEEE, reproduced with permission)

The rainfall rate data in Fig. 4.28 clearly show a higher probability of increased rainfall rates in the summer than in the winter. For temperate latitudes, this is typical of the summer rain phenomenon that occurs due mainly to convective events such as thunderstorms. Other areas may show two peaks of high rainfall rate activity (e.g. the spring and the fall [89]) or a peak that occurs in the typhoon seasons, as in Japan. In all cases, the peaks are associated with the warming effects of the Sun in association with the presence of moist air. This would suggest that a diurnal variation should also exist.

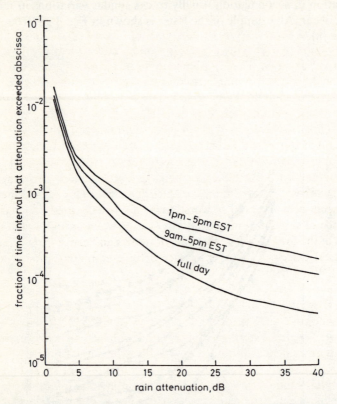

Fig. 4.29 *Cumulative distributions showing diurnal variation of 19 GHz rain attenuation*
  (*Fig. 17 of Reference 90*)
  Data are from the secondary facility
  Crawford Hill COMSTAR Experiment 19 May 1977–18 May 1978
  19·04 GHz
  38·6° Elevation
  Polarisation 21° from vertical
  (Copyright © 1982 IEEE, reproduced with permission)

(b) Diurnal variations

Earth stations that are located in regions where the high intensity rainfall is usually due to thunderstorms will show a strong diurnal variation in the path attenuation statistics. This is because the convective activity associated with thunderstorms is generated by surface warming due to solar radiation. As the day progresses, so the surface of the Earth heats up, increasing the convective activity and the probability of a shower or thunderstorm, particularly in the summer months. An example of the diurnal variation in the statistics is given in Fig. 4.29 (from Fig. 17 of Reference 90).

In Fig. 4.29, the data are split into three curves, the 'full day' data that represent the cumulative statistics, the '9 a.m. to 5 p.m.' data that represent those data for a normal working day, and the '1 p.m. to 5 p.m.' data for the afternoon period of a working day. At low attenuations (below 5 dB) there is no significant variation between the three curves but, for higher attenuations, the influence of the afternoon thunderstorms on the statistics is apparent. For this particular experiment [90], 4 p.m. local time was statistically the period when the highest path attenuations were experienced.

In other climates there may be other influences than simple convective heating that will cause a diurnal variation in the path attenuation statistics. One experiment in Japan [91] indicated that there were statistically two peaks in path attenuation that centred around 6 a.m. and 10 p.m. local time. It was speculated that this pattern of precipitation could be inherent in a coastal climate [91].

The concept of worst hour or worst day is not usually used in describing the performance of satellite communications systems even though it will influence the design of the overall link somewhat. Of more significance in determining the the statistical extremes to be encountered by a communications link is the concept of Worst Month.

4.4.2 Worst month

The worst month statistics for a given link are obtained by compiling a composite curve using, at each threshold level, the highest exceedence probability obtained in any calendar month. Fig. 4.30 illustrates the process.

In Fig. 4.30, a set of 12 curves are drawn representing statistical data for each of the calendar months in a year. From these individual monthly statistics, a composite curve can be constructed that comprises the outer envelope of the data; the worst month. In this case, no one calendar month is the worst month, the composite curve being built up of segments from months 9, 11 and 12.

Some communications services, particularly those that involve broadcasting (e.g television), require the link to operate with a specified outage in the worst month. Typically, an outage of 1% is tolerated in the worst month. The ratio between the attenuation experienced in a worst month to that exceeded in an average year depends on the probability level selected and the climate. Fig. 4.31a shows this dependence for a number of climates. (Note, however, that the climates depicted do not follow the same 15 climate descriptors given in Figs. 4.26a and b and Table 1.4).

From Fig. 4.31a, for example, if a path attenuation of 3 dB is experienced in curve C at an annual probability level of 2×10^{-4} (i.e. 0·02% of a year) then the path attenuation for 0·02% of the worst month is $4 \times 3 \text{dB} = 12 \text{dB}$. The multiplying ratio Q, which is the average annual-worst-month probability divided by the average annual probability, can be related to the average annual probability Y by Reference 93:

$$Q = AY^{-\beta} \tag{4.39}$$

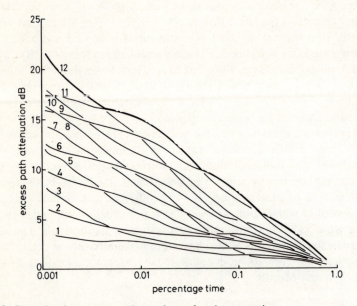

Fig. 4.30 *Example of a worst month envelope of path attenuation*
—·— Envelope of the worst month made up of curves for months 9, 11 and 12

where A and β are constants. The values of A range between 1·20 and 3·30 and those for β between 0·167 and 0·074 [92] with those for North America and Europe giving [92]

$$Q = 1·64Y^{-0·130} \tag{4.40}$$

Using percentages rather than probabilities (with $A = 3$ now rather than 1·64), Report 338 [94] inverts eqn. 4.40 in terms of the average annual-worst-month probability p_w and average annual probability p, giving

$$p = 0·29p_w^{1·15}\% \tag{4.41}$$

Note that $Q = p_w/p$ and $Y = p$. Eqns. 4.40 and 4.41 have been found to give reasonable fits to worst month data for rainfall rate measurements as well as path attenuation [95, 96] and a recent test of the model given by equation 4.39 indicated an RMS error of about 24% which was considered to be a reasonable accuracy [180]. Of the two equations above, eqn. 4.41 tends to be more widely used.

An exponential model that describes the relationship between the average monthly excess and the probability that the monthly excess exceeds this figure has been proposed [170]. Of more interest, perhaps, is the variation of the worst month data. This variation has been derived [170] and the results are shown in Fig. 4.31*b*.

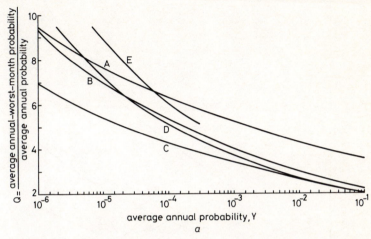

Fig. 4.31a *Ratio of annual-worst-month to average annual probability, Q, as a function of average annual probability Y(from Fig. 4 of Reference 92)*

A: Prairie + Northern ⎫
B: Central + Mountain ⎬ Rain rate, Canada [93]
C: Coastal + Great Lakes ⎭
D: Rain rate and rain attenuation, Europe [92]
E: Rain rate, Sweden
(Copyright © 1986 ITU, reproduced with permission)

An example of the use of the Figure in Reference 170, shown in Fig. 4.31*b* is, for an average worst month/year ratio Q of 6, 10% of the worst months will show a fading percentage which is at least 1·7 times the average worst-month percentage. Thus, once Q is known for a particular outage criterion, the more extreme situations can be determined directly using Fig. 4.31*b* [170].

4.4.3 Short-term characteristics
The short-term characteristics of a phenomenon describe the instantaneous variability of that phenomenon. For path attenuation, there are three short-term characteristics that are important for communications system modelling: the duration of the attenuation event or fade, the interval between successive fades, and the rate of change of attenuation.

(a) Fade duration
The measurement data to date [e.g. 98, 176] seem to indicate that the duration of fades exceeding given thresholds have a log–normal distribution. For a given path there is apparently no significant dependence on the fade depth, at least up

to a level of 20 dB [98]. This would seem to indicate that the larger time percentages, which would include the lower fade level events or high measurement frequencies, are made up of a larger number of individual events. In

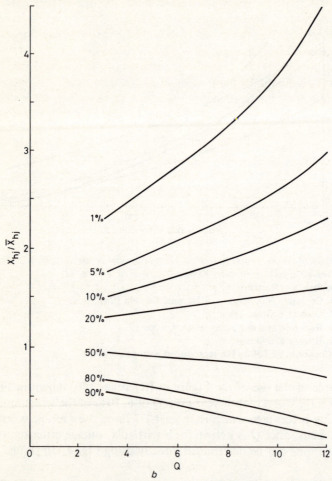

Fig. 4.31 b *Ratio of the individual worst month exceedance X_{hj}, to the average worst month exceedence $\bar{X}_{hj}$ for various percentiles as functions of Q (from Fig. 3 of Reference 170)*
The deviation of the 50% curve away from unity indicates the strong skew of the X_{hj} distribution for large values of Q
(Copyright © 1988 ITU, reproduced with permission)

essence, the average fade duration is largely independent of the fade depth over a wide range of fade levels. At small rain fade levels (less than 1 or 2 dB), equipment accuracy and other non-rain effects such as scintillation will distort the statistics. Similarly, at the extreme end of the fade margin, only one or two events will be present in a given year, and this, too, will distort the statistics.

An example of average fade duration is given in Fig. 4.32 (from Ref. 99 in Reference 98).

In Fig. 4.32, an average fade duration of approximately 5 min exists for most fade level thresholds, and this seems to be typical for most paths and climates with the exception of those regions that are subject to extremely severe and widespread events such as typhoons. It should be borne in mind, however, that the deviations possible from the average fade duration become quite large at the relatively smaller fade levels. It is not uncommon, for instance, to have a fade that exceeds 3 dB for more than an hour at frequencies above 14 GHz. For satellite–ground communciations systems designers, as well as the level and the duration of a fade being important, information on the average time between fades of a given level is also useful.

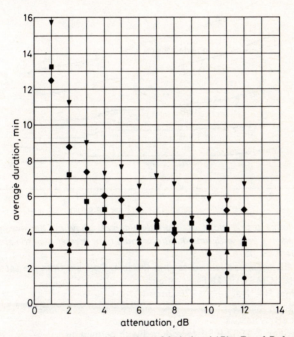

Fig. 4.32 *Average fade duration as a function of fade level* (*Fig. 7a of Reference 98*)

   ▼   Darwin, Australia   14 GHz $\theta = 60°$ (Nov. 1977–May 1979)
   ◆   Darwin, Australia   11 GHz $\theta = 60°$ (Nov. 1977–May 1979)
   ■   Innisfail, Australia   11 GHz $\theta = 45°$ (Oct. 1978–Apr. 1979)
   ▲   Melbourne, Australia 14 GHz $\theta = 45°$ (June 1980–June 1983)
   ●   Melbourne, Australia 11 GHz $\theta = 45°$ (June 1980–June 1983)

(b) Interval between successive fades

When cataloguing extreme weather statistics, meteorologists often use a term called Return Period. A 'ten-year value', for instance, implies a recorded value that is unlikely to be repeated, or return, for at least ten years. Multi-year, extreme values such as these are of interest to communications systems designers only insofar as they describe the potential variations from the annual average statistics. For economic reasons, it is the annual average statistics that are used to design the link. Of more concern than the multi-year extremes are the return periods within a year of rain events that could disrupt the performence of the link. For example, given that a 10 dB fade has occurred (causing, say, a 5 min outage) what is the average interval before the next 10 dB fade event occurs? Fig. 4.33 (from Fig. 19 of Reference 90) gives the minimum interfade interval, or return period, from a 19 GHz satellite beacon experiment.

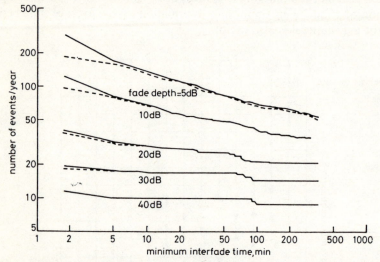

Fig. 4.33 *Cumulative distributions of interfade intervals showing the average number of intervals in a year with time durations equal to or greater than the abscissa (Fig. 19 of Reference 90)*
The 1 dB hysteresis is discussed in the text. Data are from the secondary facility.
19 GHz COMSTAR, 18.5° elevation
Crawford Hill, NJ. June 1976–June 1978
—— No hysteresis
- - - 1 dB hysteresis
(Copyright © 1982 IEEE, reproduced with permission)

In Fig. 4.33, the curves are drawn with and without a 1 dB hysteresis. The inclusion of a 1 dB hysteresis removes the effects of minor, random equipment errors and also simulates the average difference in signal power level that exists between losing lock and re-acquiring lock on a beacon signal. The significant impact of including hysteresis at the lower fade levels is noteworthy. From Fig. 4.33, the almost horizontal characteristics at the fade levels of 20, 30 and 40 dB illustrate the large return period between fades at those threshold levels. The relatively steep slope of the 5 dB thereshold curve, on the other hand, indicates a much greater variability in the return period of smaller rain fade events. In this two-year experiment [90], the median return period for 5 dB fades was 31 mins, and for 10 dB fades 46 mins. These results are typical for temperate regions and the data could be scaled with some degree of confidence to other frequencies and elevation angles (see Section 4.6.1)

(c) Rate of change of attenuation

In a similar fashion to rain fade duration satistics, the distributions of the rate of change of attenuation data appear to be log-normal with a median of about 0·1 dB/s [98]. Little difference has been observed between the positive-going (fading) and negative-going (recovering) slopes of the rate of change of attenuation for integration times of 10 s or more. However, there appears to be clear evidence that, as the fade rate increases, the difference between the fade slope and the recovery slope tends to increase with the fade slope always being greater. The difference becomes more marked as the integration time decreases below 10 s [97]. The physical explanation could be that the higher fading rates (both positive and negative) are associated with thunderstorms and that the leading edges of thunderstorms contain higher rainfall rates than the trailing edges.

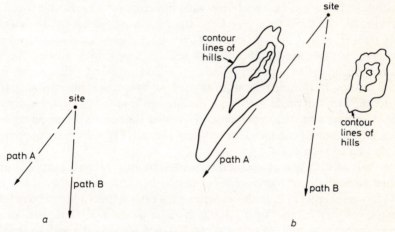

Fig. 4.34 *Examples of the sites with and without uniform topography in the area around the site*
a Uniformly flat terrain surrounding the site
b Non-uniform terrain surrounding the site
In case (*a*), path A and path B will experience similar rainfall characteristics on a statistical basis and so no significant variations will occur between their cumulative results. In case (*b*), however, significant topographic features surrounding the site will result in different rainfall characteristics along the two paths. Hilly regions usually experience more rainfall than flat regions and so the attenuation along path A should be statistically higher than that along path B for the same frequencies and elevation angles.

In most experiments reported to date, the average fade slope does not appear to depend significantly on the fade level, with a maximum fade rate of about 1 dB/s being reported in the frequency range 10–14 GHz for integration time constants on the order of 10 s. Much higher fading rates are observed with integration times below 10 s [97]. Some evidence does exist, however, that with increasing fade levels there is a trend to higher fade rates [97, 100, 171] and that there is a statistical relationship between fade slope and fade duration [101]. Both of these results are tentative at present with, if anything, the trends being too weak to permit any engineering significance to be attached to them [172].

4.4.4 Site-to-site variability

Meteorological aspects discussed in Section 1.3 indicate that small-scale variabilities in the weather patterns can occur over very small distances. These can lead to very different slant path attenuation statistics being measured from the same site using different azimuth angles, or generally uncorrelated statistics being obtained between simultaneous measurements from two sites only a few kilometres apart. These are due to azimuthal and spatial variations in the rainfall characteristics.

(a) Azimuthal variations

Even though there may be a preferred wind direction for a particular site, if the terrain around the site is flat over a large area, simultaneous slant-path measurements made from that site with different azimuth angles will show no significant differences. If there are significant variations in topography around the site, however, regions of enhanced rainfall and regions of partial shadowing from rainfall can result. If these regions of rainfall variations are some distance from the site, the azimuth angle used from the site can lead to significant variations in the slant path statistics depending on whether the measurement path passes through the region of enhanced, or reduced, rainfall. Fig. 4.34 gives a schematic presentation of such cases.

The first satellite beacon measurements to confirm the possibility of simultaneous azimuthal variations were conducted in Germany during the SIRIO and OTS compaigns [102]. In this experiment, a ridge of hills a little above 200 m in height roughly parallel to the azimuth angle to the SIRIO satellite caused appreciably more rainfall to occur along that path than on the path to the OTS satellite which was essentially over the floor of a valley. The situation was similar to that shown in Fig. 4.34*b*.

(b) Spatial variations

Small-scale variations in rainfall rate characteristics due to both topographical features of the local terrain and the limited horizontal extent of severe rainfall events can be used with advantage to reduce the net attenuation experienced along a slant path if more than one site is used simultaneously. This technique is referred to as site diversity or path diversity [103].

4.4.4.1 Site diversity

Path diversity in communications systems involves the provision of alternative propagation paths for signal transmission, with the capability to select the least impaired path when conditions warrant. For satellite communications systems, implementation of path diversity requires the deployment of two or more interconnected earth terminals at spatially separated sites; hence the use of the term 'site diversity'. A pictorial representation is shown in Fig. 4.35. This concept is based on the observation [104] that the most severe slant path impairments at frequencies above about 10 GHz are generally caused by intense rainfall occurring in individual rain cells of limited spatial extent. Deployment of multiple terminals with site separations somewhat in excess of the average horizontal dimensions of individual intense rain cells is expected to improve markedly the system availability, since simultaneous (joint) path outages are presumed to be random and infrequent for such a configuration.

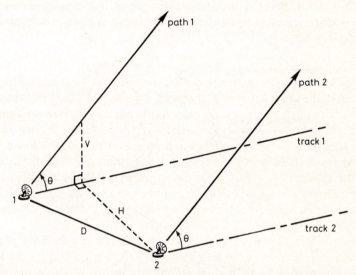

Fig. 4.35 *Dual-site diversity configuration illustrating the horizontal and vertical path separations (Fig. 3.2 of Reference 103)*
(Reproduced with permission from INTELSAT and COMSAT)

Although the meteorological aspects of site diversity (see Section 1.3.3) are somewhat more complex than this simple picture would imply, experimental data do confirm that the system performance achievable with two well-sited diversity terminals can be significantly superior to that obtainable with single-path operation. Measurements [105] and analysis [106] also indicate, however, that the additional performance benefits obtained by operation with more than two terminals is small. Implementation of additional diversity terminals to operate with a main station would also be complex and expensive. The emphasis of investigations, therefore, concentrates on dual-site (as opposed to three or more sites) diversity configurations for satellite communications.

There are several factors that may affect the site diversity performance of a particular installation. These include:

(a) Terminal separtion D
(b) Path geometry (elevation angle θ and azimuth angle ϕ)
(c) Local meteorological characteristics (rainfall rate statistics, degee of convectivity, rain cell dimensions, shapes, relative separation etc.)
(d) Frequency f
(e) Orientation of the baseline joining the sites
(f) Local topographical features.

Interrelationships exist between these parameters, which are shown pictorially in Fig. 4.36, so that isolating the dependence of diversity behaviour on any one of them is difficult. Before considering these topics, the characteristics of site diversity performance will be discussed.

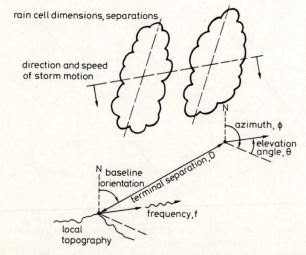

Fig. 4.36 *Pictorial representation of factors affecting site diversity (Fig. 1.1 of Reference 103)* (Reproduced with permission from INTELSAT and COMSAT)

(i) *Characteristics of site diversity peformance*
(a) *Calculation of site diversity performance*
A hypothetical, but representative, example of cumulative distributions of 11 GHz rain attenuation for two single paths (A and B), and for the diversity (joint) distribution (J), constructed from the concurrent fading records for paths A and B by selecting the lesser attenuation for each data sample, is shown in Fig. 4.37.

The mean (average) of the two values of rain attenuation at each time percentage defines the mean single-site attenuation distribution (dashed line in

Fig. 4.37), which is commonly used as a reference distribution for classifying diversity performance.

Two standard statistical approaches exist for relating the mean single-site and diversity attenuation distributions, as illustrated in Fig. 4.37. The diversity improvement factor [104], or diversity advantage [105] is defined as the ratio of the single-path p_m and diversity p_{div} time percentages for a specified rain

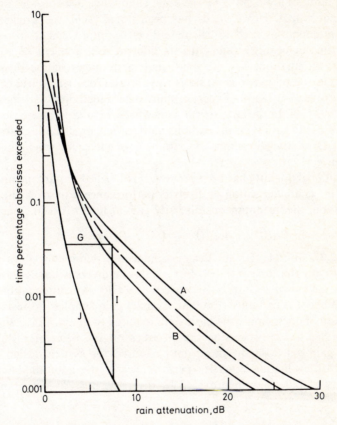

Fig. 4.37 *Typical characteristics of 11 GHz single-path (A and B) and diversity (J) rain attenuation distributions giving the definitions of diversity gain (G) and diversity advantage or improvement (I) (Fig. 1.2 of Reference 103)*
(Reproduced with permission from INTELSAT and COMSAT)

attenuation A:

$$I(A) = p_m(A)/p_{div}(A) \tag{4.42}$$

Diversity gain G [107, 108] is defined in an orthogonal sense to diversity advantage and is the difference (in decibels) between the single-path and diversity rain attenuations for a given time percentage p:

$$G(p) = A_m(p) - A_{div}(p) \tag{4.43}$$

One apparent merit of the diversity improvement factor is that it can be determined for small time percentages of the diversity distribution. Conversely, it is not defined for large single-site attenuations. For diversity gain, the regions of applicability are just reversed.

Both quantities are defined by the same cumulative distributions and are, in a sense, equivalent. However, as has been noted [106], the single-path and diversity attenuations used to calculate diversity improvement correspond to different time percentages, and thus to different measurement reliabilities. In particular, attenuations for the small time percentages of the diversity distribution may be subject to large statistical uncertainties, which will be reflected in the improvement factor. Experimentally determined diversity improvement factors [109, 110] do, in fact, reveal somewhat irregular, scattered behaviour, whereas diversity gain typically exhibits more-or-less predictable characteristics [109, 111]. Diversity gain is thus the parameter of choice for specifying diversity performance.

An additional quantity has been proposed [112] called 'instantaneous diversity gain' for the characterisation of diversity performance. For an N-site diversity configuration, the instantaneous diversity gain at time t is defined as:

$$G_i(t) = A_{max}(t) - A_{min}(t) \tag{4.44}$$

where $A_{max}(t)$ and $A_{min}(t)$ are the maximum and minimum values of the N single-path attenuations at time t. For a dual-site diversity configuration, $G_i(t)$ is simply the positive difference between the two path attenuations at any instant. While it is certainly true that diversity gain supplies no instantaneous information, it does supply the basic data required to design a diversity system. On the other hand, processing the statistics on the basis of instantaneous diversity gain will lose the basic data required for system design. What is required for evaluating diversity performance is the additional (statistical) availability of fade margin provided by diversity, not the instantaneous difference between attenuations.

(b) Reference distribution
Ideally, if the least-impaired path of a diversity configuration could always be identified and selected for communications, the cumulative distribution defined at each time percentage by the lesser of A and B in Fig. 4.37 would constitute a more reasonable reference distribution than does the mean single site distribution. Diversity gain would then be less than computed from the mean single path attenuation, which presents an apparant paradox: Why does not perfect switching between diversity paths maximise the available diversity gain?

In fact, such switching does maximise the diversity gain but the results are reflected in the diversity (joint) distribution of Fig. 4.37. The assumption of perfect switching is implicit in available data, which are processed by selecting the smallest path attenuation for each sampling interval to compile the joint attenuation distribution. Diversity combining would also achieve the goal of 'perfect switching' since the stronger of the two signals would always be used as the reference. For analogue signals, very accurate phase control is required to achieve diversity combining, and so tends to be limited to fairly narrow-band applications.

In operating systems, switching will be performed in accordance with an algorithm designed to minimise the number of switches (and the concomitant possibility of a temporary link interruption) while maintaining acceptable service [113]. Switching will typically be avoided, even if the primary link is somewhat impaired, provided the path is available with some reasonable reserve margin. This is similar to the hysteresis employed earlier in developing single site statistics (see Section 4.5.3). In any case, measurement inaccuracies in the monitoring equipment will inevitably introduce errors in establishing the least impaired path, and therefore in switching efficiency. The joint attenuation distribution achievable with perfect switching will not be attained in such circumstances. The degree to which it will be achieved is a function of the switching strategy and the measurement accuracy of the equipment.

The above considerations render the details of the reference distribution somewhat moot for many applications. In practice, the mean single-site attenuation is a convenient reference that presumably averages out some of the experimental inaccuracies and interannual variabilities of the measured statistics for the individual paths.

(ii) *Factors affecting site diversity performance*
A variety of site diversity measurements have been performed and are summarised in several References [e.g. 114–116]. Many of the parameters potentially affecting a particular site diversity installation are interrelated (e.g. baseline orientation, path geometry and local terrain) and some of these relationships will be noted.

(a) *Dependence on site separation*
The concept of site diversity is based on the assumption that propagation impairments on different paths are more or less uncorrelated for sufficient spatial separations between paths. Conversely, propagation degradations for identical paths (i.e. parallel paths with zero separation) are completely correlated. Between these limits, site (or path) separation must be a strong determinant of diversity gain. Not unexpectedly, measurements show that site separation D is the controlling factor in diversity performance for separations of less than 10 to 20 km.

The dependence of diversity gain G on site separation is illustrated in Fig. 4.38 [111], based on data at three frequencies from Ohio and New Jersey, USA. Diversity gain is observed to increase rapidly as D increases from zero until the

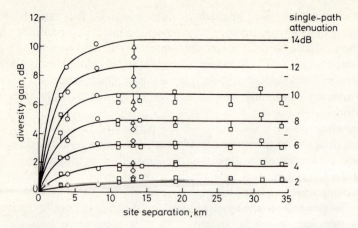

Fig. 4.38 *Dependence of diversity gain on site separation (Fig. 3.1 of Reference 103 after Fig. 1 of Reference 111)*
  ○ 15·3 GHz   ◇ 20 GHz
  □ 16 GHz   △ 30 GHz
(Copyright © 1976 IEEE, reproduced with permission from the IEEE, INTELSAT and COMSAT)

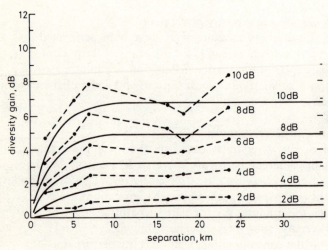

Fig. 4.39 *Comparison of diversity gain versus site separation for two sets of data from the USA [111] and Europe [118] (Fig. 3.1(a) of Reference 103)*
  —— From Fig. 3.1 (Reference 111)
  ●–––● Appleton Lab. (Reference 118)
(Reproduced with permission from INTELSAT and COMSAT)

separation exceeds 10–15 km, after which the benefits from further increases in *D* becomes small. The site separation for which 95% of the available diversity gain is achieved varies somewhat among different measurements, ranging from about 15 to about 30 km [117]. These differences may be due to influences of parameters other than site separation, such as configuration geometry, climatic differences or terrain effects. Data from an 11·6 GHz radiometric site diversity experiment in the UK [118] show similar trends to the US data. A comparison of the two sets of data is shown in Fig. 4·39 and some interesting differences are apparent between the results.

Firstly, in general, the diversity gain in the UK experiment seems to exceed that for the US experiments, which supports the argument that, for a given site separation, diversity gain decreases as frequency increases. The dependence is fairly weak, however, especially at the medium elevation angles (25–55°) used in these experiments. Secondly, the increase in diversity gain with increasing site separation seems to be significantly higher in the UK experiment than that which would be 'predicted' using the USA data for small site separtions, in agreement with other European data [131]. Thirdly, and more significantly, there is a distinct drop in the diversity gain in the UK experiment for site separations between 7 and 18 km. This occurred because two, independent, intense raincells passed through the site diversity network and caused simultaneous rain fades at the widely spaced sites.

The observed dependence of diversity gain on site separation is consistent with joint probability characteristics of point rainfall rate (e.g Fig. 1.23). Both quantities show rapid decorrelation with increasing site separation, until 'saturation' effects begin to limit the quantity to a value corresponding to somewhat less than complete decorrelation [119, 120]. Concurrently measured path attenuation and point rainfall rate statistics for diversity sites often show comparable behaviour which suggests that joint rainfall rate statistics may be used to predict site diversity performance, analagous to the predictive techniques used to convert point rainfall rate statistics to single path attenuation statistics [109].

Because the rain medium is bounded in the vertical direction, and the precipitation pattern in the horizontal plane is often banded, it is possible that separation (both vertical and horizontal) of the radio paths may be more important than the physical site separation [116]. The geometry for the perpendicular path separtion was illustrated in Fig. 4.35, which also demonstrated the interdependence of these path separations on baseline orientation, path azimuth, and path elevation angle. For a specific configuration, local terrain and climatic factors (e.g. alignment between configuration geometry and the directions of local weather fronts, or any regional anisotropy in rain structures) may also be related to path separation. Such effects are not yet fully understood. However, it has been concluded [119] from meteorological considerations that horizontal increases in path separation were more useful than equivalent vertical increases for reducing the probability of simultaneous path impairments, at least for latitudes below 60° and elevation angles above 3°.

(*b*) *Baseline orientation effects*

As noted above, meteorological considerations [119] imply that maximising the horizontal path separation (for a specified site separation) will maximise diversity gain. From Fig. 4.35, the maximum horizontal path separation is achieved by orienting the diversity baseline prependicular to the path azimuth. Data from a three-site 15·5 GHz radiometric experiment in New Jersey [121] and from a six-site network of 11·6 GHz radiometers near Slough, UK [118] had previously supported this view. A later survey [109] also concluded that the preferred baseline orientation was normal to the radio path. It has been surmised [122], however, that orienting the baseline normal to the radio path, and secondarily to the direction of travel of convective weather fronts, would be most beneficial for site diversity applications. This idea has been apparently confirmed by extensive measurements in the USA [110].

Radar simulations of diversity configurations for Wallops Island, Virginia, USA, with baselines parallel to the slant path azimuths, revealed that baselines oriented perpendicular to the predominant direction of raincell elongation were preferred [123]. Later measurements from the six-site diversity experiment mentioned earlier also revealed that the prevailing direction of movement of weather systems could affect diversity performance [124]. Other radar simulations [117, 125] and analyses [126], however, showed only a small dependence of diversity performance on baseline orientation. It therefore appears that the results are not conclusive as regards baseline orientation. This situation is probably indicative that baseline orientation is not the dominant parameter for many diversity configurations, and also that the factors potentially affecting diversity performance are sufficiently interrelated to make ascribing observed performance to a single parameter (other than site or path separation) difficult. For example, in an experiment in southern Ontario, Canada, consistently poor diversity performance was attributed to orographic rainfall induced by an escarpment parallel to the baseline [127]. Possibly this topographic effect could have been negated by re-orienting the baseline prependicular to the escarpment.

Since the probability of simultaneous path impairments must inevitably be larger for parallel baseline and path azimuths at low elevation angles (because the vertical path separations will be small), especially for small site separations, baseline orientation will be important in these cases. As the possibility of joint impairment by a single cell is high for such a configuration, the preferred baseline orientation will indeed probably be the bisector of the obtuse angle between the path azimuth and the major axis of cell anisotropy (though in general the latter information will be unknown, and may thus be ignored). For large site separations, however, variations in the joint probability are small for differing baseline orientations, as confirmed by the measured data.

(*c*) *Effects of path geometry*

The geometry of a radio path is defined by the path azimuth (ϕ) and elevation

(θ) angles. For most cases, the azimuth angle dependence is covered by baseline orientation effects since it is the relative angle between the path and the baseline that appears to be important in determining diversity gain. Azimuth effects have been shown to be significant in some cases of single-site operation [102], although the effects noted may not have been as substantial in a two-site diversity experiment.

Strong elevation angle effects on single path propagation impairments are often observed (e.g. [98]). As the elevation angle decreases, the slant path length through the troposphere rapidly increases, thereby increasing the occurrence and severity of single path impairments. A concomitant increase in the probability of joint impairments is also expected and has been substantiated by diversity measurements. The diversity gain at 11·6 GHz for an elevation angle of 6° was found to be approximately half that achieved at 30° for a 7·1 km site separation near Slough, UK [128]. Similar diversity gain behaviour was observed on 14/11 GHz paths in western Japan at a 6° elevation angle and 17 km path separation [129] and in Virginia, USA at 11° and 7·3 km separation [130]. These diversity gain curves are shown in Fig. 4.40.

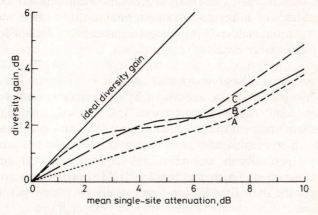

Fig. 4.40 *Diversity gain behaviour at a frequency of 12 GHz for low elevation angles (Fig. 3.3 of Reference 103)*
A: Slough, UK (D = 7·1 km, θ = 6°) [128]
B: Kurashiki, Japan (D = 17 km, θ = 6°) [129]
C: Blacksburg, USA (D = 7·3 km. θ = 11°) [130]
(Reproduced with permission from INTELSAT and COMSAT)

Interestingly, the gain curves are rather flat for single path attenuations in the range from about 3 to 6 dB, revealing minimal increases in performance over this range or even a decrease, which behaviour seems to be characteristic for low elevation angle experiments amongst other factors. The occurrence of this 'flattening' could be due to the simultaneous presence of two separate raincells over the two earth stations. It is very probable, however, that if the measurements were continued for a number of years, the flat portion would blend into the overall characteristic giving a monotonic increase in diversity gain with single-site attenuation [131]. It is therefore considered that the apparent flattening or non-monotonic characteristic seen in low elevation angle site diversity experiments to date is due to a sampling error; i.e. insufficient data to provide an accurate result. It seems clear that diversity performance degrades as the path elevation angle becomes smaller, and, from Fig. 4.40, that elevation angle may be a dominant consideration for $\theta \leqslant 15°$. For higher elevation angles, diversity performance has only a weak dependence on path elevation angle.

(d) Frequency dependence
Diversity performance is expected to decrease with increasing frequency because the impact of widespread, low rainfall rate structures on the single-path statistics will increase. These widespread phenomena are more correlated over large distances, and thus the decorreltion between paths should decrease. Early measurements of diversity gain, however, indicated that the gain appeared to be almost independent of frequency [132, 133]. A later analysis [114] of the apparent residual frequency dependence of diversity gain (after estimated site separation effects had been deleted) also yielded weak dependence. The frequency dependence of diversity gain has not been tested on low elevation angle paths, and it is probable that, for these paths, frequency effects will be important, particularly above 30 GHz.

(e) Local meteorological and topographic effects
Diversity performance can be influenced by local terrain, since topographic features often modify rainfall characteristics [134] as noted earlier in one experiment [127]. Both apparent rainfall enhancement [127] and rain 'shadow' [124] can occur. Such orographic effects can be anticipated in general, although their magnitude will be difficult to predict. Interposing significant geographical features (a hill, river etc.) between the sites should always increase the decorrelation between the sites. Establishing a pair of diversity terminals with a substantial difference in height above mean sea level [135] might permit greater than normal diversity gain to be achieved owing to the substantially shorter path through the troposphere of the higher terminal. This result may not always occur [136] and the advantage to be gained by siting one terminal up a mountain may be obviated by logistics, adverse weather and other problems. There is no doubt, however, that small height separations (akin to those in terrestrial line-of-site space diversity operations) is the preferred method at elevation angles below 3° at high latitudes [137, 138].

4.5 Corrrelation of attenuation data

In developing and testing predictive models, particularly those that are based in part on empirical data, some of the first aspects that are investigated are the degree of correlation between various experimental results and the tendency of the results towards any identifiable trend. The correlation investigations of the attenuation data can be broken down into four general categories: those that deal with a comparison of long-term data, those that compare short-term characteristics, those that compare different experimental techniques, and those that investigate differential effects. The first three are essentially to derive scaling laws so that one set of measured data may be transposed in frequency, elevation angle, polarisation etc. to any other geographical location.

4.5.1 Long-term scaling of attenuation with frequency
(a) Constant attenuation ratio

The CCIR has adopted [98] a simple, empirical frequency scaling law [139] that gives a constant attenuation ratio relating the attenuation measured at one frequency to that at another frequency for the same probability of occurrence along identical Earth–space paths. If A_1 and A_2 are the equiprobable value of attenuation, in decibels, at frequencies f_1 and f_2, in gigahertz, respectively, then the scaling law gives

$$A_1/A_2 = g(f_1)/g(f_2) \tag{4.45}$$

where

$$g(f) = (f^{1.72})/(1 + 3 \times 10^{-7} f^{3.44}) \tag{4.46}$$

The path attenuations A_1 and A_2 are those in excess of atmospheric gaseous attenuation. The expression gives reasonable results over a frequency range of about 7–50 GHz and for attenuation values of practical interest [173]. It should be noted, however, that eqn. 4.45 gives a single value that is independent of fade depth, temperature, rainfall rate, etc. Dual-frequency measurements have shown that the long-term frequency scaling ratio of attenuation changes with the fade depth, amongst other parameters. In all cases, the ratio of the attenuation observed at the higher frequency to that observed at the lower frequency decreases as the fade depth increases. An example of one such measurement is shown in fig. 4.41.

(b) Variable attenuation ratio

An attempt by Hodge [140] to introduce rain medium parameters into the long-term frequency scaling assumed a gaussian rainfall rate distribution along the path, giving

$$A_2 = \left(\frac{k_2}{k_1^{\alpha_2/\alpha_1}}\right) \pi^{1/2 \cdot (1-\alpha_2/\alpha_1)} l_0^{(1-\alpha_2/\alpha_1)} \left(\frac{\alpha_1^{\alpha_2/\alpha_1}}{\alpha_2}\right)^{1/2} A_1^{\alpha_2/\alpha_1} \text{ decibels} \tag{4.47}$$

A_2 and A_1 are the total path attenuations at frequencies f_2 and f_1, respectively, k and α are the power law coefficients for specific attenuation (see eqn. 4.7 and Table 4.1) for the frequencies f_2 and f_1, and l_0 is a parameter to scale the length of the path.

In the frequency range 10–30 GHz, $\alpha_2 \simeq \alpha_1$ and $l_0^{(1-\alpha_2/\alpha_1)} \simeq 1$, giving

$$\frac{A_2}{A_1} \simeq \frac{k_2}{k_1}\left(\frac{A_1}{k_1}\left[\frac{\alpha_1}{\pi}\right]^{1/2}\right)^{\left(\frac{\alpha_2}{\alpha_1}-1\right)} \times \left(\frac{\alpha_1}{\alpha_2}\right)^{1/2} \tag{4.48}$$

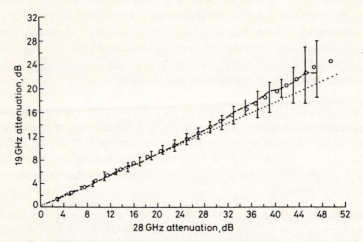

Fig. 4.41 *Relationships between measured attenuations at 28.56 GHz and 19.04 GHz (Fig. 10 of Reference 90)*

    o   Statistical relationship between pairs of attenuation values that are exceeded for the same amount of time on the attenuation distributions.

   – – – Median values of the instantaneous 19 GHz attenuation for each 2 dB increment of 28 GHz attenuation.

   I   10–90% spread in instantaneous 19 GHz attenuation.

   ···· Idealised frequency squared relation

The data are from the main facility

Path elevation angle 38·6°

Polarisations 21° from vertical

(Copyright © 1982 IEEE, reproduced with permission)

Another model that introduced a path length dependence was that due to Rue [141]. In its original form, the long-term frequency scaling prediction gave

$$A_2 = 5^{\alpha_2}k_2 D + 3k_2\left(\frac{A_1 - 5^{\alpha_1}k_1 D}{3k_1}\right)^{\alpha_2/\alpha_1} \text{ decibels} \tag{4.49}$$

A_2, A_1, α_2, α_1, k_2 and k_1 are the same parameters as before with D now the path length parameter. In eqn. 4.49

$$D = \frac{H}{\sin\theta} - \frac{3}{\cos\theta} \tag{4.50}$$

Where H is the height of the 0° Celsius isotherm (the freezing level) and θ is the path elevation angle. Over a limited range of frequencies (11·6–17·8 GHz) the Rue model, in its revised version, was found to give very good results [142].

Both the Hodge and the Rue models take account of polarisation angle and so are more accurate than the simple, constant-attenuation-ratio CCIR formulation given in eqns. 4.45 and 4.46. However, the measurement errors and year-to-year variability are much larger than the difference in prediction accuracy between the CCIR, Hodge, and Rue models. If absolute simplicity is a primary objective, the CCIR model is to be preferred.

Two recent proposals within the CCIR [173] give simple attenuation scaling formulae with frequency that incorporate a measure of polarisation and attenuation variability. The first, due to Fedi [174], uses the parameters k and α (see Table 4.1) and the second, due to Boithias, uses an empirical formulation.

(i) *Fedi k and α method*

In this method, the attenuation at frequency f_2 is related to that at frequency f_1 by

$$A_2 = 4k_2 \times \left(\frac{A_1}{4k_1} \right)^{\alpha_2/\alpha_1} \text{ decibels} \tag{4.51}$$

This is equivalent to assuming a constant path length of 4 km. The results are fairly good below a frequency of about 60 GHz [173].

(ii) *Boithias method*

In this method, an empirical fit was made to all the available data. The attenuation at a frequency f_2 is related to that at frequency f_1 by

$$A_2 = A_1 \times \left(\frac{\phi_2}{\phi_1} \right)^{1 - H(\phi_1, \phi_2, A_1)} \text{ decibels} \tag{4.52}$$

where

$$\phi_{(f)} = \frac{f^2}{1 + 10^{-4} f^2}$$

and

$$H(\phi_1, \phi_2, A_1) = 1 \cdot 12 \times 10^{-3} \left(\frac{\phi_2}{\phi_1} \right)^{0 \cdot 5} (\phi_1 A_1)^{0 \cdot 55}$$

with the frequency f expressed in gigahertz.

Eqns. 4.45, 4.46, 4.51 and/or 4.52 can provide the essential long-term scaling information for relatively close pairs of frequencies (e.g. 14/11 GHz and 30/20 GHz) without recourse to path length information, although a path length of 4 km is inherent in eqn. 4.51.

There are some reservations about using a long-term frequency scaling method that employs path length relationships when the frequencies are very different, e.g. 11 GHz and 30 GHz. At 11 GHz, the parts of the rain medium that are significant are much less extensive than those that are significant at 30 GHz.

The effective path lengths being compared will therefore, in general, be different. An example that highlights this and the possible error in using a single frequency scaling ratio over all attenuation levels is shown in Fig. 4.42 (from Fig. 10 of Reference 143).

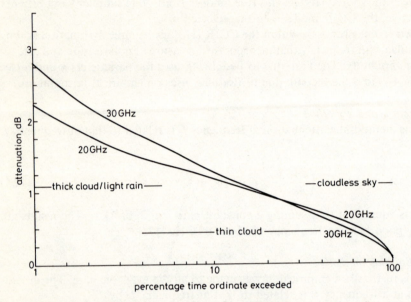

Fig. 4.42 *Cumulative excess attenuation data at 20 and 30 GHz measured by radiometers over a three-year period. (Fig. 10 of Reference 143)*
Martlesham Heath
Elevation angle = 29·9°
Period: 1 Oct. 1978–30 Sept. 1981
Recording times (h):
20 GHz   23346·8 (88.8%)
30 GHz   23257·1 (88·4%)
Linear polarisation (11·8° from vertical)

In Fig. 4.42, the cumulative statistics of 30 GHz and 20 GHz slant path attenuations cross over at about the 20% time level. For percentage time levels above 20%, atmospheric humidity causes more attenuation at 20 GHz than at 30 GHz due to the proximity of the 22 GHz water vapour resonance line. In a similar vein, very light rain will cause appreciable attenuation at 30 GHz while its effect will be insignificant at 10 GHz. The effect is even more marked in short-term frequency scaling.

4.5.2 Short-term frequency scaling
Radiowave attenuation through rain is a function of many parameters including dropsize distribution, rainfall rate and temperature. Even if all of these parameters are held constant except one, variations in that one parameter can

cause significant effects both in predicting the path attenuation and the ratio between attenuations. An example is shown in Fig. 4.43 [144] where the ratio of 14/11 GHz specific attenuation has been calculated using the power law relationship, a Laws and Parsons dropsize distribution, and values of k and α given in Reference 12.

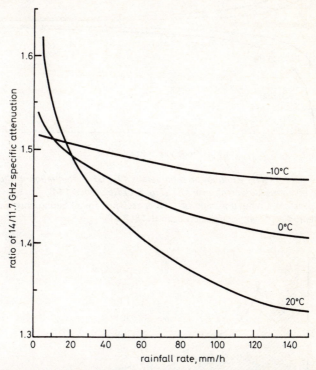

Fig. 4.43 *Ratio of 14/11·7 GHz specific attenuation for various temperatures versus rainfall rate using the Laws and Parsons distribution (Fig. 6 of Reference 144)*
(Reproduced with permission from INTELSAT)

From Fig. 4.43, it can be seen that, for a given rainfall rate, the specific attenuation ratio can vary appreciably. Equally, a ratio of 1·425 can be obtained for a rainfall rate of 50 mm/h or 100 mm/h depending on the temperature of the rain. While the fluctuations between total path attenuation ratios will be less than those measured over a kilometre (the specific attenuation), large variations have been observed in total path attenuation ratios.

One element of the variation in attenuation ratio is experimental error. Fig. 4.44 shows the error bounds in the frequency scaling ratio if an error E (in decibels), occurs on an 11·7 GHz downlink and a constant ratio of 1·45 is assumed between the total path attenuations of 14·0 and 11·7 GHz.

Clearly, equipment error is a major contributor in being able either to

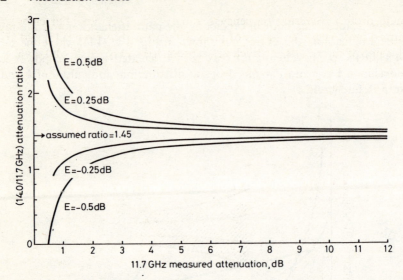

Fig. 4.44 *(14·0/11·7 GHz) Attenuation ratio versus 11·7 GHz attenuation in the presence of the given measurement error (E) (Fig. 8 of Reference 144)*
A constant attenuation ratio of 1·45 is assumed for all fading conditions
(Reproduced with permission from INTELSAT)

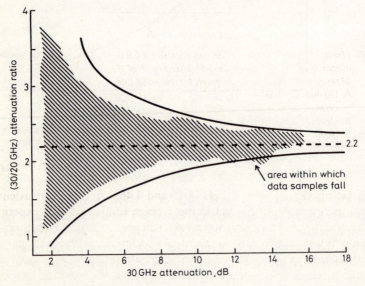

Fig. 4.45 *Combined scattergram for 20 rain events (1318 min of data) (Fig. 11 of Reference 144 after Fig. 6 of Reference 145)*
Error Bounds given by the solid line assume:
±0·4 dB at 20 GHz
±0·6 dB at 30 GHz
(Reproduced with permission from INTELSAT)

estimate or to verify the frequency scaling ratio. This potential error was confirmed in two excellent series of measurements with ATS–6 [145] and OTS [146], some of the results of which are reproduced in Figs. 4.45 and 4.46.

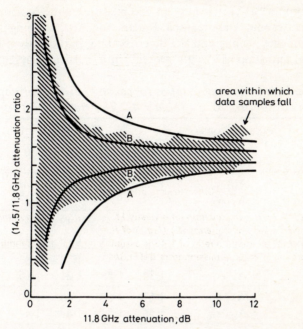

Fig. 4.46 *(14·5/11·8 GHz) Attenuation ratio versus 11·8 GHz attenuation for seven major events (Fig. 13 of Reference 144 after Fig. 7 of Reference 146)*
Measurement error bounds given by the solid lines assume:
A: Up-link ± 1 dB
   down-link ± 0·5 dB
B: Up-link ± 0·4 dB
   down-link ± 0·2 dB
(Reproduced with permission from INTELSAT)

Three factors can be seen in Figs. 4.45 and 4.46. Firstly, as the attenuation increases, there is a definite trend in the average ratio; secondly, experimental errors, even very small experimental errors, can give large scaling ratio errors at low attenuations; and thirdly, some of the scaling ratio fluctuations at high attenuations cannot be explained by experimental errors. This would suggest that, while a reasonably stable long-term frequency scaling ratio can be used with reasonable accuracy, a short-term scaling ratio (necessary for uplink power control) is quite complex to model and is open to many sources of errors [142, 147]. This is particularly true when scaling between polarisations at two different frequencies. Table 4.6 (from Table 4.4.2.2 of Reference 148) illustrates the potential errors.

4.5.3 Correlation between experimental techniques

Comparisons between results obtained using different measurement techniques have generally shown a fairly good correlation between the long-term, or statistical, data. Scaling from radiometer data to obtain equivalent satellite beacon data and vice versa is now an accepted technique provided the limitations of each technique are realised. Radars, too, are becoming accepted for providing frequency scaling and site diversity data, particularly if the radar is dual-polarised and supported with 'ground truth' measurement equipment.

Table 4.6 *Predicted attenuation ratios for scaling between orthogonal linear polarisations at 0° C (Table 7.4.2.2 of Reference 148)*

Attenuation ratio	Polarisations H = horizontal V = vertical	Rain rate (mm/h)					
		5	10	25	50	100	150
$A_{14\cdot455}$	A_H/A_V	1·77	1·76	1·75	1·74	1·74	1·73
$A_{11\cdot786}$	A_V/A_H	1·39	1·34	1·27	1·22	1·18	1·15
(30° elevation)	A_V/A_V	1·57	1·53	1·49	1·45	1·42	1·40
	A_H/A_H	1·57	1·54	1·50	1·47	1·44	1·42
$A_{14\cdot455}$	A_H/A_V	1·83	1·83	1·83	1·83	1·83	1·83
$A_{11\cdot786}$	A_V/A_H	1·34	1·28	1·21	1·15	1·10	1·07
(10° elevation)	A_V/A_V	1·57	1·53	1·48	1·44	1·40	1·38
	A_H/A_H	1·56	1·53	1·49	1·46	1·43	1·41
A_{30}	A_H/A_V	2·77	2·70	2·59	2·52	2·44	2·40
A_{20}	A_V/A_H	2·16	2·05	1·89	1·78	1·68	1·63
(30° elevation)	A_V/A_V	2·42	2·35	2·22	2·13	2·04	1·99
	A_H/A_H	2·47	2·36	2·21	2·11	2·01	1·95
A_{30}	A_H/A_V	2·85	2·79	2·70	2·63	2·56	2·52
A_{20}	A_V/A_H	2·06	1·94	1·77	1·66	1·56	1·50
(10° elevation)	A_V/A_V	2·39	2·31	2·19	2·10	2·01	1·97
	A_H/A_H	2·45	2·34	2·18	2·08	1·98	1·92

The problems occur, however, when short-term comparisons are attempted. Figs. 4.47*a*, *b* and *c* illustrate the difficulties in correlating instantaneous radiometer data with those obtained using the same frequency and antenna along the same slant path to the ATS-6 satellite at a frequency of 30 GHz (from Reference 35).

In Fig. 4.47*a*, the instantaneous radiometer samples are plotted against the corresponding instantaneous satellite beacon measurements. Although there is a general agreement, the spread is quite large with almost a random scatter between the upper and lower bounds. In Fig. 4.47*b*, the scatter is no longer random with two clear parts to the storm. In both of these events, the characteristics of the propagation medium changed during the event, giving markedly different effective medium temperatures. Since a constant effective medium temperature was assumed, the radiometer-predicted attenuations showed appreciable variations from the directly measured satellite beacon data.

Fig. 4.47*c* apparently illustrates a completely different effect from that seen in Fig. 4.47*a* and *b*. Here, an intense raincell within the first Fresnel zone caused substantial attenuation to the beacon signal. The averaging effect of the radiometer beam, however, caused the peak attenuation to be completely underestimated. If all the radiometer data are compared with the satellite beacon data, however, a surprisingly good result is obtained. Fig. 4.48 compares the long-term data obtained in the same experiment that produced the results given in Figs. 4.47*a*, *b* and *c*. This is further confirmation that near-field effects are not important in satellite beacon attenuation measurements [47].

4.5.4 *Differential effects*
Differential effects due to rain can occur in time, frequency and polarisation. The last, the differential amplitude effects at the same frequency but in orthogonal polarisations, is a feature of depolarisation and is discussed in Chapter 5. The first two can cause ranging errors and dispersion effects, respectively.

(*a*) *Ranging errors*
The phase variations on a link are due to the variations in the real part of the bulk refractive index. In clear sky, it was shown (see Section 3.2.7) that the variations in relative humidity caused the major changes in apparent range. Rain, with its inherently higher bulk refractive index, will cause additional ranging errors [149]. Figs. 4.5 and 4.6 showed the variations of the bulk refractive index of rain with frequency. The real part peaked at about 10 GHz and the effect translates to a peak in the phase and group delay in the same frequency range as is shown in Fig. 4.49 [152].

As can be seen from Fig. 4.49, the resulting delay variations due to rain are generally measured in hundredths of nanoseconds. While this can cause errors in ranging, it is well within the guard times of Time Division Multiple Access (TDMA) systems [80]. Most ranging measurements for determining satellite ephemeris data are conducted in clear sky conditions and so range errors due to rain can be avoided. In addition, when range/delay variations start approaching TDMA system limits, the total link attenuation will be such as already to have exceeded the fade margin of the link. If instead of a narrow band signal as is used for ranging, a wide band signal is transmitted, the differential effects across the band of frequencies can cause dispersion.

(*b*) *Dispersion effects*
From Fig 4.49, it can be seen that the rate of change of phase delay with

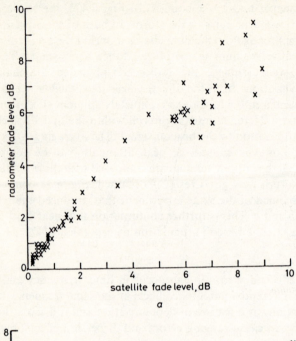

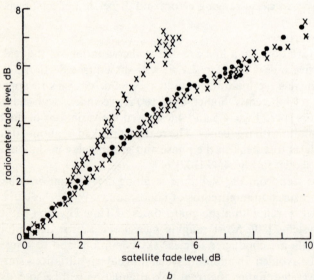

Fig. 4.47 *Comparison of satellite beacon and radiometer fade level for an event (Figs 17–19 of Reference 35)*
  (*a*) 19 November 1975 (Langley site)
  (*b*) 15 August 1975 (Winkfield site)
  ● Occurred after fade maximum
  × Occurred before fade maximum
  (*c*) 16 July 1976 (satellite and radiometer results at Langley site)
  (Reproduced with permission from ESA)

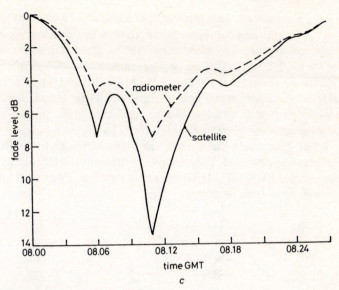

Fig. 4.47 *Continued*

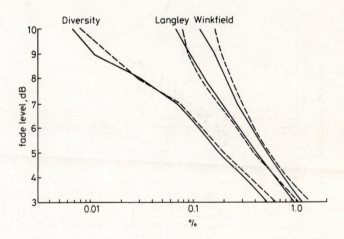

Fig. 4.48 *Cumulative statistics of simultaneous observations (from Fig. 15 of Reference 35)*
--- Radiometer
— Satellite
Observation hours: 1742 h 55 m
Observation period: 1 July 1975–2 Aug. 1976
Elevation angle: 22°,
Frequency: 30 GHz,
Site separation: 12·3 km
(Reproduced with permission from ESA)

frequency is at a maximum between the frequencies of 10 and 50 GHz. Any wideband system operating through intense rain in this frequency range will experience maximum phase variations across the bandwidth. A series of experiments using the COMSTAR satellites [90] measured both differential amplitude and phase across bandwidths of 264 MHz, 528 MHz and 9·5 GHz at a frequency of 28·56 GHz. No evidence was found for any significant amplitude or phase dispersion other than that which could be attributed soley to the frequency dependence of microwave propagation in rain. That is, no frequency selective dispersion effect that could have been due, for instance, to multipath or resonance effects was observed. This confirms a theoretical study carried out for the frequency range 10–30 GHz [150]. Some unexpected differential amplitude effects, however, have been observed.

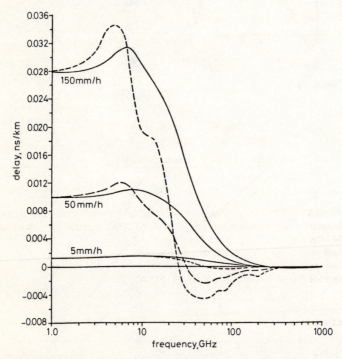

Fig. 4.49 *Group delay and phase delay versus frequency through a 1 km rain path (from Reference 152)*
——— phase delay
‑ ‑ ‑ group delay
(Reproduced with permission from Dr. D. V. Rogers, COMSAT Labs.)

Fig. 4.50 shows the range of the differential attenuation data samples obtained across a bandwidth of just over 500 MHz at a frequency of 11·6 GHz during one storm. The apparent hysteresis effect is probably explained by completely different particle sizes and concentrations being present in different periods of the storm. The effect is the same as was observed by a radiometer in Fig. 4.47*b* where the change in effective medium temperature caused the two extremes of the characteristic to be observed. The variation in the differential attenuation amounted in most cases to less than 10% of the mean carrier attenuation measured and so should not be system limiting.

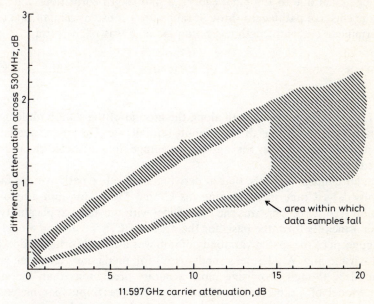

Fig. 4.50 *Differential attenuation in between ±265 MHz sidebands versus 11·6 GHz carrier attenuation measured in August 1979 from the SIRIO satellite (Fig. 14 of Reference 144 after Fig. 50 of Reference 151)*
Upper side-band 11·863 GHz
Lower side-band 11·331 GHz
(Reproduced with permission from INTELSAT)

4.6 Attenuation prediction models

4.6.1 Single-site prediction models
Since rain is the principal attenuating medium on slant paths, the first step in any attenuation prediction process is obtaining either the measured rainfall rate statistics for the site in question or the accurate prediction of the same. In the absence of measured rainfall rate statistics, the relatively simple Rice–Holmberg model for computing rainfall rate statistics on a world-wide basis is now the accepted input model for slant-path attenuation predictions (see eqn. 1.18). The difficult step in going from point rainfall rate statistics to slant path attenuation statistics has produced many competing models.

A first step in predicting slant path attenuation statistics was taken by Dutton and Dougherty [153, 154] who divided up the rainfall rate statistics into three regions, below 5 mm/h, between 5 and 30 mm/h, and above 30 mm/h, and calculated the specific attenuation using a power law equation similar to eqn. 4.7. To obtain the pathlength in the rain medium, some assumptions were made with regard to the storm height and the vertical dependence of the liquid water content.

A later model due to Lin [155] evoked a pathlength correction factor with which the effective pathlength through rain with a constant rainfall rate could be determined. The pathlength correction factor F was of the form

$$F = \frac{1}{1 + \dfrac{L}{LR_s}} \tag{4.53}$$

where L was the projected length along the ground of the slant path length to a height of 4 km and R_s was the five-minute rainfall rate. The limited availability of five-minute rainfall rates for a given percentage time restricted the utility of this prediction method.

A similar concept to Lin's, that of deriving an effective path average factor, was utilised by Crane in developing his Global Model for rain attenuation prediction [156, 157]. This was the first model with universal applicability and its basic principles form the basis for the present CCIR prediction model [98].

The concept of the spatial variability of rain was used in Morita's model [158] and in Misme and Waldteufel's model [159] for predicting rain attenuation. Both models, though promising more accurate predictions than the simple Global Model of Crane [156], utilised more complex input parameters and somewhat difficult computational procedures. Neither model has been used as widely as the Crane model.

The critical feature of a slant path rain attenuation prediction model is how to create the effective pathlength over which the rainfall rate can be quantified and the attenuation calculated. The aim is to produce an effective pathlength that has a physical basis and so is amenable to modification between different climates and/or frequencies if this is found to be necessary. Some models, like Lin's and Crane's, assume a constant rainfall rate along the path and vary the length of the path to achieve the correct value of attenuation. This variation is done by means of reduction factors. Other models, like that due to Stutzman and Dishman [160] and its refinement [161], and a later model due to Crane [162], introduce a spatial variation in the rainfall rate itself either by means of an exponential decay [160] or by forming two components to the rainfall structure [162].

Comparative analyses of the various rain attenuation prediction models have been carried out [e.g. 163–165] and it appears that, if weight is given to those attenuation measurements that have been conducted for periods in excess of two years when comparing measured results to predictions, the CCIR model [98] is generally to be preferred, both from its inherent simplicity and its reasonable accuracy — at least for frequencies of about 30 GHz or below. The procedure for the CCIR model is set out in step-by-step form in Section 2.2.1.1 of Reference 98 and is duplicated below.

"To calculate the long-term statistics of the slant-path rain attenuation at a given location, the following parameters are required [166]:

$R_{0.01}$ = point rainfall rate for the location for 0·01% of an average year, mm/h

h_s = height above mean sea level of the earth station, km

θ = elevation angle, deg.

ϕ = latitude of the earth station, deg.

f = frequency, GHz

The proposed method consists of the following steps 1–7 to predict the attenuation exceeded for 0·01% of the time, and step 8 for other time percentages.

Step 1: The rain height h_R is calculated from the latitude of the station (assuming $h_R \simeq h_{FR}$, following Section 4.4.3 in Report 563):

$$h_R(\text{km}) = \begin{cases} 4\cdot0 & 0 < \phi < 36° \\ 4\cdot0 - 0\cdot075(\phi - 36) & \phi \geqslant 36° \end{cases} \tag{4.54}$$

A considerable deviation from the above value may be expected if the important rainy season is very different from the summer season.

Step 2: For $\theta \geqslant 5°$ the slant path length L_S below the rain height is obtained from:

$$L_S = \frac{(h_R - h_S)}{\sin \theta} \text{ kilometres} \tag{4.55}$$

For $\theta < 5°$ a more accurate formula should be used:

$$L_S = \left[\frac{2(h_R - h_S)}{\left(\sin^2 \theta + \dfrac{2(h_R - h_S)}{R_e}\right)^{1/2} + \sin \theta} \right] \text{ kilometres} \tag{4.56}$$

Step 3: The horizontal projection L_G of the slant path is found from (see Fig. 4.51)

$$L_G = L_S \cos \theta \text{ km} \tag{4.57}$$

Step 4: The reduction factor $r_{0.01}$ for 0·01% of the time can be calculated from

$$r_{0.01} = \frac{1}{1 + 0\cdot045 L_G} \tag{4.58}$$

Step 5: Obtain the rain intensity $R_{0.01}$ exceeded for 0·01% of an average year (with an integration time of 1 min). If this information cannot be obtained from local data sources, an estimate can be obtained from the maps of rain climate given in Section 4.2 of Report 563. (Note that these data are reproduced in Table 1.4 and Figs. 1.29.)

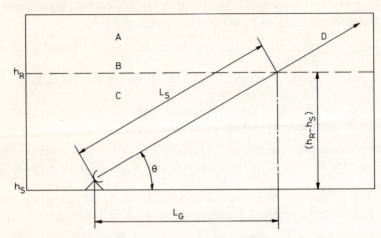

Fig. 4.51 *Schematic presentation of an Earth-space path giving the parameters to be input into the attenuation prediction process (Fig. 1 of Reference 98)*
A: Frozen precipitation
B: Rain height
C: Liquid precipitation
D: Earth–space path
(Copyright © 1986 ITU, reproduced with permission)

Step 6: Obtain the specific attenuation γ_R using the frequency-dependent coefficients given in Table I of report 721 (note that this Table is reproduced as Table 4.1) and the rainfall rate $R_{0.01}$ determined from step 5, by using

$$\gamma_R = k(R_{0.01})^{\alpha} \text{ decibels/km} \tag{4.59}$$

Alternatively, Figs 1 or 2 of Report 721 may be used (Note that these Figures are reproduced as Figs. 4.26a and b.)
Step 7: The attenuation exceeded for 0·01% of an average year may be obtained from:

$$A_{0.01} = \gamma_R L_S r_{0.01} \text{ decibels} \tag{4.60}$$

Step 8: The attenuation to be exceeded for other percentages of an average year, in the range 0·001% to 1·0%, may be estimated from the attenuation to be exceeded for 0·01% for an average year by using

$$\frac{A_p}{A_{0.01}} = 0 \cdot 12 p^{-(0 \cdot 546 + 0 \cdot 043 \log p)} \tag{4.61}$$

This interpolation formula has been determined to give factors of 0·12, 0·38, 1 and 2·14 for 1%, 0·1%, 0·01% and 0·001%, respectively."

The slant path attenuation prediction method of the CCIR reproduced above has been found to over-predict in two general cases: when the path elevation angle is low (below 10°) and when the site in question is in a high rainfall rate, tropical region. A method [165] has been proposed that attempts to remedy these defects by making the pathlength reduction factor rain-rate dependent. Using the terminology of the CCIR prediction method above and combining eqns. 4.55 and 4.56, the reduction factor $r_{0.01}$ by the proposed method can be expressed as

$$r_{0.01} = \frac{1}{1 + L_S \cos \theta / L_0} \tag{4.62}$$

Note that, instead of 0·045 (which is the reciprocal of 22·5 km, the fixed terrestrial pathlength to which the path averaged rainfall rate is referenced) a general pathlength L_0, called the characteristic length of the raincell size for a rainfall rate at 0·01% of the time, is introduced. The characteristic length L_0 is calculated from

$$L_0 = a \times e^{-b \times R_{0.01}} \text{ kilometres} \tag{4.63}$$

For frequencies between 10 and 20 GHz, values of $a = 35$ and $b = 0·015$ give the best results in terms of the mean error.

The above modification to the general CCIR attenuation prediction method provides an empirical fit to the observations that, at low latitudes, the height of the rain h_{FR} can be significantly different from the height of the freezing layer (h_{FS}), and, at low elevation angles, the propagation path can emerge from the side of a rain cell without ever passing through the freezing layer. The latter aspect is receiving considerable attention, particularly with efforts to model the shape of typical rain cells [e.g. 175].

4.6.2 Site diversity prediction models

Unlike the single-site slant path attenuation prediction procedure, there is no general agreement on a site diversity prediction model. This is probably due to the additional range of parameters that have to be factored in (see Section 4.5.4). A number of models have been proposed [103] but none has yet proved to be significantly better than the revised model due to Hodge [114]. In this, the diversity gain G is given by the product of four diversity gain functions that are related to distance (D in kilometres), frequency (f in gigahertz), elevation angle (θ in degrees), and baseline orientation (ϕ in degrees). The baseline orientation is measured with respect to the azimuth direction of the propgation path, chosen so that $\phi \leqslant 90°$. The resultant diversity gain G is

$$G = G_{(D)} \cdot G_{(f)} \cdot G_{(\theta)} \cdot G_{(\phi)} \text{ decibels} \tag{4.64}$$

where $G_{(D)} = a(1 - e^{-bD})$
$G_{(f)} = 1·64 e^{-0·025f}$
$G_{(\theta)} = 0·00492\, \theta + 0·834$
$G_{(\phi)} = 0·00177\, \phi + 0·887$

with

$$a = 0.64A - 1.6(1 - e^{-0.11A})$$
$$b = 0.585(1 - e^{-0.098A})$$

and

$$A = \text{single-path fade depth, dB}$$

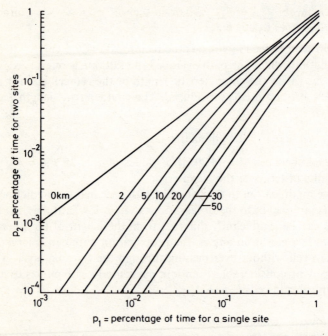

Fig. 4.52 *Relationship between percentages of time with (p_2) and without (p_1) diversity for the same attenuation on Earth–satellite paths (Fig. 4 of Annex I to Reference 178)* (Copyright © 1988 ITU, reproduced with permission)

For those cases where the ratio of the probabilities is required rather than the diversity gain, the general trends of the data have been set out (see Fig. 4.52 from Fig. 4 of Reference 178) and a general formula fitted to the curves which gives

$$p_2 = \frac{p_1^2(1 + \beta^2)}{p_1 + 100\,\beta^2} \qquad (4.65)$$

where p_2 = joint probability for two diversity sites, %
       p_1 = single-site probability, %

and

$$\beta^2 = 10^{-4} \times d^{1.33} \text{ for } d > \text{approximately 2 km}$$

The diversity improvement factor I is given by [178]

$$I = \frac{p_1}{p_2} = \frac{1}{(1 + \beta^2)}\left(1 + \frac{100\beta^2}{p_1}\right)$$ (4.66)

There is a mathematical relationship that appears to exist between I and G (see Annex of Reference 178) and so it should be possible to convert diversity gain to diversity advantage, and vice versa.

While the two models above give generally acceptable results, they both lack a method for inputting climate variations such as rainfall rate characteristics and convectivity factor.

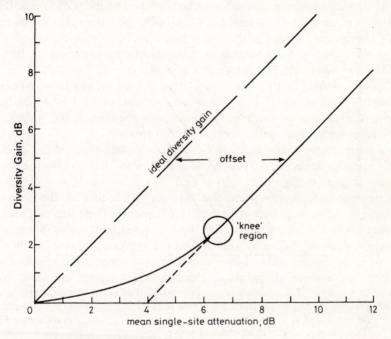

Fig. 4.53 *Characteristic diversity gain behaviour (Fig. 4.1 of Reference 103 after Fig. 1 of Reference 179)*
(Copyright © 1982 IEE, reprinted with permission from the IEE, INTELSAT and COMSAT)

One approach to try and remedy this defect proposed [103, 116] a simple model that postulated an offset between the diversity gain characteristics and the ideal diversity gain, plus a 'knee region (see Fig. 4.53). The offset was given by the single-path attenuation for 0·3% of the time, as it was considered that rainfall over the two sites would be well correlated at this time percentage, and the 'knee' region by the single-path attenuation at 25 mm/h, as it was considered that the rainfall over the two sites would be essentially decorrelated at this level of rainfall rate. The simplicity of the model led to wide variations in prediction error, although the trend of the results was fairly good; however, subsequent revisions to the model [167] have improved the performance greatly. At present, however, the most accurate models are those recommended by the CCIR. Either the Hodge model (eqn. 4.62) or the probability curve fit (eqn. 4.63) should be used depending on whether the required output is diversity gain (in decibels) or joint probability (in percentage time), respectively.

4.7 System impact

In most digital communications systems, there will be a predicted Bit Error Rate (*BER*) for a given Carrier-to-noise (*C/N*). Usually the latter is converted to a ratio of Energy per bit (E_b) to the Noise power per unit bandwidth or Noise power density (N_0) in order to provide a standardised comparison. A typical curve of E_b/N_0 versus *BER* is shown in Fig. 4.54.

In Fig. 4.54, a *BER* of 10^{-6} has been selected as the level at which an outage is deemed to have occurred. In a like manner, a level of 10^{-8} *BER* has been chosen as the nominal performance goal. Between these levels the system can operate acceptably, albeit with a degraded performance.

For a given system bandwidth, the E_b/N_b ratio can be converted to *C/N* and the link budget calculated. A synopsis of the elements that go into the initial procedure is given in Fig. 4.55.

The performance objectives and the clear-sky effects that set the clear-sky *C/N* (see Fig. 4.55) have been discussed in Chapter 3. These are particularly important for small earth stations with limited operating margins. Sometimes equipment-caused degradations [168] can lead to appreciable variations in *C/N* and must be taken into account over and above any propagation effects. Such variations are antenna de-pointing, thermal effects ageing etc., and were reviewed in Section 3.7.3. A schematic of the impact of propagation parameters on earth station design is given in Fig. 4.56.

4.7.1 Uplink fade margin

The availability objectives, which set the threshold *C/N*, are principally determined by rain fading statistics at frequencies above 10 GHz. The uplink fade margin will simply be determined by the predicted rain attenuation at the desired percentage time at the frequency, polarisation, site and path elevation angle in question. For frequencies well below 10 GHz, scintillation due to the ionosphere could be the major amplitude effect on the margin (see Chapter 2). The fade margin predicted for the site must be equal to or less than the net margin obtained in the link budget calculations (see Table 1.6). If the anticipated rain fade exceeds the net margin set aside in the link budget, the system will experience more outage time than is permitted within the availability specification. The downlink fade margin is calculated in a different way.

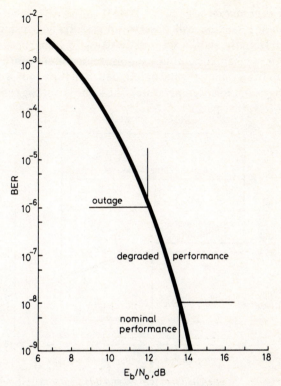

Fig. 4.54 *Typical uncoded coherent QPSK performance including implementation losses*
(*Fig. 1 of Reference 168*)
(Copyright © 1985 IEE reproduced with permission)

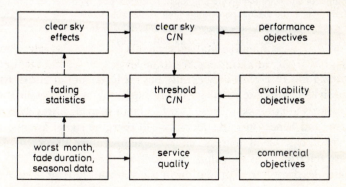

Fig. 4.55 *Synopsis of an earth station link margin calculation procedure (from Fig. 4 of Reference 168)*
(Copyright © 1985 IEE, reproduced with permission)

4.7.2 Downlink degradation

The receiver on the satellite will observe an essentially constant, high temperature emitted from the Earth except in the case of a global beam antenna (see Fig. 4.14). For this reason, the additional noise temperature contribution due to thermal emission from rain in the the uplink path will form only a small fraction of the receiving system noise temperature. The same is not the case for an earth station looking towards space.

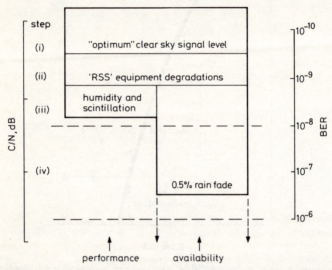

Fig. 4.56 *Schematic of the impact of propagation parameters on the determination of earth station design (Fig. 3 of Reference 168)*Steps I–IV show the order in which the various elements are assessed. First the average (optimum) clear-sky path loss is calculated to give the clear-sky signal level. Next, all the equipment error sources that contribute to changing the level of the received signal are aggregated together in a route-sum-square (RSS) procedure. Thirdly, the seasonal and diurnal impact of humidity and scintillation are predicted, followed, last, by the prediction of the rain fade level at the desired percentage time. A percentage time of 0·5% is shown in the Figure; this relates to the one way outage level for a business service
(Copyright © 1985, IEE, reproduced with permission)

An earth station receiver will generally have a lower noise temperature [80] than an equivalent satellite receiver. In addition, the antenna will usually be directed towards a 'cool' sky. These two facts will provide a low system noise temperature, and hence, in rain, not only will there be signal attenuation but there will be a significant increase in noise temperature observed. The two together are referred to as downlink degradation.

To obtain the downlink degradation in a given rain fade, it is necessary to calculate the system noise temperature in both clear-sky and in the rain fade. The calculation of the system noise temperature T_{sys} is usually referred to a point behind the antenna just prior to the receiver. In Fig. 4.57, the modified antenna temperature, $\sigma_f T_A$, is the noise temperature measured in the reference plane PP' where the system noise temperature is calculated. T_A will be made up of all the noise contributions incident on the antenna including that due to the thermal emission of the rain, but reduced by the feed transmissivity, σ_f.

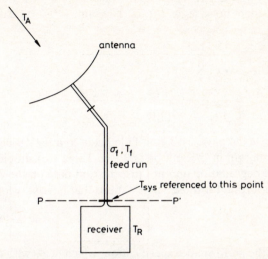

Fig. 4.57 *Simple schematic of an earth station giving the principal components for deter-mining the system noise temperature T_{sys}*
Incident upon the plane *P–P'*, at which point the system noise temperature is calculated, is the modified antenna temperature $\sigma_f T_A$. Plane *P–P'* is usually taken to be the flange connecting the feed run to the receiver. The physical temperature of the feed run is T_f while T_R is the equivalent noise temperature of the receiver. The fractional transmissivity of the feed run is σ_f (see Section 4.1.4)

In general

$$T_{\text{sys}} = T_R + (1 - \sigma_f)T_f + \sigma_f T_A \text{ degrees K} \tag{4.67}$$

where T_f, T_R and σ_f are as noted in Fig. 4.57.
In clear-sky, the only attenuation experienced along the slant path to the satellite will be gaseous attenuation A_g (dB), while, in rain, the signal will experience a combined gaseous and rain attenuation, given by A (dB). If T_m (K) is the physical temperature of the rain medium and T_c (K) is the background sky noise due to cosmic radiation, then

$$T_A|_{\text{clear sky}} = T_m(1 - 10^{-A_g/10}) + T_c \times 10^{-A_g/10} \text{ degrees K} \tag{4.68}$$

$$T_A|_{\text{rain}} = T_m(1 - 10^{-A/10}) + T_c \times 10^{-A/10} \text{ degrees K} \tag{4.69}$$

Note the similarity to eqns. 4.14 and 4.18 and remembering that $10^{-x/10} = e^{-x/4.34}$.
The downlink degradation *DND* is then given by

$$DND = A + 10 \log (T_{\text{sys}}|_{\text{rain}}/T_{\text{sys}}|_{\text{clear sky}}) \text{ decibels} \tag{4.70}$$

To illustrate the effect, letting $A_g = 0.5\,\text{dB}$, $T_c = 2.7\,\text{K}$, $T_m = 280\,\text{K}$, $A = 5.0\,\text{dB}$, $\sigma_f = 0.95$, $T_f = 280\,\text{K}$, and $T_R = 200\,\text{K}$ gives

$$T_A|_{\text{clear sky}} = 32.9\,\text{K} \tag{4.71a}$$

$$T_A|_{\text{rain}} = 192.3\,\text{K} \tag{4.71b}$$

whence

$$T_{\text{sys}}|_{\text{clear sky}} = 245.3\,\text{K} \tag{4.72a}$$

$$T_{\text{sys}}|_{\text{rain}} = 396.7\,\text{K} \tag{4.72b}$$

From this

$$DND = 5 + 2.1 = 7.1\,\text{dB} \tag{4.73}$$

Note that, in this example a 5 dB rain fade leads to a 7·1 dB downlink degradation. The net margin in the link budget (see Table 1.6) should always allow for the downlink degradation and not just the rain fade on the downlink. This is particularly important for those earth stations that have very low system noise temperatures.

4.7.3 Service quality

Over and above the performance specification, which deals with the standard to be met for a high proportion of the time, and the availability objective, which sets the outage time allowed in an average year or worst month, there are usually other criteria by which a customer will judge the quality of service. It is no good, for instance, telling a customer that the link meets the outage criteria if, two times out of ten, he fails to establish a connection due to rain fades. It may be important, therefore, to establish diurnal, and even monthly and seasonal traffice patterns, and see if the peak traffic requirements in terms of timing correlate with the peaks in the rain activity.

An example of the correlation in traffic and rain activity could be a clearing bank which, at close of business, wishes to transfer the day's records to head office via a satellite link but, since the bank is located in a region that has a high incidence of afternoon and early evening thunderstorms, there is a higher than average probability that a rain outage will occur just when the transfer of records takes place. To provide more information with regard to service quality, therefore, it is necessary to investigate much more than the simple rain fade statistics. The impact of rain on a satellite communications system will be a complex mix of cumulative rain attenuation statistics, seasonal variations, diurnal characteristics, inter-fade gaps, joint outage probabilities and return periods. The first step in establishing reliable data on all these parameters will be to obtain accurate information on all the rain climates and the structure of the rain itself. A noteworthy start on the former has been attempted within the ESA propagation programme [169]. Techniques for overcoming rain attenuation are detailed in Chapter 7.

4.8 References

1 MIE, G.: 'Beitrage zur optik truber medien, speziell kolloidaler metallosungen', *Ann. Phys.*, 1908, **25**, pp. 377–445
2 BARRETT, E. C., and MARTIN, D. W.: 'The use of satellite data in rainfall monitoring' (Academic Press, NY, USA, 1981)
3 FRASER, R. S.: 'Interaction mechanisms-within the atmosphere ', JANZA, F. J. (Ed.): in 'Manual of remote sensing. Vol. I: Theory, Instruments and Techniques' (American Society of Photogrammetry, Falls Church, VA, USA) pp. 181–123
4 RYDE, J. W.: 'The attenuation and radar echoes produced at centimetre wavelengths by various meteorological phenomena' in 'Meteorological factors in radio-wave propagation'. Report of conference 8 April 1946 at Royal Instituation (Physical Society, London, 1946) pp. 169–189
5 OLSEN, R. L., ROGERS, D. V., and HODGE, D. B.: 'The aR^b relation in the calculation of rain attenuation', *IEEE Trans.*, 1978, **AP-26**, pp. 318–329
6 Recommendations and Reports of the CCIR, XVIth Plenary Assembly, Dubrovnik, 1986, Volume V (Propagation in non-ionized media); Report 721: 'Attenuation and scattering by rain and other atmospheric particles'.
7 LAWS, J. O., and PARSONS, D. A.: 'The relation of rain drop-sizes to intensity', *Trans. Amer. Geophys. Union*, 1943, **24**, pp. 432–460
8 GUNN, R., and KINZER, G. D.: 'The terminal velocities of fall for water droplets in stagnant air', *J. Meteorology*, 1949, **6**, pp. 243–248
9 RAY, R. S.: Broadband complex refractive indeces of ice and water', *Appled Optics*, 1972, **6**, pp. 1836–1844
10 FEDI, F.: 'Attenuation due to rain on a terrestrial path', *Alta Frequenza*, 1979, **66**, pp. 167–184
11 MAGGIORI, D.: 'Computed transmission through rain in the 1–400 GHz frequency range for spherical and elliptical raindrops and any polarization; FVB Rept. 1C379' *Alta Frequenza*, 1981, **L**, pp. 262–273
12 ROGERS, D. V., and OLSEN, R. L.: 'Calculations of radiowave attenuation due to rain at frequencies up to 1000 GHz'. CRC Report No. 1299, 1976, Dept. of Communications (Canada), Communications Research Centre, Ottawa
13 FLOCK, W. L.: 'Propagation effects on satellite systems at frequencies below 10 GHz'. NASA reference publication 1108, 1983
14 ZUFFEREY, C. H.: 'A study of rain effects on electromagnetic waves in the 1–600 GHz range', M.S. Thesis, 1972, Dept. of Electrical Engineering, University of Colorado, Boulder, USA (reprinted in 1979)
15 CRANE, R. K.: 'Propagation phenomena affecting satellite communication systems operating in the centimeter and millimeter wavelength bands', *Proc. IEEE*, 1971, **59**, pp. 173–188
16 ROGERS, D. V., and OLSEN, R. L.: 'Multiple scattering in coherent radiowave propagation through rain', *Comsat Tech. Rev.*, 1983, **13**, pp. 385–402
17 BRUSSARD, G.: 'Radiometry: A useful prediction tool?', ESA publication SP-1071, 1985
18 HATA, M., DOI, S., and KONDO, N.: 'Complementary use of laser-beam and millimetric-wave propagations', Proceedings ISAP Japan, 1985, Vol. III, pp. 1099–1022
19 SHKAROFSKY, I. P., and MOODY, H. J.: 'Performance characteristics of antennas for direct broadcasting satellite systems including effects of rain depolarization', *RCA Rev.*, 1976, **37**, pp. 279–319
20 FREENY. E. E. and GABBE, J. D.: 'A statistical description of intense rainfall', *Bell Syst. Tech. J.*, 1969, **48**, pp. 1789–1851
21 FANG, D. J.: 'A new way of estimating microwave attenuation over a slant propagation path based on raingauge data', *IEEE Trans.*, 1976, **AP-24**, pp. 381–384
22 MAWIRA, A., NEESEN, J., and ZELDERS, F.: 'Estimation of the effective spatial extent of rain showers from measurements by a radiometer and a raingauge network'. International Conference on Antennas and Propagation (ICAP 81), 1981, IEE Conf. Publ. 195, pp. 133–137

23 ALLNUTT, J. E.: 'Prediction of microwave slant path attenuation from point rainfall rate measurements', *Electron. Lett.*, 1977, **13**, pp. 376–378

24 SEMPLAK, R. A. and TURIN, R. H.: 'Some measurements of attenuation by rainfall at 18·5 GHz', *Bell Syst. Tech. J.*, 1969, **48**, pp. 1767–1787

25 NORBURY, J. R. and WHITE, W. J. K.: 'A rapid response raingauge', *J. Phys. E.*, 1971, **4**, pp. 601–602

26 WANG, T-i., KUMAR, P. N., and FANG, D. J.: 'Laser rain gauge: near-field effect', *Applied Optics*, 1983, **22**, pp. 4008–4012

27 THIRLWELL, J., and EMERSON, D. J.: 'Rain rate statistics at Martlesham Heath (Jan. 1979–Dec. 1981) and their dependence on raingauge integration time'. British Telecom Research Labs. Research & Technology Executive Internal Memorandum TA6/009/85, 1985.

28 HARDEN, B. N., NORBURY, J. R., and WHITE, W. J. K.: 'Estimation of attenuation by rain on terrestrial radio links in the UK at frequencies from 10 to 100 GHz', *IEE J. Microwaves, Optics & Acoustics.*, 1978, **2**, pp. 97–104

29 LIN, S. H.: 'Dependence of rain-rate distribution on raingauge integration time', *Bell Syst. Tech. J.*, 1976, **55**, pp. 135–141

30 TIURI, M. E.: 'Radio telescope receivers' in KRAUS, J. D. (Ed.) 'Radio astonomy' (McGraw-Hill, NY, 1966)

31 YOKOI, H., YAMADA, M., and SATOH, T.: 'Atmospheric attenuation and scintillation of microwaves from outer space', *Astronomical Society of Japan*, 1970, **22**, pp. 511–524

32 Recommendations and Reports of the CCIR, XVIth Plenary Assembly, Dubrovnik, 1986, Volume V (Propagation in non-ionized media); Report 720-1: 'Radio emission from natural sources in the frequency range above about 50 MHz'.

33 HOGG, D. C., and CHU, T-S.: 'The role of rain in satellite communications', *Proc. IEEE*, 1975, **63**, pp. 1308–1330

34 DICKE, R. H.: 'The measurement of thermal radiation at microwave frequencies', *Rev. Sci. Instrum.*, 1946, **17**, pp. 268–275

35 ALLNUTT, J. E., and SHUTIE, P. F.: 'Slant path attenuation and space diversity results at 30 GHz using radiometer and satellite beacon receivers'. Proceedings of the Final ATS-6 Experimenters Meeting held at ESTEC 'ATS-6 propagation experiments in Europe', 1977, **SP-131**, pp. 69–78

36 SMITH, E. K., and NJOKU, E. G.: 'The microwave noise environment at a geostationary satellite caused by the brightness of the Earth', *Radio Science*, 1985, **20**, pp. 318–323

37 CHANDRASEKHAR, S.: 'Radiative Transfer' (Dover publications, NY, 1949)

38 BRUSSAARD, G.: 'Radiative transfer in size-limited bodies', Proc. URSI Commission F Symposium, 1983, ESA **SP-194**, pp. 371–377

39 ZAVODY, A. M.: 'Effect of scattering by rain on radiometer measurements at millimetre wavelengths', *Proc. IEE*, 1974, **121**, pp. 257–263

40 MAWIRA, A.: 'Microwave thermal emission of rain', *Electron. Lett.*, 1981, **17**, pp. 162–163

41 ALTSHULER, E. E., FALCONE, V. J., and WULFSBERG, K. N.: 'Atmospheric effects on propagation at millimeter wavelengths', *IEEE Spectrum*, 1968, **5**, 83–90

42 ALLNUTT, J. E., and UPTON, S. A. J.: 'Results of a 12 GHz radiometric experiment in Hong Kong', *Electron. Lett.*, 1985, **21**, pp. 1217–1219

43 GARDNER, F. M.: Phaselock techniques' (John Wiley, 1966)

44 ALLNUTT, J. E. and GOODYER, J. E.: 'Design of receiving stations for satellite-to-ground propagation research at frequencies above 10 GHz', *IEE J. Microwaves, Optics & Acoustics*, 1977, **1**, pp. 157–164

45 AGRAWAL, B. N.: 'Design of geosynchronous spacecraft' (Prentice Hall, 1986)

46 PRATT, T., and BOSTIAN, C. W.: 'Satellite communications' (John Wiley, 1986)

47 HAWORTH, D. P., McEWAN, N. J., and WATSON, P. A.: 'Effect of rain in the near field of an antenna', *Electron. Lett.*, 1978, **14**, pp. 94–96

48 ARNOLD, H. W., COX, D. C., and HOFFMAN, H. H.: 'Antenna beamwidth independence of measured rain attenuation on a 28 GHz Earth-space path', *IEEE Trans.*, 1982, **AP-30**, pp. 165–168

49 BRUSSAARD, G.: 'Variability of atmospheric noise temperature in 11–14 GHz bands due to water vapour and clouds', *Electron. Lett.*, 1981, **17**, pp. 20–22

50 HALL, M. P. M.: 'Effects of the troposphere on radio communications' (Peter Perigrinus, 1979)

51 J. JOSS, SCHRAM, K., THAMS, J. C., and WALDVOGEL, A.: 'On the quantitative determination of precipitation by radar'. Wissenschaftliche Mitteilung no. 63, Zurich, Eidgenossische kommission zum stadium der hagelbildung und der Hagelabwehr

52 MARSHALL, J. S., and PALMER, W. M. K.: 'The distribution of raindrops with size', *J. Meteorology*, 1948, **5**, pp. 165–166

53 Recommendations and Reports of the CCIR, XVth. Plenary Assembly, Geneva, 1982, Volume V (Propagation in non-ionized media); Report 563-2: 'Radiometeorology data'.

54 WICKERTS, S.: 'Dropsize distribution in rain', FOA rapport C 20438-E1(E2), 1982 Forsvarets Forskningsanstalt, Huvwdavdelning, 2, 102 54 Stockholm, Sweden

55 JONES, D. M. A.: '3 cm and 10 cm wavelength radiation back-scatter from rain', Proc. 5th Radar Weather Conference, 1985, pp. 281–285

56 McCORMICK, K. S. M.: 'A comparison of precipitation attenuation and radar backscatter along Earth-space paths', *IEEE Trans.*, 1972, **AP-20**, pp. 747–754

57 STRICKLAND, J. I.: 'The measurement of slant path attenuation using radar, radiometers, and satellite beacon', IUCRM Colloquium Proceedings, 1973, pp. III.6.1–III.6.9

58 YAMADA, M., OGAWA, A., FURUTA, O., and YOKOI, H.: 'Measurement of rain attenuation by dual-frequency radar'. International symposium on Antennas and Propagation, 1978, pp. 469–472

59 LIGTHART, L. P., NIEUWKERK, L. R., and DISSANAYAKE, A. W.: 'Radar study of the melting layer'. ESA Report ESA CR(P), 1974

60 RUST, W. D., and DOVIAK, R. J.: 'Radar research on thunderstorms and lightning', *Nature.*, 1982, **287**, pp. 461–468

61 McCORMICK, G. C., and HENDRY, A.: 'Principles of radar discrimination of the polarization properties of precipitation'. *Radio Science*, 1975, **10**, pp. 421–434

62 HENDRY, A., and ANTAR, Y. M. M.: 'Precipitation particle identification with centimeter wavelength dual-polarized radars', *Radio Science*, 1984, **19**, pp. 115–122

63 GODDARD, J. W. F., and CHERRY, S. M.: 'New developments with the Chilbolton dual-polarised radar: 5th International Conference on Antennas and Propagation (ICAP 87), IEE Conf. Publ. 274, 1987, Vol. 2 pp. 325–327

64 STAPOR, D. P. and PRATT, T.: 'A generalised analysis of dual-polarization radar measurements in rain', *Radio Science*, 1984, **19**, pp. 90–98

65 SELIGA, T. A., and BRINGI, V. N.: 'Potential use of radar reflectivity measurements at orthogonal polarizations for measuring precipitation', *J. Appl. Met.*, 1976, **15**, pp. 69–76

66 PRATT, T., STUTZMAN, W. L., BOSTIAN, C. W., POLLARD, K. J., and PORTER, R. E.: 'The prediction of slant path attenuation and depolarization from multiple polarization radar measurements', 5th International Conference on Antennas and Propagation (ICAP 87), IEE Conf. Publ. 274, 1987, Vol. 2, pp. 6–10

67 ULBRICH, C. W., and ATLAS, D.: 'Assessment of the contribution of differential polarization to improved rainfall measurements', *Radio Science*, 1984, **19**, pp. 49–57

68 GOLDHIRSH, J.: 'A review on the application of non-attenuating frequency radars for estimating rain attenuation and space diversity performance', *IEEE Trans.*, 1979, **GE-17**, pp. 218–239

69 GUNN, K. L. S., and EAST, T. W. R.: 'The microwave properties of precipitation particles', *J. Rev. Meteor. Soc.*, 1954, **80**, p. 522

70 PIERCE, J. R., and KOMPFNER, R.: 'Transoceanic communications by means of satellites', *Proc. IRE.*, 1959, p. 372

71 HOGG, D. C., and SEMPLAK, R. A.: 'The effect of rain and water vapor on sky noise at centimeter wavelengths', *Bell Syst. Tech. J.*, 1961, **40**, pp. 1331–1348

72 WILSON, R. W.: 'Sun tracker measurements of attenuation by rain at 16 and 30 GHz', *Bell Syst. Tech. J.*, 1969, **48**, pp. 1383–1404

73   DAVIES, P. G., and LANE, J. A.: 'Statistics of tropospheric attenuation at 19 GHz from observations of solar noise', *Electron. Lett.*, 1970, **6**, pp. 522–523

74   WRIXON, G. T.: 'Measurements of atmospheric attenuation on an Earth-space path at 90 GHz using a sun tracker', *Bell Syst. Tech. J.*, 1971, **50**, pp. 103–114

75   CCIR Data bank: Study Period 1982–86. Doc. 5/378 (Rev. 1), ITU, 2 Rue Varembé 1211, Geneva 20, Switzerland

76   'Influence of the atmosphere on radiopropagation on satellite-earth paths at frequencies above 10 GHz'. COST project 205 Final Report, 1985, Commission of the European Comminities, EUR 9923 EN

77   ATLAS, D., KERKER, M., and HITSCHFELD, W.: 'Scattering and attenuation by non-spherical atmospheric particles', *J. Atmos. Terr. Phys.*, 1953, **3**, pp. 108–119

78   STRICKLAND, J. I.: 'The measurement of slant path attenuation using radar, radiometers, and satellite beacons', *J. Rech. Atmos.*, 1974, **8**, pp. 347–358

79   'Special Issue on Multiparameter Radar Measurements of Precipitation', *Radio Science*, 1984, Vol. 19

80   MIYA, K. (Ed.): 'Satellite communications technology (second edition), (KDD Engineering & Consulting Inc., Tokyo (English Language Edition, 1985)

81   YOKOI, H., YAMADA, M., and OGAWA, A.: 'Measurement of precipitation attenuation for satellite communications at low elevation angles', *J. Recherches Atmosphérique*, 1974, **8**, pp. 329–338

82   IPPOLITO, L. J.: 'Millimeter wave propagation measurements from the Applications Technology Satellite (ATS-V)', *IEEE Trans.*, 1970, **AP-18**, pp. 535–552

83   IPPOLITO, L. J., (Ed.): '20 and 30 GHz millimeter wave measurements with the ATS-6 satellite'. NASA Technical Note TN D-8197, 1976

84   Proceedings of the Final ATS-6 Experimenters Meeting held at ESTEC 'ATS-6 propagation experiments in Europe', 1977, SP-131

85   Recommendations and Reports of the CCIR, XVIth Plenary Assembly, Dubrovnik, 1986, Volume V (Propagation in non-ionized media); Report 724-2: 'Propagation data required for the evaluation of coordination distance in the frequency range 1 to 40 GHz'

86   HANSSON, L.: Private communication following discussions on: HANSSON, L., and DAVIDSON, C.: 'Final report: OTS propagation experiment in Stockholm'. Swedish Telecommunications Administration, Radio Dept., Marbackagatan 11, S-123 86 FARSTA, Sweden, 1983

87   POIARES BAPTISTA, J. P. V., ZHANG, Z. W., and McEWAN, N. J.: 'Stability of rain-rate cumulative distributions', *Electron. Lett.*, 1986, **22**, pp. 350–352

88   GOLDHIRSH, J.: 'Slant path fade and rain-rate statistics associated with the COMSTAR beacon at 28·56 GHz from Wallops Island, Virginia over a three-year period', *IEEE Trans.*, 1982, **AP-30**, pp. 191–198

89   ALLNUTT, J. E.: 'Low elevation angle propagation measurements in the 6/4 GHz and 14/11 GHz bands', IEE Conf. Publ. 248, ICAP 85, 1985 pp. 62–66

90   COX, D. C., and ARNOLD, H. W.: 'Results from the 19 and 28 GHz COMSTAR satellite propagation experiments at Crawford Hill', *Proc. IEEE*, 1982, **70**, pp. 458–488

91   FUJITA, M., SHINOZUKA, T., IHARA, T., FURUHAMA, Y., and INUKI, H., ETS-II Experiments Part IV: characteristics of millimeter and centimeter wavelength propagation', *IEEE Trans.*, 1980, **AES-16**, pp. 581–589

92   Recommendations and Reports of the CCIR XVIth Plenary Assembly, Dubrovnik, 1986, Volume V (Propagation in non-ionized media); Report 723-2: 'Worst month statistics'

93   SEGAL, B.: 'The Estimation of worst-month precipitation attenuation probabilities in microwave system design', *Ann. Telecommun.*, 1980, **35**, pp. 429–433

94   Recommendations and Reports of the CCIR XVIth Plenary Assembly, Dubrovnik, 1986, Volume V (Propagation in non-ionized media); Report 338-5: 'Propagation data and prediction methods required for line-of-sight radio-relay systems'

95   YON, K. M., STUTZMAN, W. L., and BOSTIAN, C. W.: 'Worst-month rain attenuation and XPD statistics for satellite paths at 12 GHz', *Electron. Lett.*, 1984, **20**, pp. 646–647

96 DINTELMANN, F.: 'Worst-month statistics', *Electron. Lett.*, 1984, **20**, pp. 890–892

97 HEWITT, M. T., EMERSON, D., RABONE, D. C., and THORN, R. W.: 'Fade rate and fade duration statistics from the OTS slant-path propagation experiment'. British Telecom Research Laboratories, Research & Technology Executive Internal Memorandum No. 5004/84 Issue 1, 1985

98 Recommendations and Reports of the CCIR XVIth Plenary Assembly, Dubrovnik, 1986, Volume V (Propagation in non-ionized media); Report 564-3: 'Propagation data and prediction methods required for Earth-space telecommunications systems'

99 FLAVIN, R. K.: 'Rain attenuation considerations for satellite paths in Australia', *Austral. Telecomm. Res.*, 1982, **16**, pp. 11–24

100 MATRICCIANI, E.: 'Rate of change of signal attenuation from SIRIO at 11·6 GHz', *Electron. Lett.*, 1981, **17**, pp. 139–141

101 DINTELMANN, F.: 'Analysis of 11 GHz slant-path fade duration and fade slope', *Electron. Lett.*, 1981, **17**, pp. 267–268

102 RUCKER, F.: 'Simultaneous propagation measurements in the 12 GHz band on the SIRIO and OTS satellite links', URSI Comm. F Open Symposium, 1980, Lennoxville, PQ, Canada, pp. 4.1.1–4.1.5

103 ALLNUTT, J. E., and ROGERS, D. V.: 'Aspects of site diversity modelling for satellite communications systems', Technical Memorandum IOD-E-84-22, Oct. 1984, INTELSAT, 3400 International Drive, Washington, DC 20008-3096, USA

104 HOGG, D. C.: 'Path diversity in propagation of millimeter waves through rain', *IEEE Trans.*, 1967, **AP-15**, pp. 410–415

105 WILSON, R. W., and MAMMEL, W. L.: 'Results from a three-radiometer path-diversity experiment', IEE Conf. Publ. 98, 1973, pp. 23–27

106 HODGE, D. B.: 'Path diversity for reception of satellite signals', *J. Recherches Atmos.*, 1974, **8**, pp. 443–449

107 ALTMAN, F. J., and SICHAK, W.: 'Simplified diversity communications system for beyond-the-horizon links', *IRE Trans.*, 1956, **CS-4**, pp. 50–55

108 HODGE, D. B.: 'The characteristics of millimeter wavelength satellite-to-ground space-diversity links', IEE Conf. Publ. 98, 1973, pp. 28–32

109 ALLNUTT, J. E.: 'Nature of space diversity in microwave communications via geostationary satellites: A review', *Proc. IEE*, 1978, **125**, pp. 369–376

110 LIN, S. H., BERGAMANN, H. J., and PARSLEY, M. V.: 'Rain attenuation on Earth-satellite paths-Summary of 10-year experiments and studies', *Bell Syst. Tech. J.*, 1980, **59**, pp. 183–228

111 HODGE, D. B.: 'An empirical relationship for path diversity gain', *IEEE Trans.*, 1976, **AP-24**, pp. 250–251

112 TOWNER, G. C., BOSTIAN, C. W., STUTZMANN, W. L., and PRATT, T.: 'Instantaneous diversity gain in 10–30 GHz satellite systems', *IEEE Trans.*, 1984, **AP-32**, pp. 206–208

113 WALLACE, R. G., and CARR, J. L.: 'Site diversity system operation study-Final report'. Tech. Report 2130, ORI Inc., Silver Spring, Maryland, USA, 1982

114 HODGE, D. B.: 'An improved model of diversity gain on Earth-space paths', *Radio Science*, 1982, **17**, pp. 1393–1399

115 IPPOLITO, L. J., KAUL, R. D., and WALLACE, R. G.: 'Propagation effects handbook for satellite systems design', NASA Ref. Publ. 1082(03), NASA, Washington, DC, USA, 1983

116 ROGERS, D. V., and ALLNUTT, J. E.: 'Evaluation of a site diversity model for satellite communications systems', *IEE Proc. F*, 1984, **131**, pp. 501–506

117 HODGE, D. B.: 'Path diversity for Earth-space communications links', *Radio Science*, 1978, **13**, pp. 481–487

118 HALL, J. E., and ALLNUTT, J. E.: 'Results of site diversity tests applicable to 12 GHz satellite communications', IEE Conf. Publ. 126, 1975, pp. 156–162

119 ROGERS, R. R.: 'Statistical rainstorm models: their theoretical and physical foundations', *IEEE Trans.*, 1976, **AP-24**, pp. 547–566

120 BARBALISCIA, F., and PARABONI, A.: 'Joint statistics of rain intensity in eight Italian locations for satellite communications networks', *Electron. Lett.*, 1982, **18**, pp. 118-119

121 GRAY, D. A.: 'Earth-space path diversity: dependence on base-line orientation', IEEE G-AP Int. Symp., 1973, University of Colorado, Boulder, Colorado, USA, pp. 366–369

122 HOGG, D. C., and CHU, T. S.: 'The role of rain in satellite communications', *Proc. IEEE*, 1975 pp. 1308–1331

123 GOLDHIRSH, J.: 'Earth-space path attenuation statistics influenced by orientation of rain cells', Proc. 17th Conf. on Radar Meteorology, Seattle, Washing, USA (American Meteorological Society, 1976) pp. 85–90

124 MacKENZIE, E. C., and ALLNUTT, J. E.: 'Effect of squall-line direction on space-diversity improvement obtainable with millimeter-wave satellite radio communications systems', *Electron. Lett.*, 1977, **13**, pp. 571–573

125 FERGUSSON, A., and ROGERS, R. R.: 'Joint statistics of rain attenuation on terrestrial and Earth-space propagation paths', *Radio Science*, 1978, **13**, pp. 471–479

126 MASS, J: 'Diversity and baseline orientations', *IEEE Trans.*, 1979, **AP-27**, pp. 27–30

127 STRICKLAND, J. I.: 'Radiometric measurement of site diversity improvement at two Canadian locations', URSI Commission F open Symposium, 1977, La Baule, France (late paper)

128 ALLNUTT, J. E.: 'Variation of attenuation and space diversity with elevation angle on 12 GHz satellite-to-ground radio paths', *Electron. Lett.*, 1977, **13**, pp. 346–347

129 ROGERS, D. V.: 'Diversity and single-site radiometric measurements of 12 GHz rain attenuation in different climates', IEE Conf. Publ. 195, 1981, pp. 118–123

130 TOWNER, G. C., MARSHALL, R. E., STUTSMAN, W. L., BOSTIAN, C. W., PRATT, T., MANUS, E. A., and WILEY, P. H.: 'Initial results from the VPI & SU SIRIO diversity experiment', *Radio Science*, 1982, **17**, pp. 1489–1494

131 WITTERNIG, N., RANDEU, W. L., RIEDLER, W., and KUBISTA, E.: '3-years analysis report (1980–1983), 1987, Final Report of INTELSAT contract IS-900: 12 GHz quadruple-site radiometer diversity experiment, Institut fur Angewandte Systemtechnik in der Forschungsgesellschaft Joanneum, Inffeldgasse 12, A-8010 Graz, Austria

132 GOLDHIRSH, J., and ROBISON, F. L.: 'Attenuation and space diversity statistics calculated from radar reflectivity data of rain', *IEEE Trans.*, 1975, **AP-23**, pp. 221–227

133 HODGE, D. B., THEOBALD, D. M., and TAYLOR, R. C.: 'ATS-6 millimeter wavelength propagation experiment'. Report 3863-6, ElectoScience Laboratory, Ohio State University, Columbus, Ohio, USA, 1976

134 HARROLD, T. W., and AUSTIN, P. M.: 'The structure of precipitation systems-a review', *J. Recherches Atmos.*, 1974, **8**, pp. 41–57

135 OTSU, Y., KOBAYASHI, T., SHINOZUKU, T., IHARA, T., and AOYAMA, S.-i.: 'Measurement of rain attenuation at 35 GHz along the slant paths over two sites with a height difference of 3 km', *J. Radio Research Labs. (Japan)*, 1978, **25**, pp. 1–21

136 MISME, P., and WALDTEUFEL, P.: 'Affaiblisements calcules pour des liaisons Terresatéllite en France', *Ann. Telecommun.*, 1982, **37**, pp. 325–333

137 MIMIS, V., and SMALLEY, A.: Low elevation angle site diversity satellite communications for the Canadian Arctic', IEEE International Conf. on Communications, 1982, pp. 4A.4.1–4A.4.5

138 GUTTERBURG, O.: 'Measurements of atmospheric effects on satellite links at very low elevation angles'. AGARD EPP Symposium on Characteristics of the lower atmosphere influencing radiowave propagation, 1983, Spatind, Norway, pp. 5–1 to 5–19

139 BOITHIAS, L., and BATTESTI, J.: 'Au sujet de la dependence en frequence de l'affaiblissement due a la pluie', *Ann. Telecomm.*, 1981, **36**, pp. 483

140 HODGE, D. B.: 'Frequency scaling of rain attenuation', *IEEE Trans.*, 1977, **65**, pp. 446–447

141 RUE, O.: 'Radio wave propagation at frequencies above 10 GHz: New formulas for rain attenuation', *TELE*, 1980, **1**, pp. 11–17

142 'Project COST 205: Frequency and polarisation scaling of rain attenuation', *Alta Frequenza*, 1985, **LIV**, pp. 157–181

143 THORN, R. W.: 'Long-term attenuation statistics at 12, 14, 20 and 30 GHz on a 30 degree slant-path in the UK', British Telecom Research Labs, Research & Technology Executive Internal Memorandum, R6/014/85(L), 1984

144 ALLNUTT, J. E.: 'Correlation between up-link and down-link signal attenuation along the same satellite-ground radio path', Tech. Memo. IOD-P-81-01, INTELSAT, 1981

145 HOWELL, R. G., THIRLWELL, J., BELL, R. R., GOLFIN, N. G., BALANCE, J. W., and MacMILLAN, R. H.: '20 and 30 GHz attenuation measurements using the ATS-6 satellite', ESA SP-131 ATS-6 Propagation Experiments in Europe, 1977, pp. 55-68

146 THIRLWELL, J., and HOWELL, R. G.: 'OTS and radiometric slant-path measurements at Martlesham Heath', URSI Open Symposium, 1980, Lennoxville, Canada, pp. 4.3.1–4.3.9

147 HOLT, A. R., McGUINESS, R., and EVANS, B. G.: 'Frequency scaling propagation parameters using dual-polarisation radar results', *Radio Science*, 1984, **19**, pp. 222–230

148 THIRLWELL, J.: 'Frequency scaling of slant-path attenuation', British Telecom Research Labs., Research & Technology, Executive Internal Memorandum R6/002/83, 1982

149 NUSPL, P. P., DAVIES, N. G., and OLSEN, R. L.: 'Ranging and synchronisation accuracies in a regional TDMA experiment', Proc. 3rd International Conference on Digital satellite Communications, 1975, Kyoto, Japan, pp. 292–300

150 STUTZMAN, W. L., PRATT, T., IMRICH, D. M., SCALES, W. A., and BOSTIAN, C. W.: 'Dispersion in the 10-30 GHz frequency range: Atmospheric effects and their impact on digital satellite communications', *IEEE Trans.*, 1986, **COM-34**, pp. 307–310

151 DINTELMANN, F., and RUCKER, F.: '11 GHz propagation measurements on satellite links in the Federal Republic of Germany', AGARD 26th. Symposium of Electromagnetic Wave Propagation, 1980, London, pp. 18–1 to 18–9

152 ROGERS, D. V., and OLSEN, R. L.: 'Delay and its relation to attenuation in radiowave propagation through rain', Abstract in USNC/URSI National Radio Science Meeting, 1975, University of Colorado, pp. 142–143; Fig. 4.49 supplied by private correspondence from Dr. D. V. Rogers, COMSAT Laboratories, Clarksburg, Maryland, 20871, USA

153 DUTTON, E. J., and DOUGHERTY, H. T.: 'Modeling the effects of cloud and rain upon satellite-to-ground system performance', OT Report 73–5, Office of Telecommunications, Boulder, Colorado, USA, 1973

154 DUTTON, E. J.: 'Earth–space attenuation prediction procedure at 4 to 16 GHz', OT Report 77-123, Office of Telecommunications, Boulder, Colorado, USA, 1977

155 LIN, S. H.: 'Empirical rain attenuation model for Earth–satellite paths', *IEEE Trans.*, 1979, **COM-27**, pp. 812–817

156 CRANE, R. K.: 'A global model for rain attenuation prediction', EASCON 1978 Record, IEEE Publication 78 CH 1354–4 AES, 1978, pp. 391–395

157 CRANE, R. K.: 'Prediction of attenuation by rain', *IEEE Trans.*, 1980, **COM-28**, pp. 1717–1733

158 MORITA, K.: 'Estimation methods for propagation characteristics on Earth-to-space links in microwave and millimeter wavebands', *Rev. ECL NTT Japan*, 1980, **28**, pp. 459–471

159 MISME, P., and WALDTEUFEL, P.: 'A model for attenuation by precipitation on a microwave Earth-to-space link', *Radio Science*, 1980, **15**, pp. 655–665

160 STUTZMAN, W. L., and DISHMAN, W. K.: 'A simple model for the estimation of rain-induced attenuation along Earth-to-space paths at millimeter wavelengths', *Radio Science.*, 1982, **17**, pp. 1465–1476

161 STUTZMAN, W. L., and YON, K. M.: 'A simple rain attenuation model for Earth-space radio links operating at 10–35 GHz', *Radio Science*, 1986, **21**, pp. 65–72

162 CRANE, R. K.: 'A two component rain model for the prediction of attenuation statistics', *Radio Science*, 1982, **17**, pp. 1371–1388

163 IPPOLITO, L. J.: 'Rain attenuation prediction for communications satellite systems', AIAA 10th Communications Satellite Systems Conference, 1984, pp. 319–326

164 MACCHIARELLA, G.: 'Assessment of various models for the prediction of the outage time on the Earth-to-space link due to excess rain attenuation', *Ibid.*, pp. 332–335

165   KARASAWA, Y., YASUNAGA, M., YAMADA, M., and ARBESSER-RASTBURG, B.: 'An improved prediction method for rain attenuation in satellite communications operating at 10–20 GHz', *Radio Science*, 1987, **22**, pp. 1053–1062

166   FEDI, F.: 'A simple method for predicting rain attenuation statistics on terrestrial and Earth-Space paths', RIF: 1B1081, 1981, Fondazione Ugo Bordoni, Via Baldassarre Castiglione, 59, 00142 Roma, Italy

167   MAAS, J.: 'A simulation study of rain attenuation and diversity effects on satellite links', *COMSAT Tech. Rev.*, 1987, **17**, pp. 159–188

168   ALLNUTT, J. E., and ARBESSER-RASTBURG, B.: 'Low elevation angle propagation modelling consideration for the INTELSAT Business Service', IEE Conf. Publ. 248, 1985, pp. 57–61

169   WATSON, P. A., GUNES, M., POTTER, B. A., SATHIASEELAN, V., and LEITAO, J.: 'Development of a climatic map of rainfall attenuation for Europe', Report 327, Univerity of Bradford, Bradford, UK, 1982; (Final Report for the European Space Agency under ESTEC contract 4162/79/NL (DG(SC))

170   Conclusions of the Interim Meeting of Study Group 5 (Propagation in non-ionized media), Geneva, 11–26 April 1988, Document 5/204; Report 723-2 (MOD I): 'Worst month statistics'.

171   WEBBER, R. V., and SCHLESAK, J. J.: 'Fade rates at 13 GHz on Earth-space paths', *Ann. Telecommun.*, 1986, **41**, pp. 562–567

172   Interim Report on INTELSAT Contract INTEL-238 [Amendment No, 3]: 'Analysis of dynamic features of long-term INTELSAT V and OTS propagation data'.

173   Conclusions of the Interim Meeting of Study Group 5 (Propagation in non-ionized media), Geneva, 11–26 April 1988, Document 5/204; Report 721-2 (Mod I): 'Attenuation and scattering by rain and other atmospheric particles'.

174   FEDI, F.: 'Rain attenuation on Earth–satellite links: a prediction method based on joint rainfall intensity', *Ann. des Telecom.*, 1981, **36**, pp. 73–77

175   BUNÉ, P. A. M., HERBEN, M. H. A. J., and DIJK, J.: 'Rain rate profiles obtained from the dynamic behaviour of the attenuation of microwave radio signals', *Radio Science.*, 1988, **23**, pp. 13–22

176   OYINLOYE, J. O.: 'Characteristisc of the non-ionised media and radio wave propagation in equatorial areas-A review', *Telecommun. J.*, 1988, **55**, pp. 115–129

177   BOITHIAS, L., BATTESTI, J., and ROORYCK, M.: 'Prediction of the improvement factor due to diversity reception on microwave links', MICROCOLL, Budapest, 1986

178   Conclusions of the Interim Meeting of Study Group 5 (Propagation in non-ionized media), Geneva 11–26 April 1988, Document 5/204; Report 564-3 (MOD I): 'Propagation data and prediction methods required for Earth-space telecommunications systems'.

179   ALLNUTT, J. E., and ROGERS, D. V.: 'A novel method for predicting site diversity gain on satellite-to-ground radio paths', *Electronics Letters*, 1982, **18**, pp. 233–235

180   CASIRAGHI, E., and PARABONI, A.: Assessment of CCIR worst-month prediction method for rain attenuation', *Ibidem*, 1989, **25**, pp. 82–83

Chapter 5

Depolarisation effects

5.1 Introduction

The rapid increase in the demand for telecommunications capacity, combined with the pressures to conserve the bandwidth used as much as possible, led to the concept of frequency re-use. On a satellite, this can be achieved in two ways: via spatial isolation or via polarisation isolation. Fig. 5.1 illustrates the two concepts.

In Fig. 5.1a, the spatial separation of the two beams combined with the narrow beamwidths of the antennas allows the signal transmitted in one beam to be discriminated from any residual signal that may be inadvertently received from the other beam. In Fig. 5.1b, no such spatial isolation is available and the only factor which permits the signal in one beam to be discriminated from the potentially interfering signal in the other beam is the polarisation isolation that exists between the two beams.

The polarisation isolation of a wave refers to the degree of non-randomness in the orientation of the electric vector. A completely unpolarised wave is one with no detectable orientation sense of the electric vector. As noted in Section 1.2.4, there is one general case of polarisation sense, that of elliptical polarisation, with linear polarisation and circular polarisation being special cases of elliptical polarisation. For circular polarisation, the two senses are Right Hand Circular Polarisation (RHCP) and Left Hand Circular Polarisation (LHCP). A wave is RHCP if the sense of rotation of the electric field corresponds to the natural curl of the fingers of the right hand when the right thumb is pointed along the propagation direction [1]. For LHCP, the same definition applies but with the left hand used instead of the right. For linear polarisation, it is common to use vertical and horizontal as the two orthogonal reference axes.

A perfectly polarised wave will have no component in the orthogonal sense. There are only two states for a perfectly polarised wave: linear or circular. Fig. 1.7 showed how an elliptically polarised wave can be decomposed into two orthogonal circularly polarised waves. Fig. 5.2 illustrates a similar decomposition of a linearly polarised wave into two orthogonal linear vectors.

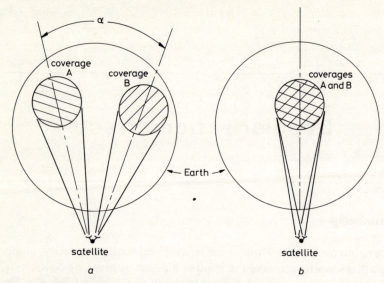

Fig. 5.1 *Illustration of the two techniques of frequency re-use*
  a Spatial isolation
  b Polarisation isolation
  In both cases, identical frequency bands are used in both coverages. In (*a*), the angle
  α between the two beams provides spatial isolation, while in (*b*), the use of
  orthogonal polarisations in the two contiguous coverages provides the isolation.

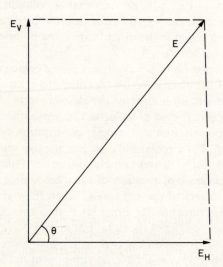

Fig. 5.2 *Decomposition of a linearly polarised vector into two orthogonal components*
  Vector *E* is oriented at a general angle, θ, to the horizontal (*H*) direction and at
  (90° − θ) to the vertical (*V*) direction, giving
  $E_H = E \cos \theta$
  $E_V = E \sin \theta$

A wave that is not perfectly polarised will have a component in the orthogonal sense. The energy in the wanted sense of polarisation is usually referred to as the co-polarised component while the energy in the orthogonal sense is called the cross-polarised component. The difference between the two senses of polarisation is called the cross-polarisation discrimination (XPD). The XPD of a signal can be reduced by: (*a*) the imperfections of the antennas [2], (*b*) the mutual pointing errors of the antennas and (*c*) the intervening propagation medium. The accurate prediction of depolarisation due to the various sources is important in assessing link margins and interference criteria. In this Chapter, only the depolarisation due to hydrometeors will be considered; depolarisation due to Faraday rotation of the electric vector in the ionosphere is dealt with in Chapter 2.

5.2 Basic hydrometeor depolarisation considerations

5.2.1 Medium anisotropy: Differential effects
In Fig. 5.3, a schematic presentation is given of two perfectly polarised signals, one linearly vertical and the other RHCP, entering a propagation medium and exiting on the other side following depolarisation. The linearly polarised signal now has components in both orthogonal linear senses and the circularly polarised signal has components in both orthogonal circular senses. The XPD of both incident signals has been reduced owing to the depolarisation.

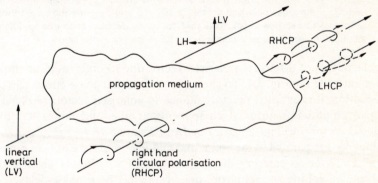

Fig. 5.3 *Schematic presentation of the depolarisation of radiowave signals by a propagation medium*

Depolarisation occurs due to the anisotropy of the propagation medium. If the medium (e.g. rain) is composed of symmetrical particles (in this example, perfectly spherical raindrops), no signal depolarisation would occur. In light rain or fog, this is generally the case, such conditions only causing signal attenuation through absorption and/or scattering. As noted in Section 1.3.5(*b*), however, the shape of the raindrops distorts owing to hydrodynamic forces. The drops, as well as becoming non-symmetrical in shape, also tend to be tilted away from the local horizontal and vertical axes of symmetry owing to wind gusting in heavy rainstorms. As will be seen later, a linearly polarised signal from a satellite is also not usually aligned with the local vertical and horizontal axes of symmetry [3]. This tilt of the incident electric vector away from the axes of symmetry of the raindrop will cause signal depolarisation. The depolarisation will be caused by differential attenuation and differential phase effects between the two axes of symmetry of the raindrop.

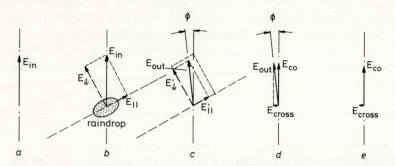

Fig. 5.4 *Illustration of depolarisation due to differential attenuation between the major and minor axes of symmetry of the raindrop*

In (*a*), the incident electric vector E_{in} is vertically polarised with no component in the horizontal sense. E_{in} is resolved into $E_{\parallel}$ and $E_{\perp}$, parallel and perpendicular to the major axis of the raindrops, respectively, in (*b*). $E_{\parallel}$ will be attenuated more than $E_{\perp}$ and the resultant signal E_{out} shown in (*c*) is tilted ϕ degrees away from the vertical. E_{out} is resolved back into the original polarisation axes giving E_{co}, the co-polarised vector, and E_{cross}, the cross-polarised vector, in (*d*). E_{co} and E_{cross} are the two orthogonal output vectors shown in (*e*). Note that, following depolarisation due to differential attenuation, there is now a component of E in the horizontal sense.

Depolarisation of a linearly polarised vector due to differential attenuation effects is illustrated in Fig. 5.4. To show how differential phase effects can cause depolarisation, it is easiest to picture a linearly polarised vector, resolved into two equal amplitude sinusoidal components at plus and minus 45° to the orthogonal vector, incident upon a raindrop. The two vectors resolved at plus and minus 45° are parallel to the major and minor axes of the elliptical raindrop, respectively. Prior to entering the raindrop, both 45° vectors are exactly in phase, but, on exiting the raindrop, one vector has been phase-delayed with respect to the other. As with differential amplitude effects, the differential phase effects between the principal axes of symmetry of the raindrop have caused a tilt of the original vector away from its former orientation. This is illustrated in Fig. 5.5.

The vectorial addition of the two vectors, one subjected to differential attenuation effects and the other to differential phase effects, will yield the resultant electric vector exiting the raindrop. A perfectly circularly polarised wave, since it is made up of two orthogonal, equal-amplitude, linearly polarised vectors with a 90° phase difference between them, will be affected in much the same way by differential phase and amplitude effects as are linearly polarised waves. The differential amplitude effects will cause the two linearly polarised vectors to exit the rain medium with different amplitudes; the differential phase effects will cause the phase between the two vectors to change from an exact 90°. Either of the differential effects, or a combination of the two, will cause a circularly polarised wave to become elliptically polarised, which, in turn, can be resolved into two circularly polarised waves with orthogonal senses.

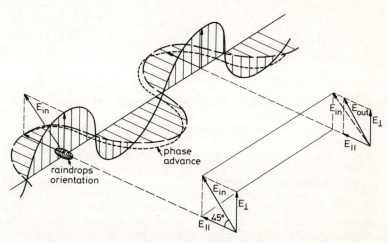

Fig. 5.5 *Illustration of differential phase effects causing depolarisation*
The incident vector E_{in}, is at 45° to the principal axes of the ensemble of ellipsoidal raindrops and the resolved vectors with respect to the major axes, $E_\perp$ and $E_\parallel$, are exactly in phase. On exiting the ensemble of raindrops, the peak magnitudes of $E_\perp$ and $E_\parallel$ are unaltered but the vectors are no longer in phase. The differential phase between them will cause an effective tilt of the resultant electric vector E_{out} with respect to the incident vector E_{in}. If E_{out} is resolved back into the co-ordinate axes of E_{in}, there will be a co-polarised vector parallel to E_{in} and a cross-polarised vector perpendicular to E_{in}.

A perfectly polarised, linear vector will suffer no depolarisation if it is exactly aligned with either of the principal axes of symmetry of a raindrop since no orthogonal component of the vector exists to be resolved prior to entering the raindrop, and hence no differential effects will be observed on exit. The signal will suffer attenuation and phase shifts, however, but there will be no effective rotation of the vector away from its original orientation. As the incident orientation of the perfectly polarised linear vector moves away from the axes of symmetry, so the magnitude of the orthogonally resolved component of the incident vector increases, and with it the differential amplitude and phase effects. These reach a maximum when the incident signal is at an angle of 45° with respect to the principal axes of symmetry. Since this corresponds to the case of circularly polarised signals, it follows that the depolarisation of a linearly polarised signal will always be less than that of a circularly polarised signal except at an angle of 45° with respect to the principal axes of symmetry of the raindrop. Fig. 5.6 illustrates this schematically.

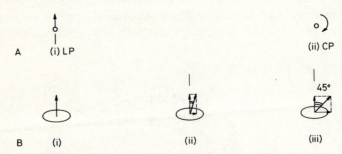

Fig. 5.6 *Schematic presentation of the effect of orientation of the electric vector with respect to the principal axes of the raindrops*
A The raindrop here is perfectly spherical and so the orientation of the electric vector is immaterial.
B In (i), the vector is aligned with the minor axis of symmetry. Assuming perfect-polarisation of the linear vector, no orthogonal component exists. With no component in the orthogonal axis, no differential effects are observed. The same would be the case if the perfectly polarised signal were aligned with the major axis of symmetry. In (ii), a small orthogonal component can be resolved into the major axis because the perfectly polarised vector is tilted away from the minor axis by a small amount. In (iii), the tilt angle is 45° and the two components resolved into the major and minor axes of symmetry are of equal amplitude leading to maximum differential effects. A circularly polarised wave has two equal-amplitude orthogonal components, as in case (iii) here, and so the depolarisation of an LP wave with a tilt angle of 45° will be the same as for a CP wave.

The amount of differential attenuation and differential phase experienced by a signal passing through an ensemble of raindrops will depend upon a number of factors, including frequency and rainfall rate. The maximum effects will be observed when the raindrops present the largest cross-sectional area towards the signal; i.e. the major axis of the ellipsoidal raindrop is perpendicular to the direction of propagation (see Fig. 5.7).

If the raindrops are assumed to be aligned with their major axis of symmetry at 90° to the direction of propagation, the specific differential attenuation and differential phase shift with frequency at 20° Celsius with a Laws and Parsons drop-size distribution are as given in Figures 5.8a and b, respectively [4]. Note that differential phase effects dominate at frequencies below 10 GHz and that differential attenuation dominates at frequencies above 20 GHz. Between 10 and 20 GHz there is a region of cross-over between the dominance of the two differential effects.

In general, the raindrops will not be aligned with their major axes exactly at 90° to the direction of propagation nor will one of the axes of symmetry of the raindrops be parallel to the polarisation orientiation. There will usually be some misalignment of the raindrop axes of symmetry, both with respect to the local horizontal (the canting angle) and with respect to the radiowave signal polarisation orientation (the tilt angle).

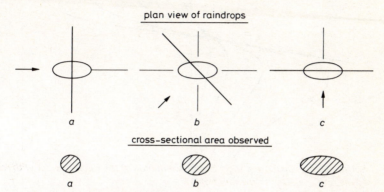

Fig. 5.7 *Illustration of the different apparent assymmetrical shapes of the same raindrop when viewed from different directions.*
In the three plan-view sketches, the direction of propagation is perpendicular to the minor axis of symmetry in (*a*) and perpendicular to the major axis of symmetry in (*c*); in (*b*) it is at an intermediate angle between the two. In the projections of the cross-sectional areas, the ellipticity (or assymetry) of the areas can be seen to increase from (*a*) to (*c*), thus yielding greater differential effects.

5.2.2 Tilting and canting angles
(a) Tilt angle
Linearly polarised transmissions from geostationary communications satellites usually have the orientation of the electric vector specified with respect to the equator. A horizontal polarisation is one that has the electric vector parallel to the equator while a vertical polarisation is perpendicular to the equator at the sub-satellite point (see Fig. 5.9)

If the satellite is transmitting with a vertical polarisation, only earth stations that lie on the meridian will receive signals with the polarisation vector aligned with the local vertical. At a general point P the polarisation orientation will have tilted away from the local vertical by an angle τ. As the point P moves further away from the meridian, so the angle τ increases until, when P is on the equator, $\tau = 90°$. The means that the polarisation vector at the satellite appears to be 90° different from that at the earth station; i.e. a vertical polarisation at the satellite (by definition at right angles to the equator) is received at an orientation parallel to the local horizontal at the earth station. The general equation for the tilt angle τ is [5].

$$\tau = \arctan (\tan \alpha / \sin \beta) \text{ degrees} \qquad (5.1)$$

where α = earth station latitude (positive for the northern hemisphere and negative for the southern hemisphere)

  β = satellite longitude minus the earth station longitude, with longitude expressed in degrees east.

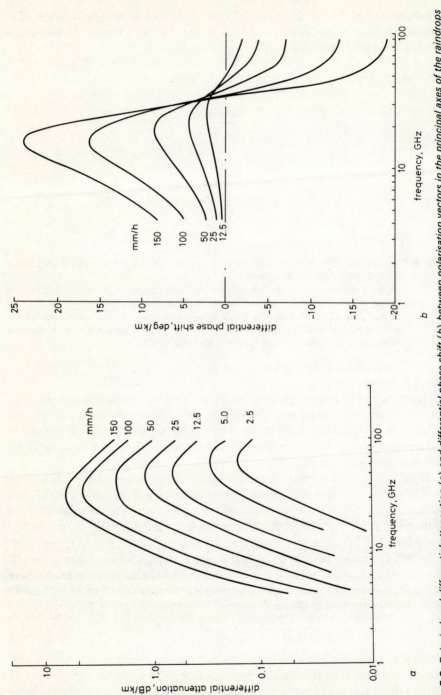

Fig. 5.8 *Rain-induced differential attenaution (a) and differential phase shift (b) between polarisation vectors in the principal axes of the raindrops*

(curve designations/notes from Figs. 1 and 2 of Reference 4)

(*b*) *Canting angle*

A raindrop falling in stagnant air will, if large enough, distort as is shown in Fig. 1.26. The major axis of symmetry, sometimes called the axis of rotational symmetry since the cross-section is almost circular in a plane perpendicular to this axis, will be horizontal. If the air mass is moving horizontally, but the wind speed is constant with height, the drops will still align themselves with their major axes of symmetry horizontal. Windspeed, except in gusts or downbursts,

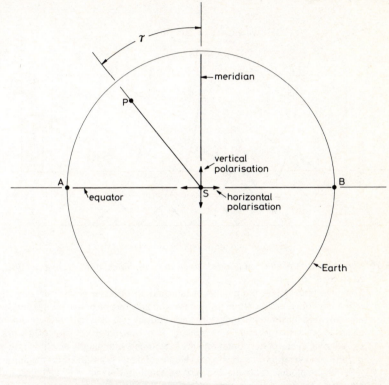

Fig. 5.9 *Definition of horizontal and vertical orientations of linear polarisation with respect to the satellite*

The point S is the sub-satellite point directly below the geostationary satellite on the equator. The meridian is the longitude that passes through S. Points A and B are at oposite ends of the equator at the extremes of the visibility to the satellite (i.e. lowest operational elevation angle at each station). P is a general point on the Earth at which point the electric vector will be tilted at τ degrees from the local horizontal or vertical, depending whether the original linearly polarised vector from the satellite was horizontally or vertically polarised, respectively.

tends to decrease approximately exponentially with altitude, however. As the drop falls, it will encounter air moving at a different speed and so the drops will tend to cant away from the horizontal.

An illustration of the wind forces on drops falling from an altitude of 3 km is shown in Fig. 5.10.

The drops initially fall vertically but soon encounter air that is slowing down relative to the 3 km air speed. The horizontal velocity of the drops will decrease at an ever increasing rate until it reaches zero at the ground (in the absence of gusts). As the drops fall, they will tend to orientate their major axes of symmetry normal to the net aerodynamic force and so will cant out of the horizontal. This is illustrated in Fig. 5.11.

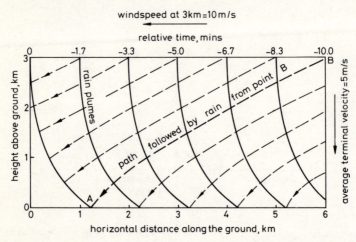

Fig. 5.10 *Illustration of the effect of wind shear on the path traced out by rain as it falls (from Reference 10)*

The solid lines depict the plumes of rain that an observer would see viewing the rainstorm from one side. Normally there would only be one or two plumes but several are shown to illustrate the trajectory traced out by rain as it falls. A 3 km rain height is assumed with the raincloud moving at a horizontal velocity of 10 m/s. An exponential decay of horizontal velocity is assumed with height, reaching zero at the ground. An average vertical terminal velocity of 5 m/s is assumed for the rain drops.

While the plumes show the apparent trajectory of the rain, the *real* trajectory is as shown by the dashed lines. One such trace, BA is highlighted. At a time −10 mins relative to the observer, the rain reaching the ground at point A about 8 mins later starts out at point B.

Notes:

(i) As the rain nears the ground the slope of the *real* fall (dashed lines) increases as the apparent slope of the rain in the plume decreases (solid line)

(ii) The observer to the side would assume that the rain reaches the ground going left to right-down the plume when in actual fact it goes from right to left down the dashed line

(Reproduced by permission from Prof. G. Brussard)

In the same way that clouds will have their major axes aligned with the direction of the wind, so the raindrops will tend to be 'streamlined' with their narrow cross-section pointing into the horizontal wind component. In deriving the possible canting angles that could exist, Brussaard [6] assumed that the raindrops would not only face into the direction of the wind but would cant their major axes away from the horizontal so that it was aligned parallel to the local wind flow around the drop.

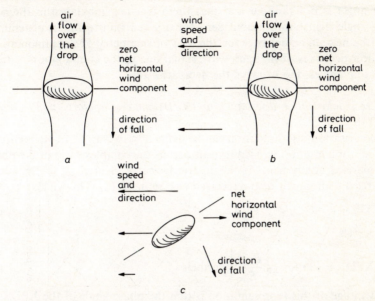

Fig. 5.11 *Relative alignment and canting angle of a raindrop in various wind conditions*
a Stagnant air
b Wind velocity constant with height
c Wind shear conditions: velocity decreasing with height

  In (*a*), the forces around the drop have stabilised since the drop is at its terminal velocity and the air is stagnant.

  In (*b*), the forces around the drop have also stabilised once the drop has accelerated to the speed of the horizontal wind, which is constant with height at this point. In both (*a*) and (*b*) it is gravity that is causing the only net wind flow over the raindrop and, since the raindrop will, in a steady state, always orientate the major axis of symmetry at right-angles to the aerodynamic force, the drops will fall without any canting angle.

  In (*c*), a situation that is more typical occurs with a horizontal wind shear with height (see Fig. 5.10). There will be two wind components, a vertical one due to gravity and a net horizontal component in the *opposite* direction to the wind (i.e. the horizontal wind tends to slow down as altitude decreases giving a decelerating horizontal component). The direction of fall will be the vector sum of the two forces, and the raindrop, since it aligns its major axis to be perpendicular to the net aerodynamic force, will 'cant' out of the horizontal. Since windshear forces tend to, on the average, increase as the altitude decreases, canting angles will increase as the drop falls closer to the ground.

An interesting feature of Brussaard's theory [6] is that the canting angles will differ for different equivalent raindrop radii. This follows from the different terminal velocities of raindrops of various sizes and hence the different wind shear gradients encountered per unit time during the fall. Canting angles can be both negative and positive with respect to a given co-ordinate system. It is immaterial whether the raindrops are canted downwards 5° (say) or upwards 5° since the net depolarisation effect is the same. What is important is that the canting angle is not random and that there is a net imbalance in the mean canting angle distribution about zero for a given rainstorm. In general, the greater the mis-alignment between the axes of symmetry of the raindrops and the polarisation axes of the radiowave signal due to the combination of tilting and canting angles, the worse is the depolarisation.

5.2.3 Cross-polarisation discrimination (XPD) and cross-polarisation isolation (XPI)

Cross-polarisation discrimination can be defined in two ways. In Reference 7 it is defined, for a mono-polarised transmitter, as the complex ratio of the phasor cross-polarised component of the received electric field E_{cross} to the phasor co-polarised component of the received electric field E_{co}. This yields

$$\text{XPD} = E_{cross}/E_{co} \tag{5.2}$$

or in decibels

$$\text{XPD} = 20 \log_{10} \left| \frac{E_{cross}}{E_{co}} \right| \text{ decibels} \tag{5.3}$$

The advantage of this representation is that the phase angle of the XPD is the same as the relative phase of the cross-polarised signal measured with respect to the co-polarised signal. The disadvantage of this presentation is that it yields negative values for XPD and can lead to confusion in defining what an 'increase' in XPD actually means. The CCIR have adopted a presentation format that is the inverse of eqn. 5.3 namely

$$\text{XPD} = 20 \log_{10} \left| \frac{E_{co}}{E_{cross}} \right| \text{ decibels} \tag{5.4}$$

Eqn. 5.4 yields positive values of XPD for all normal cases of propagation conditions and will be used as the basis for further discussion in this Chapter.

In general, a receiver will detect one polarisation sense and it is the isolation of signals in this polarisation sense from those transmitted in the orthogonal sense that is important. This is the cross-polarisation isolation (XPI). Fig. 5.12 illustrates the definitions of XPD and XPI.

By definition, from Fig. 5.12

$$\text{XPD} = 20 \log_{10} \frac{ac}{ax} \text{ decibels} \tag{5.5}$$

$$\text{XPI} = 20 \log_{10} \frac{ac}{bx} \text{ decibels} \tag{5.6}$$

For a rain medium, it can be shown [8] that XPD and XPI are equivalent and measurements [9] have confirmed this. The latter, since they included the freezing level and ice crystal effects, lend credence to the large number of XPD measurements undertaken worldwide.

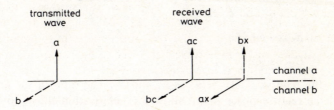

Fig. 5.12 *Terms used to define the difference between XPD and XPI*
ac and *bc* are the co-polarised components of signals transmitted in channels *a* and *b*, respectively. *ax* and *bx* are the cross-polarised components of signals transmitted in channels *a* and *b*, respectively.

A dual-polarised receiver that is designed to receive orthogonal channels simultaneously will detect both the wanted, or co-polarised, signals *ac* and *bc* and the unwanted, or cross-polarised signals *bx* and *ax*. The isolation in channel *a* will be the ratio of *ac/bx*. In general, experiments use only mono-polarised transmissions and so what is measured is the discrimination ratio *ac/ax* or *bc/bx*.

5.3 Measurement techniques

If the differential attenuation and differential phase induced by the propagation medium can be measured, the cross-polarisation discrimination can be obtained immediately from

$$XPD = 20 \log_{10} \left| \frac{e^{-(\alpha + j\beta)} + 1}{e^{-(\alpha + j\beta)} - 1} \right| \text{ decibels} \tag{5.7}$$

where α = differential attenuation, nepers
β = differential phase, rad

The measurement of differential phase implies a coherent detection system and so incoherent receivers, such as radiometers, will not be sufficient to deduce XPD. Coherent receivers can be of the direct or indirect type and a number of techniques can be employed that are generally variations on a theme. Before describing these, it is worthwhile to review some basic theory.

5.3.1 Basic theory
Fig. 5.13, from Fig. 1 of Reference 11, shows the general polarisation ellipse with a tilt angle τ and angles ε and γ as indicated. Note that the axial ratio r (see Section 1.2.4) is positive for LHCP and negative for RHCP by definition [11].

The polarisation state of a wave can be characterised by [11]:

(i) Shape of the ellipse (i.e. the axial ratio r)

(ii) Orientation of the ellipse (i.e. the tilt angle τ)
(iii) Sense of rotation of the electric field vector (i.e. sign of r)
(iv) Energy in the wave

In Fig. 5.13, the instantaneous electric field $\mathscr{E}$ can be decomposed into two orthogonal vectors $\mathscr{E}_x$ and $\mathscr{E}_y$ given by:

$$\mathscr{E}_x = E_1 \cos \omega t \tag{5.8a}$$

$$\mathscr{E}_y = E_2 \cos (\omega t + \delta) \tag{5.8b}$$

where δ is the relative phase between $\mathscr{E}_x$ and $\mathscr{E}_y$. The ratio of the amplitudes E_2/E_1 and the relative phase δ will allow the polarisation ellipse to be reconstructed. Their detection, or inference from other measurements, can be by both direct or indirect means.

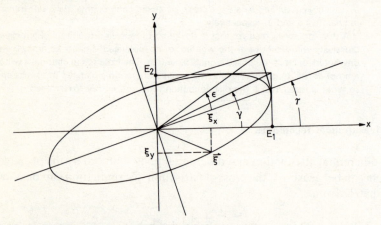

Fig. 5.13 *Polarisation ellipse (from Fig. 1 of Reference 11)*
A general vector $\mathscr{E}$ will have components $\mathscr{E}_x$ and $\mathscr{E}_y$ in the x and y axes, respectively. The maximum values that $\mathscr{E}_x$ and $\mathscr{E}_y$ can take are E_1 and E_2, respectively.
    Note that $\varepsilon = \cot^{-1}(a/b) = \cot^{-1}(r)$, where r is the axial ratio, and that $\gamma = \tan^{-1}(E_2/E_1)$. The angle that the major axis of the polarisation ellipse makes with the horizontal (x) axis, the tilt angle, is given by τ.

5.3.2 Direct measurements

Direct XPD measurements on Earth–space paths involve the detection of a coherent signal either transmitted from a satellite as a beacon or transponded by a satellite. The latter involves the transmission of a carrier signal to the satellite, usually from the earth station that will eventually detect the transponded signal. This is referred to as a looped-back set-up.

There are four general methods of obtaining XPD by direct measurements [11]:

(a) *Polarisation-pattern method*: In this, a linearly polarised antenna is continuously rotated about an axis along the propagation direction. The maximum and minimum levels detected, corresponding to the major and minor axes of the polarisation ellipse, plus the relative orientations of the receiving antenna to the horizontal axis when detecting the maximum and minimum values, will yield the polarisation ellipse but not the sense of the polarisation.

(b) *Linear-component method*: The two orthogonal, linearly polarised components E_1 and E_2 plus the relative phase between them, δ, are measured in this method which yields the polarisation ellipse directly.

(c) *Circular-component method*: This is the same as (b) but with circularly polarised feed elements used instead of linearly polarised feed elements.

(d) *Multiple-component method*: to obviate the need to measure the relative phase, a series of four power measurements need to be made from a possible six. These comprise one set of two measurements plus one measurement each from the other two sets of measurements of the following: two orthogonal linear polarisations; two orthogonal linear polarisations oriented at 45° with respect to the first set; and LHCP and RHCP.

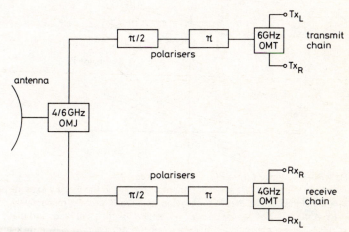

Fig. 5.14 *Schematic of an earth-station feed network incorporating rotating polarisers (from Fig. 2.7 of Reference 13)*
The $\pi/2$ and π polariseres are independently rotatable in both the transmit and receive chain in order to optimise the polarisation purity of the system.

Of these four direct detection methods, (b) and (c) are most commonly used. Method (a) unnecessarily complicates the detection of the signal, particularly if the incoming signal is highly polarised (i.e. almost no power in the cross-polarised sense), leading to a loss of lock and the necessity to re-acquire signal lock twice during each revolution of the antenna. Method (d) is again very complicated, this time in the antenna feed requirements.

The linear-component method, (b), can be used to measure the XPD of a linearly polarised signal or a circularly polarised signal, while the circular-

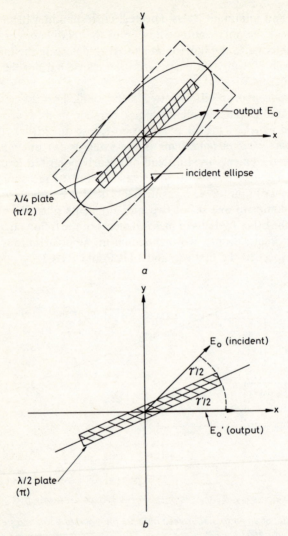

Fig. 5.15 *Schematic representation of the action of cascaded polarisers (after Bryant, from Fig. 3.3 of Reference 13)*

In (a), the quarter-wave plate of the polariser is aligned with the major axis of the incident polarisation ellipse. The resultant output vector E_0 is then passed to the half-wave polariser. In (b), the half-wave polariser plate is aligned so that it bisects the angle between E_0 and the desired orientation (in this case, the x axis). The final output of the cascaded $\pi/2$ and π polarisers is E_0' in the x axis.

component method, (c), can only be used to measure the XPD of a circularly polarised signal.

To convert a circularly polarised signal into a linearly polarised signal requires the use of a polariser [12]. A polariser consists essentially of a dielectric plate inside a waveguide section that introduces a given phase delay, usually 90° ($\pi/2$ rad) or 180° (π rad). The waveguide section is rotatable so that the polariser plate can be aligned to any position. Usually two polarisers are cascaded together in an antenna feed network as shown in Fig. 5.14 [13].

The orthomode junction (OMJ) separates the transmitted signals from the received signals like a diplexer, 6 and 4 GHz in Fig. 5.14, respectively, and the orthomode transducer (OMT) separates the orthogonal polarisations, in this case RHCP and LHCP. Usually, the quarter-wave polariser plate $\pi/2$ is aligned so that it is parallel with the major axis of the incoming polarisation ellipse. This will ensure that, at the output of the polariser, there is a linearly polarised signal. This linearly polarised signal is then input to the half-wave polariser π, which, if it is aligned halfway between the incoming orientation and the desired output orientation, will yield a linearly polarised signal at the desired port of the OMT. Fig. 5.15 illustrates the action of the polariser [13].

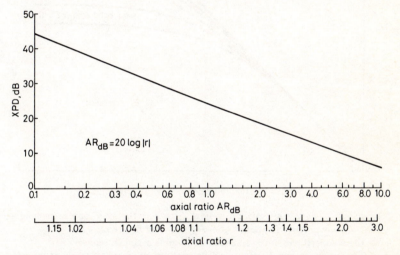

Fig. 5.16 *XPD versus axial ratio of an elliptically polarised wave (from Fig. 4.2–4 of Reference 1)*

The measurement of XPD is more difficult than the measurement of only attenuation since the antennas (both spacecraft and earth station) introduce a non-negligible cross-polarised component themselves. Remembering from eqns. 1.6 and 1.8 that

$$\text{XPD} = 20 \log \frac{(r + 1)}{(r - 1)} \text{ decibels} \qquad (5.9)$$

where r is the axial ratio, the clear-sky XPD of an antenna can immediately be found from eqn. 5.9. In Fig. 5.16 [1], XPD is plotted against both r and AR_{dB}. $AR_{dB} = 20 \log |r|$ and is another method commonly used to express antenna axial ratio.

The axial ratio of an antenna will set a lower bound for the isolation that is measurable or attainable in a link. Fig. 5.17 [1] gives a family of curves of actual XPD versus the detected XPD or isolation with AR_{dB} as parameter.

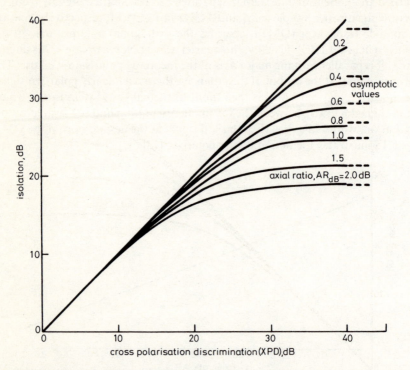

Fig. 5.17 *Family of curves of isolation against XPD, with the axial ratio as parameter, for a circularly polarised system (from Fig. 4.2–5 of Reference 1)*
Co-polarised and cross-polarised axial ratios assumed equal
The curves denote the measured isolation for an incoming signal of the given XPD by an antenna with the given axial ratio. Note that the axial ratio sets the measurable isolation limit independently of the polarisation purity of the incoming signal.

In Fig. 5.17, if the axial ratio is perfect (i.e. 0 dB for a circularly polarised antenna), the medium induced XPD will be reflected one-for-one as a change in the link isolation, or measured XPD. As the antenna axial ratio degrades, however, an ever decreasing asymptotic, limiting value of measurable isolation is imposed by the antenna axial ratio. The difference between the 0 dB axial ratio line and the actual axial ratio curve gives the potential measurement error in detecting the true XPD value. Fig. 5.18 [1] shows the same set of curves for a linearly polarised antenna. Note that, for a linearly polarised antenna, the axial ratio is infinity if it is perfectly polarised.

The linearly polarised antenna can additionally be rotated out of the required orientation of the incoming linearly polarised signal leading to a mis-alignment error. This is shown in Fig. 5.19 [1].

Since no antenna is perfect, there will always be a residual cross-polarised vector introduced. In Fig. 5.20, the clear-sky co-polarised (E_{co}) and the cross-polarised (E_{cross}) vectors of an antenna system are oriented with a rain-induced cross-polarisation vector E_R, shown at the tip of E_{cross}.

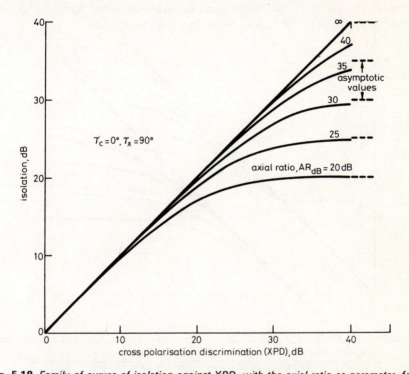

Fig. 5.18 *Family of curves of isolation against XPD, with the axial ratio as parameter, for a linearly polarised system (from Fig. 4.2–6 of Reference 1)*
Co-polarised and cross-polarised axial ratios assumed equal.
In this example there is no misalignment between the axis of the antenna feed and the incoming signal orientation.

The rain vector can describe a complete circle around the tip of the vector E_{cross} depending on the phase of the rain-induced cross-polarisation E_R with respect to the clear-sky cross-polarised vector E_{cross} introduced by the antenna. If the rain-induced vector is equal in magnitude to the clear-sky cross-polarised vector, then the measured XPD can vary between $-6\,dB$ and $+\infty$ from the actual value. For example, if the clear-sky XPD $= 20 \log (E_{co}/E_{cross}) = 30\,dB$ and $|E_{cross}| = |E_R|$, then the measured XPD can lie anywhere between $24\,dB$ and $\infty\,dB$.

In an attempt to elliminate the residual clear-sky component introduced by the antenna, two techniques can be employed: (*a*) software cancellation and (*b*) static (hardware) cancellation:

(*a*) *Software cancellation*: If a coherent detection system similar to that shown in Fig. 5.21 [14] is used, the phase of the cross-polarised output with respect to the co-polarised signal can be measured directly. A knowledge of the amplitude and the phase of the clear-sky cross-polarised signal will permit the measured XPD in rain to be adjusted in the analysis.

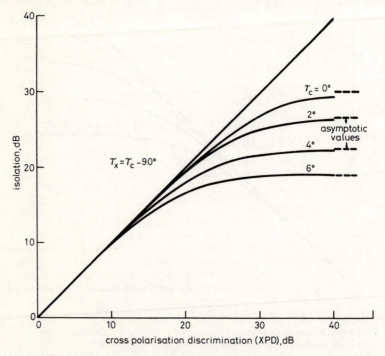

Fig. 5.19 *Family of curves of isolation against XPD, with the misalignment τ_c between the feed and the incoming signal orientation as parameter, for a linearly polarised system (from Fig. 4.2–7 of Reference 1)*
In this example, the axial ratio of the antenna is 30 dB ($=AR_{dB}$).

(*b*) *Static cancellation*: If a cross-coupling network can be introduced between the co- and the cross-polarised channels prior to the receivers, as shown in Fig. 5.22, then the insertion of E_{cross} in anti-phase to the clear-sky residual cross-polarised signal introduced by the antenna will effectively eliminate all the cross-polarised components (including those due to the spacecraft antenna and the mutual mis-alignment of the spacecraft and earth station antennas).

In general, static cancellation is very effective provided that there are no significant diurnal clear-sky variations introduced by the satellite and/or the earth station due to thermal effects, tracking errors or other causes. If there are such large diurnal changes, it is probably simpler to obtain the diurnal 'signatures' of such variations in clear-sky conditions and then subtract these effects from those obtained during rain or ice depolarisation events using software techniques.

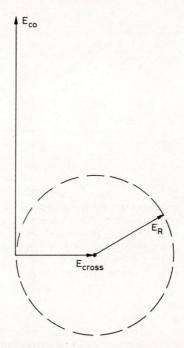

Fig. 5.20 *Illustration of the effect the clear-sky residual XPD of the antenna has on the measurement accuracy*
The clear-sky residual XPD of the antenna is due to the non-perfect axial ratio (see eqn. 5.9). The vectorial sum of E_{cross} and E_R, together with E_{co}, will give the XPD in rain. The vector E_R is the cross-polarised component induced by the rain. Depending on the relative phase between E_{cross} and E_R, the overall XPD can fluctuate by large amounts. In the Figure, a worst-case situation is shown with $|E_{cross}| = |E_R|$, which can lead to overall XPD values of infinity when the relative phase between E_{cross} and E_R is 180°.

5.3.3 Indirect measurements

The only indirect technique that has been used to date involves a dual-polarised radar [e.g. 16]. The principal difficulties with using a dual-polarised radar are: (i) there are not enough independent measurement parameters to characterise explicitly the XPD of the propagation medium [17]; and (ii) the 'dead zone' of the radar will prevent XPD events from being measured close to the receiving

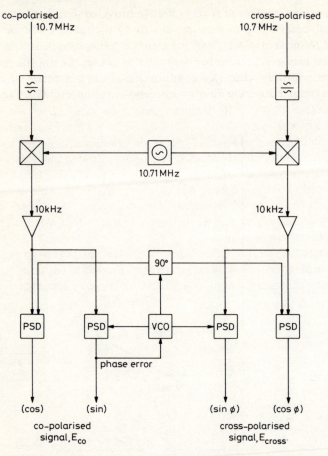

Fig. 5.21 *Example of a coherent detection circuit for the reception of a signal in both the co-and cross-polarised channels (from Fig. 8 of Reference 14)*

The stronger, co-polarised signal is used to detect the weaker, cross-polarised signal coherently since they are both derived from the same source. For a strong, co-polarised signal, the sine of the co-polarised signal will be zero. There will be a phase difference ϕ between the co- and cross-polarised signals, however, with $\phi = \tan^{-1}$ (sin ϕ/cos ϕ). The magnitude of E_{cross} is found by squaring both cross-polarised outputs, adding them together, and square-rooting the sum

(Copyright © 1977 IEE, reproduced with permission)

site. The effect of (i) above is that at least two parameters have to be assumed (usually the average raindrop canting angle and its distribution) and that of (ii) above means that only relatively distant rain/ice events that are outside the radar 'dead-zone' can be examined accurately, leading to a loss of statistical data. Neverthless, some success has been claimed [18] for **XPD** measurements using dual-polarised radars.

5.4 Experimental results

5.4.1 Identifying the problem

The significance of depolarisation due to the propagation medium was not realised until relatively recently. For frequencies up to and including, UHF, the major signal impairments on Earth–space paths, with the exception of Faraday rotation, were not related to depolarisation phenomena. Even on terrestrial microwave systems in the bands below 10 GHz, the major impairment was multi-path fading (i.e. signal attenuation due to the destructive interference of reflected rays), particularly on long hops. The introduction of frequency-re-use terrestrial microwave systems in the bands below 10 GHz led to the identification of depolarisation effects, but these were generally due to multi-path phenomena which caused 'in-band' distortion [e.g. 19]. The inherently frequency-selective nature of destructive interference due to multipath effects would lead to impairments over only a relatively small segment of the channel, or band; hence the term 'in-band' distortion. Rain effects tend to be relatively constant over bandwidths of a few percent and this led to the term 'flat fading' being given to rain attenuation on terrestrial microwave systems.

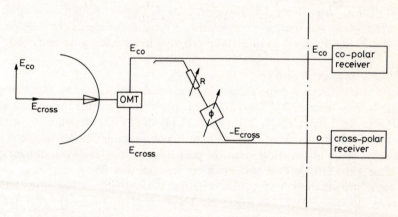

Fig. 5.22 *Static cross-polarisation cancellation (from Reference 15)*
The clear sky, residual, cross-polarised component E_{cross}, introduced because of the imperfect antennas etc., is cancelled exactly by coupling in a signal from the co-polarised channel that is of equal amplitude and 180° in antiphase
R signifies a resistive element and ϕ a phase element, both of which can be adjusted in value.

When frequencies above 10 GHz were introduced on terrestrial microwave systems, the hop-length between repeaters was reduced because of the higher levels of attenuation expected. This, in turn, reduced the incidence of multi-path effects and, when these systems became dual-polarised in order to increase the communications capacity, rain depolarisation was recognised as a significant propagation impairment for the first time. Early rain depolarisation experiments on terrestrial paths [e.g. 20, 21] quickly identified that the theory of the period was inadequate to describe all of the effects, one of which was clearly the contribution due to the off-axis performance of the antennas [2]. The shape of the raindrop was also investigated and, for large raindrops, the Pruppacher and Pitter model [22] was proposed and is now generally accepted. A number of theoretical approaches were made to describe the depolarisation phenomena fully; in particular a remarkable series of calculations by Oguchi, reviewed in [23].

The need to verify the theoretical predictions of the impairments that might potentially impede the full utilisation of dual-polarised frequency re-use techniques on communications satellite systems in the 6/4 GHz bands stimulated the first experimental depolarisation measurements in 1972 on satellite-to-ground paths [24]. These were conducted for INTELSAT by COMSAT Laboratories and, as with earlier terrestrial measurements, highlighted the difficulties in matching theory with measured results, particularly when taking isolated events. This, and other initial experiments conducted by INTELSAT and other organisations, were of an 'event' nature and did not utilise equipment developed expressly for depolarisation studies; the results were therefore not meaningful on a statistical basis. They did, however, provide experience in mounting such experiments that was used when satellites carrying experimental beacon packages specifically designed to undertake propagation studies were launched.

5.4.2 Early slant-path results

The first satellite designed specifically for slant-path depolarisation experiments was the NASA spacecraft ATS–6 and some preliminary results for a frequency of 20 GHz were reported [4, 25]. Fig. 5.23 [4] shows one depolarising event.

In Fig. 5.23, it can be seen that there is considerable scatter in the XPD value for a given attenuation value, in particular at low levels of attenuation. The clear sky XPD of the measurement system was 26 dB and it is clear that some propagation medium cancellation is occurring at low attenuation values as illustrated earlier in Fig. 5.20. The lack of instantaneous correlation between XPD and attenuation was also observed in Reference 25 but, unlike Reference 4, unexpectedly low values of XPD were also observed during very low attenuation events (see Fig. 5.24 [25]).

When ATS-6 was drifted over to 35° E in 1975 to allow propagation experiments to be undertaken in Europe, the initial results appeared to follow theory quite well (see Fig. 5.25 [26]).

In the event shown in Fig. 5.25, two theoretical curves are shown for effective average canting angles of 15° and 25°. Since the linearly polarised signals from AT6-6 were tilted of the order of 25° from the horizontal at the receiving site due to the location of the satellite with respect to the longitude and latitude of the earth station, a canting angle of 25° signifies an effective canting angle of 0° to the local horizontal. The data from Fig. 5.25 appeared to show the raindrops'

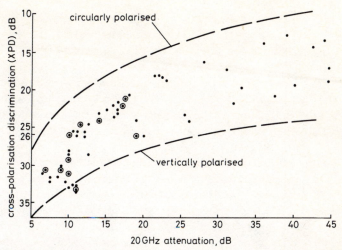

Fig. 5.23 *Measured cross-polarisation discrimination (XPD) versus attenuation on a satellite-to-ground link from ATS-6 at a frequency of 20 GHz (from Fig. 35 of Reference 4)*
⊙ Multiple point
The two broken lines indicate the theoretical limits of XPD for a given attenuation for two cases: circularly polarised signals (worst case) and linearly vertically polarised signals (best case).
Note the apparent improvement (i.e. high XPD) for some values of XPD as the attenuation increases from 5 to 10 dB, contrary to theory and probably due to the medium-induced depolarisation cancelling the residual XPD of the antenna. The incident polarisation is oriented 20° from the plane containing the local vertical
(Copyright © 1974 AT & T, reproduced with permission)

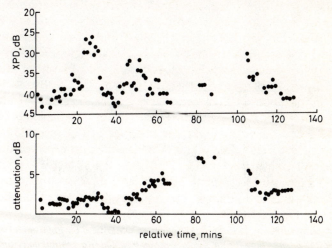

Fig. 5.24 *Measured cross-polarisation discrimination (XPD) and signal attenuation plotted against time for the same event on a satellite-to-ground link from ATS-6 at a frequency of 20 GHz (from Fig. 3 of Reference 25)*
Note between the relative times 20 and 40 the stable, low value of attenuation occurring at the same time that significant depolarisation is observed
(Copyright © 1975 IEEE, reproduced with permission)

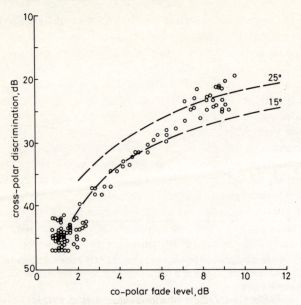

Fig. 5.25 *Measured cross-polarisation discrimination versus co-polarised signal attenuation on a satellite-to-ground link from ATS-6 at a frequency of 30 GHz (from Fig. 2 of Reference 26)*
The broken lines refer to theoretical curves of depolarisation versus attenuation for the given canting angles
(Copyright © 1977 URSI, reproduced with permission)

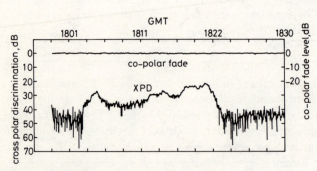

Fig. 5.26 *Measured cross-polarisation discrimination (XPD) and co-polar signal attenuation plotted against time for the same event on a satellite-to-ground link from ATS-6 at a frequency of 30 GHz (from Fig. 4 of Reference 27)*
Note that, throughout the event, the co-polar fade level (signal attenuation) is virtually zero while the cross-polarisation discrimination changes from a clear-sky residual XPD value of about 45 dB to almost 20 dB
(Copyright © 1977 ESA, reproduced with permission)

average canting angle flattening out to zero as the rain intensity increased. No significant cancellation effects were observed. Soon afterwards, so-called anomalous depolarisation events were observed, one being shown in Fig. 5.26 [27].

Early speculation had suggested dry snow [25] or ice crystals [28] effects were causing the degraded XPD in the absence of significant attenuation. It was not until range-gated radars were available to sample the slant path simultaneously with the beacon receivers that conclusive evidence was gathered [29, 30] confirming that the anomalous depolarisation was due to ice crystal effects. A review of ice crystal depolarisation theory and experiment is contained in Reference 31.

The fact that ice crystals could cause significant depolarisation indicated an orientation effect was present that was aligning the crystals (whether they were shaped in the form of 'needles' or 'plates' [32]) in much the same way that raindrops are affected by aerodynamic forces (see Fig. 5.11). Ice crystals, however, are generally very light and so tend to fall very slowly. Some general characteristics of ice crystals are given in Table 5.1.

Table 5.1 *Nature of ice clouds and crystals*

Form around dust-particle nuclei
Shape influenced by temperature
● $-25°C$ mainly needles
● $-25°C$ to $-9°C$ mainly plates
Cirrus clouds: can exist for an indefinite period
Cumulo–nimbus: regular life cycle [formation, growth
    (by sublimation), falling, melting]
Sizes: $0.1–1$ mm long
Concentrations: $10^3–10^6$ particles/m³ ($\sim$ vol. of 2×10^{-6} m³)
Tops of thunderstorms have much higher concentrations
    than do cirrus clouds

The wind shear effects that cause the canting and longitudinal aligning of raindrops are probably insufficient to cause the degree of alignment that must be present to cause the severely reduced values of XPD observed in ice-crystal depolarising events. A secondary alignment mechanism was therefore suspected to be present and a number of investigators believed the mechanism to be electrostatic forces.

Significant field strengths build up in clouds; in general, the larger the cloud the higher the field strengths. Even cirrus clouds have electric fields associated with them and, once the ice crystals have been aligned into a horizontal plane by the aerodynamic forces, the relatively weaker electrostatic forces can now rotate the ice crystals in an essentially frictionless environment so that the long axes are mutually parallel. This is illustrated schematically in Fig. 5.27

Fairly strong evidence that electrostatic forces caused a net alignment of the ice crystals was demonstrated when a number of experimenters saw rapid changes in XPD associated with lightning bolts. An example of these rapid changes is shown in Fig. 5.28 (from Fig. 1 of Reference 33).

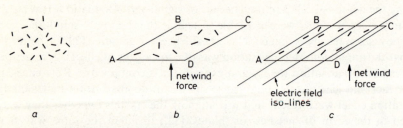

Fig. 5.27 *Aerodynamic and Electrostatic alignment of ice particles in clouds*
  a Randomly oriented ice crystal cloud prior to starting to fall.
  b Ice crystals generally oriented in the plane ABCD by the net wind force generated either by their fall or by an up-draft.
  c Ice crystals starting to align themselves parallel to the prevailing electrostatic field lines following a general orientation into plane ABCD by aerodynamic forces.
  In (a) the ice crystals are randomly oriented. As they start to fall, the aerodynamic forces (wind shear) tend to align them in a plane perpendicular to the net wind force. The presence of an electric field will cause a degree of alignment to occur of the major axes within the plane at right angles to the net wind force due to the polarity of the ice crystals. It should be noted that the ice crystals will tend to oscillate about the preferred 'electric' orientation unless significant damping forces are present [33]. Only a small net alignment, however, will yield significant depolarising effects [33].

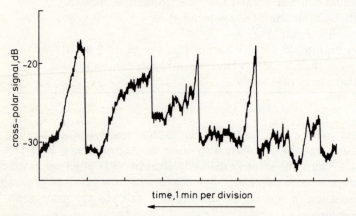

Fig. 5.28 *Rapid changes in XPD level in a thunderstorm (from Fig. 1 of Reference 33)* (Reproduced with permission from Nature Vol. 266, pp. 703–704, Copyright 1977 Macmillan Magazines Ltd.)

The rapid change in XPD can be in either direction, i.e. improving or degrading XPD. In Fig. 5.28 only degradations are shown. A field probe in the same experiment [33] showed simultaneous changes in field strength with

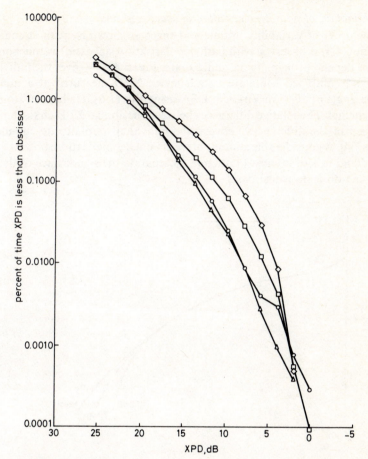

Fig. 5.29 *Variability from year to year of annual XPD statistics (from Fig. 5 of Reference 34)*
    □ Combined
    ○ 1979
    △ 1980
    ◇ 1981
The data were taken at an elevation angle of 10·7° using a circularly polarised
11·6 GHz beacon from the SIRIO satellite
(Copyright © 1986 IEEE, reproduced with permission)

changes in XPD, further supporting the theory that the rapid alignment of ice crystals required the presence of electrostatic fields. Although the XPD levels observed in 'ice crystal' events could be of the same order as XPD levels produced by severe rain events, the relative statistical significance of ice crystal depolarisation compared with rain depolarisation required long-term measurements.

5.4.3 *Variability of path depolarisation in space and time*

The same order of variability in time and space as shown by path attenuation
(see Section 4.4) is observed with path depolarisation since the two phenomena
appear to depend principally on rainfall rate. Fig. 5.29 [34] gives the individual
annual and combined cumulative statistics of XPD measured at a nominal
elevation angle of 10·7° in Virginia, USA using the 11·6 GHz beacon from the
SIRIO satellite. The relative differences between the annual XPD characteristics
are echoed in the differences between the individual rainfall rate cumulative
statistics [34]. A considerable amount of ice crystal depolarisation, however, was
observed and it is of interest to attempt to separate the significance of the two
phenomena on a statistical basis.

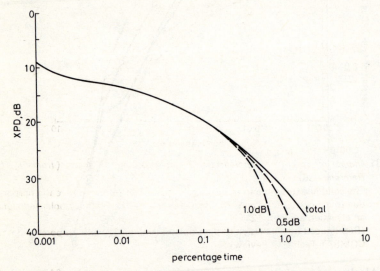

Fig. 5.30 *Impact of ice crystal depolarisation in an experiment in Indonesia (from Fig. 5 of*
Reference 35)
The solid line gives the total XPD results, inclusive of all rain and ice events. The
dashed lines are the XPD statistics that would result if events in which the attenuation exceeded the given values were excluded. The XPD data refer to the downlink, circularly polarised carrier signal with a frequency of 3·7 GHz while the attenuation is that of the looped-back carrier (uplink 5·925 GHz, downlink 3·7 GHz). The
elevation angle was 38°
(Reprinted with permission from INTELSAT, after Fig. 1 of [55], © 1982 IEE,
reproduced with permission)

(a) *Ice crystal depolarisation: statistical significance*

The first attempt to quantify the significance of ice crystal depolarisation in the
6/4 GHz bands utilised data acquired in an experiment conducted in Indonesia
[35]. The depolarisation data were separated into those that occurred in the
absence of significant attenuation and those that were accompanied by significant attenuation. This criterion was used since ice crystal depolarisation has

been observed to occur in events with very low attenuation [29]. Fig. 5.30 shows the results using attenuation thresholds of 0·5 and 1·0 dB [35].

In Fig. 5.30 it can be seen that there does not appear to be any statistical significance in ice crystal depolarisation below a percentage time of 0·1%. Data obtained at approximately the same frequency in an experiment in Alaska [36], however, showed measurable amounts of ice crystal depolarisation to be present at all percentage times (see Fig. 5.31 [36]).

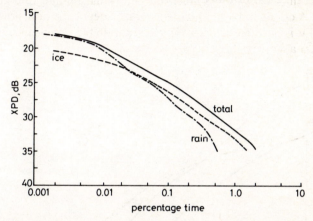

Fig. 5.31 *Impact of ice crystal depolarisation in an experiment in alaska (from Fig. 14 of Reference 36)*
The solid line gives the total XPD while the two broken lines give the proportions due to ice crystals and rain. The data are for a 4 GHz, circularly polarised signal at an elevation angle of 12°
(Reproduced by permission of the Communications Satellite Corporation from the COMSAT Technical Review)

The Alaskan experiment was conducted at an elevation angle of 12° while that in Indonesia was at an elevation angle of 38°. An initial attempt was made to see if elevation angle could be uniquely isolated as the principal parameter affecting the incidence of ice crystal depolarisation without much success [37]. Increasingly, however, depolarisation measurements at low elevation angles have followed the same trend as shown in Fig. 5.31, including those at frequencies in the 14/11 GHz band (see Fig. 5.32).

The relatively constant offset between the rain and ice crystal XPD statistics shown in Figs. 5.31 and 5.32 appears to be a feature of low elevation angle measurements and it would therefore seem that, as the elevation angle decreases, so the incidence of occurrence of ice crystal depolarisation tends to increase.

Ice clouds can be fairly extensive, although relatively thin, and the longer path length through the cloud that results with a lowering of the elevation angle will tend to increase the depolarisaiton effect on the signal [40]. The bulk of the significant ice crystal depolarisation events tend to show relatively small canting angles [42], indicating that the severe electrical effects associated with rapid and large changes in depolarisation (and canting angle) are statistically rare. One distribution of canting angles for ice crystal depolarisation effects is shown in Fig. 5.33 (from Fig. 5 of Reference 42).

On balance, therefore, the impact of ice crystal depolarisation on a statistical basis is insignificant, being far outweighed in importance in terms of depolarisation by rain depolarisation [37, 39, 40]. The relative significance of rain and ice crystal depolarisation is illustrated further in discussions on correlation effects in later Sections of this Chapter.

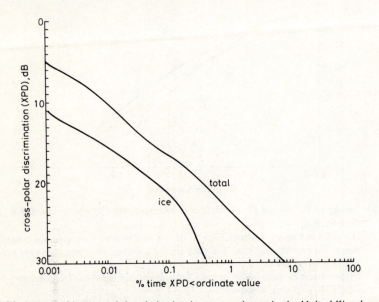

Fig. 5.32 *Impact of ice crystal depolarisation in an experiment in the United Kingdom (from Figs. 8.13 and 8.15 of Reference 38)*
The 'total' line refers to all events while the 'ice' line refers to all those events where the path attenuation was less than 1 dB. The data are for linear polarisation (approximately 56° to the local horizontal) at a frequency of 14.3 GHz.

(b) Seasonal characteristics

Since the severe depolarisation events are usually associated with heavy rain events (although ice crystals are usually present above rain, their effects are generally masked by the heavy rainfall below), the seasonal characteristics of XPD statistics will typically follow those of the rainfall statistics. That is, if the high rainfall rate events occur in the summer season, so will the majority of the severe XPD events [76]. Since ice crystal depolarisation can occur in the absence of rain, and appears also to be associated with the formation of cold fronts [40], the seasonal characteristics of XPD statistics can be less extreme than those of path attenuation, sometimes being spread out more evenly throughout the year [40].

(c) Diurnal characteristics

In most climates, the heavy rain events are associated with atmospheric warming and the diurnal incidence of severe XPD events will follow the same trend. For temperate climates, this will show a peak of events in the late afternoon and early evening associated with the occurrence of thunderstorms. In the same way as with path attenuation statistics, there is a need to quantify the extreme behaviour within an average year and this is usually expressed in terms of the Worst Month.

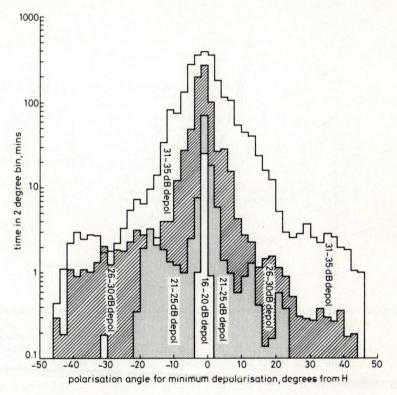

Fig. 5.33 *Canting angle distributions with XPD as parameter (from Fig. 5 of Reference 42)*
Crawford Hill 19 GHz
Elev. 38·6°
Co-polarisation attenuation < 1·5 dB
The data are shown in terms of the angles yielding the minimum depolarisation, i.e. the apparent canting angle of the particles. The data are for ice crystal depolarisation events, which, for these data, assumes that the path attenuation is less than 1·5 dB
(Copyright © 1980 IEEE, reproduced with permission)

5.4.4 Worst month

The same method of building up worst month statistics from annual data used in Section 4.4.2 for path attenuation is employed for XPD and is derived from the CCIR recommendation [43]. It is only recently that worst month data for XPD measurements have been published, and the two sets of worst month data to date [44, 45] generally support the CCIR recommendation. Fig. 5.34 gives one of the data sets [44]. The spread of the measured data around the prediction in Fig. 5.34 could be a feature of the experimental equipment and the quantity of the data used to formulate the curve as much as any real difference between the results and the CCIR model.

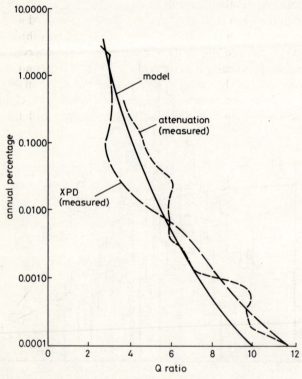

Fig. 5.34 *Average annual worst-month ratio for path attenuation and XPD (from Fig. 1 of Reference 44)*
The data were from measurements at an elevation angle of 10·7° and a frequency of 11·6 GHz using the circularly polarised SIRIO satellite beacon. The model curve is that of Report 723 in Vol V of the CCIR Green Books
(Copyright © 1984 IEE, reproduced with permission)

5.4.5 Short-term characteristics

There are, as with path attenuation, three short-term XPD phenomena that are of interest to system designers: the duration of depolarising events, the interval between successive events at the same threshold level, and the rate-of-change of depolarisation.

(a) Duration of depolarising events

The evidence built up in the previous Section that points towards rain depolarisation being statistically more significant than ice crystal depolarisation on Earth–space paths would tend to indicate that the same distribution of depolarisation events would be observed as is seen in the case of path attenuation event duration statistics. As noted in Section 4.4.3(*a*), the distribution of path attenuation events followed a generally log–normal pattern. For depolarisation, however, polarisation sense, polarisation vector orientation (if linear), and canting angle play a greater part than in path attenuation.

Significant differences in path depolarisation can be observed with changes in polarisation sense and with the orientation of the vector. In a three-year experiment with the OTS satellite [46], marked differences were observed between the event duration for the linearly polarised beacon (11·5 GHz) and the circularly polarised carrier signal (11·8 GHz). The data are presented in Fig. 5.35 [47]. There was little difference in the shape of the curves with changes in the threshold value chosen for sampling the event duration and the percentage distribution.

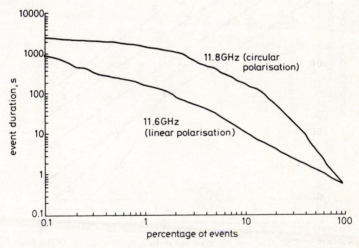

Fig. 5.35 *XPD event duration: Comparison between circular and linear polarisation (from Figs. 2.2 and 2.3 of Reference 47)*
The data cover three years of an experiment with the OTS satellite. The minimum sampling time was 0·5 s and the XPD threshold was set at 30 dB. The elevation angle was 29·9° with the linearly polarised vector making an angle of 11·8° with respect to the horizontal.

It was speculated [47] that the orientation of the linear vector with respect to the local horizontal (11·8° in this case) caused much smaller XPD values to be observed for a given rainfall rate than for the circularly polarised signal. Since the raindrop canting angle does not affect the depolarisation for circular polarisation, only the degree of alignment of the canting angle being significant, it would appear from Fig. 5.35 that, whenever there is rain present, there is a marked tendency to have some degree of alignment of the canting angles [47]. This would cause an 'inflation' of the number of events in the middle of the event duration range as is seen in Fig. 5.35. Data taken in Japan using one of the circularly polarised beacons from an INTELSAT V satellite (11·452 GHz) show similar trends. The data presentation in Fig. 5.36 is different from that in Fig. 5.35 but, for the severe depolarising events with XPD values of 5 or 10 dB, a clear log–normal trend can be seen. For the events with smaller depolarisation (15 to 20 dB), the same down-turn of the data can be seen at long duration times

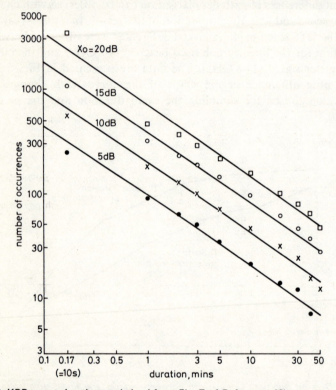

Fig. 5.36 *XPD event duration statistics (from Fig. 7 of Reference 48)*
Yamaguchi
11·452 GHz (CP)
Jan.–Dec. 1983
Elev. angle; 6·5°
The data thresholds were 5, 10, 15 and 20 dB as shown
(Copyright © 1985 IECE, reproduced with permission)

The relatively high proportion of depolarising events in the middle time ranges (about 1–3 min) compared to that expected for a log–normal distribution has also been observed in the USA at 11·6 GHz [34] and Hong Kong at 4 GHz [49], both with circular polarisation.

(*b*) *Interval between successive depolarising events*
The more severe the depolarising event, the less frequently it is likely to occur, and therefore the greater will be the interval between such events. This mirrors the results of the return interval between successive path attenuation events at a given threshold. Unlike path attenuation measurements, however, the measuring equipment plays a very significant part in the low level events. An antenna with an inherently poor XPD will tend to observe as many enhancements as degradations in small depolarising events and so the effects of the propagation medium will be masked by equipment-induced effects at low levels of XPD. At the severe levels of depolarisation, however, the XPD data, both event duration and return period, will tend to follow the rainfall rate statistics' trends.

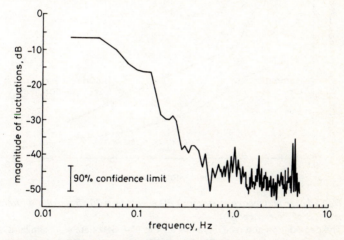

Fig. 5.37 *Spectrum of the downlink (3·7 GHz) depolarisation in severe events in an experiment in Hong Kong (from Fig. 9 of Reference 37)*
The data are for a 3 min period during a severe rain depolarisation event
(Copyright © 1984 John Wiley and Sons Ltd., reprinted with permission)

(*c*) *Rate of change of depolarisation*
A spectral analysis of a number of severe rain depolarisation events at 4 GHz [50] showed that the bulk of the energy was contained well below 1 Hz. One typical result [37] is shown in Fig. 5.37

With depolarisation, particularly for linearly polarised signals, there is the possibility of fairly rapid changes in XPD due to either slight changes in the average canting angle of the propagation medium for a system that has its linear vector orientation close to the average canting angle of the medium or to lightning effects. Measurements undertaken for both linearly and circularly polarised signals [47] again showed slightly different trends for the two polarisation senses although the recovery rates tended to be about the same as the degradation rates. Some of the data are shown in Fig. 5.38 [47].

5.4.6 Site-to-site variability

Variations in precipitation-induced depolarisation will tend to follow the variations in the principle meteorological parameter that causes depolarisation, namely rain. The more severe the rain climatic zone is in terms of rainfall rate, the more severe the measured depolarisation is likely to be on a statistical basis. Site-to-site variability is therefore to be expected and Figs. 5.39 and 5.40 give two examples (from References 51 and 37, respectively).

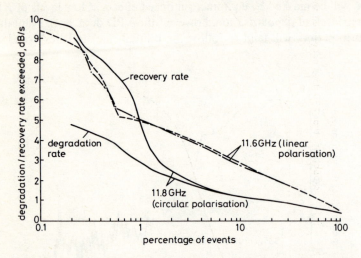

Fig. 5.38 *Degradation and recovery rates for XPD events for a 30 dB threshold (from Figs. 2.2 and 2.3 of Reference 47)*
The degradation and recovery rates for the 11·6 GHz data were similar over most of the data, while for the 11·8 GHz data, the two rates differ below about 1%. This difference is probably systemmatic; the circularly polarised beacon on OTS exhibited large diurnal variations of a regular nature due, it is thought, to thermal effects in the spacecraft antenna. Later measurements on the circularly polarised INTELSAT V beacons showed no such disparity in degradation and recovery rates.

The variations in Figs. 5.39 and 5.40 also include those associated with elevation angle and seasonal variations. There could also be smaller-scale local variations induced by azimuth and spatial variations.

(a) Azimuth variations

In a similar manner to path attenuation, geographic/orographic features that tend to enhance or reduce the likelihood of rain will naturally affect the likelihood of depolarisation occurring the same way. Depolarisation can also be induced by ice crystals at high altitude and there is some evidence [31] that air blowing from the sea has a higher likelihood of containing a higher proportion

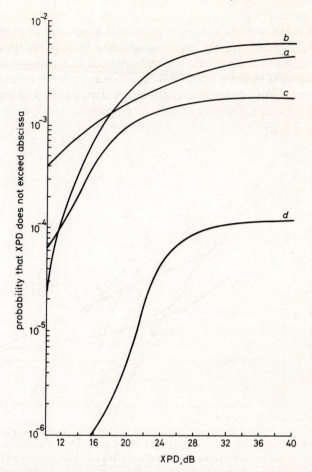

Fig. 5.39 *Illustration of site-to-site variability of XPD statistics within one country (Canada)*
(from Fig. 2 of Reference 51)

The data were taken using the 11·6 GHz beacon (circularly polarised) from the
Canadian satellite HERMES.

Site	Observation period
a St John's	173 days
b Halifax	173 days
c Toronto	195 days
d Vancouver	110 days

of large ice crystals. Paths, particularly low elevation angle paths, whose
azimuthal direction is principally over the sea are therefore likely to suffer a
greater degree of ice crystal depolarisation than those with their paths mainly
over land.

(b) *Spatial variations*

The separation of two earth stations by several kilometres, in a configuration known as site or path diversity, will generally allow one of the sites to be clear of any severe precipitation effects during heavy rain or thundershower activity. A growing amount of measured data is being accumulated both to quantify and to model the diversity improvement gained by operating paired sites in such a

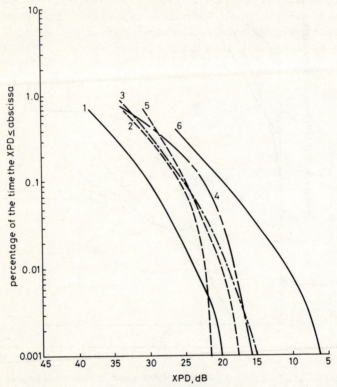

Fig. 5.40 *Illustration of site-to-site variability of XPD statistics on a world-wide basis*
The data were all taken using INTELSAT satellites at a frequency of about 4 GHz (circular polarisation) with approximately one year of data in each set. The countries were:

(1) Lario   : Italy
(2) Ibaraki   : Japan
(3) Jatiluhur : Indonesia
(4) Taipei   : Taiwan
(5) Sitka   : Alaska (USA)
(6) Honk Kong : Hong Kong
(Copyright © 1984 John Wiley and Sons Ltd., reproduced with permission)

configuration from a path attenuation viewpoint (see Section 4.4.4(b)). A similar improvement was expected for depolarisation and a few experiments have been conducted to assess the degree of improvement [40, 48, 52].

In Fig. 5.41, the XPD data are presented for a 12 GHz site diversity experiment conducted in Denmark. The 'predicted' curve shown is one deduced from an attenuation path diversity model. As with single-site XPD measurements, the earth station equipment can mask low-level XPD data. In this experiment, while the Albertsland site had a clear-sky XPD of 35 dB, that at Lyngby was as poor as 25 dB for major parts of the time. A similar experiment in the USA [40], using a site separation of 7·3 km, gave similar differences between the statistics measured at the two sites. Nevertheless, significant improvement in XPD performance can be observed. A third experiment, at approximately the same frequency but using much larger site separations, was conducted in Japan [52].

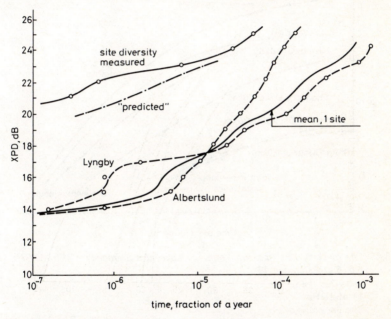

Fig. 5.41 *Site diversity XPD measurements in Denmark at a frequency of 11·8 GHz (from Fig. 1 of Reference 52)*
The elevation angle was about 28°, the polarisation sense circular, and the site separation 15 km. Although the data were taken over a period of less than one year, they are considered to be representative of the general site diversity peformance. Note the large differences in single-site XPD statistics, particularly in clear-sky
(Copyright © 1982 IEE, reproduced with permission)

In the Japanese experiment, the much greater severity of the climate can be noted from the single-site data (see Fig. 5.42), the XPD actually going negative at low percentage times. While there is an improvement in the joint XPD measured with the use of site diversity, the low elevation angle and the occurrence of widespread typhoon-type weather have greatly reduced the apparent improvement obtainable for a given percentage time.

5.5 Correlation of XPD data

When measured XPD data exist at one frequency for a particular site configuration, it is often of importance to be able to scale these data to another frequency. Two timescales are of relevance here: long-term frequency scaling (to obtain the average annual statistics) and short-term frequency scaling (to obtain the spread in the instantaneous correlation between XPD values measured simultaneously at two frequencies along the same path).

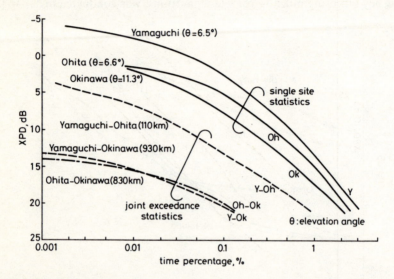

Fig. 5.42 *Site diversity XPD measurements in Japan at a frequency of 11·452 GHz (from Fig. 5 of Reference 48)*
Jan.–Dec. 1983
11·452 GHz
Circular polarisation
The elevation angle was about 6° and it is instructive to note the differences between these data and those in Fig. 5.41
(Copyright © 1985 IECE, reproduced with permission)

5.5.1 Long-term frequency scaling
When the precipitation particle is small with respect to the wavelength of the radiowave signal, the traditional Rayleigh scattering formulations can be applied. For frequencies below about 30 GHz, this will imply a frequency-squared relationship, i.e.

$$XPD_{(f_1)} = XPD_{(f_2)} - 20 \log (f_1/f_2) \text{ decibels} \qquad (5.10)$$

where f_1 and f_2 are the two frequencies in question and the XPD values are those for the same percentage times of an average year, i.e. the equi-probable values.

A relationship of the form given in eqn. 5.10 assumes a negligible contribution due to differential attenuation effects. Statistical results relating the XPD measured between 14·5 and 11·8 GHz seemed to bear out this simple relation [53] and Fig. 5.43 reproduces some of the data from that experiment [53].

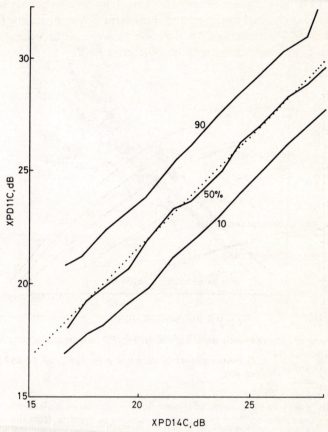

Fig. 5.43 *Long-term frequency correlation of XPD statistics: spread of instantaneous values*
(from Fig. 16 of Reference 53)
OTS 1979 + 1980
—— beacons
· · · · theory
The theory curve is a simple $(f)^2$ ratio and the '90' and '10' curves relate to the envelope containing 90% and 10% of the instantaneous data. Circular polarisation is used at both frequencies
(Copyright © 1984 American Institute of Aeronautics and Astronautics, reproduced with permission)

The long-term, frequency-squared relationship of Fig. 5.43 has been generally confirmed, even for much higher frequency separations [39]. The point to note from Fig. 5.43 is that, although there was a significant spread in the frequency scaling relationship, the data do not appear to show that less depolarisation occurred at the higher frequency than at the lower frequency. Other data [54] which compared depolarisation occurring at the same two frequencies gave strikingly different results in some events. The results from six events are shown in Fig. 5.44 [54].

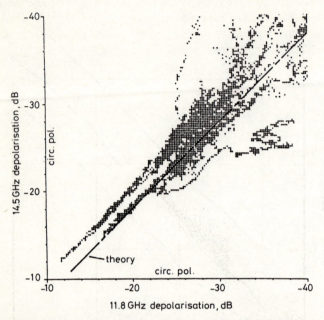

Fig. 5.44 *Trend of long-term frequency correlation of XPD: anomalous scaling ratio (from Reference 54)*
0·5 s samples
6 Events
Elevation = 29·9°
The 'theory' line is a simple $(f)^2$ ratio. Some points are multiple points; no attempt has been made to place any statistical envelope to the data. Note the large spread in the data at low XPD values (owing possibly to ice crystal and propagation medium differential cancelling effects at the two frequencies) and the two clear trends at low XPD values, one of which is closer to an $(f)^{-2}$ ratio than an $(f)^2$ scaling ratio. Circular polarisation was used
(Copyright © 1980 URSI, reproduced with permission)

There appear to be two quite different trends in the data shown in Fig. 5.44, one with an f^2 frequency correlation and the other with an f^{-2} frequency correlation. The f^{-2} correlation has yet to be explained satisfactorily.

5.5.2 Short-term frequency scaling
The data shown in Figs. 5.43 and 5.44 were essentially scatter plots of instantaneous XPD measurements made at two frequencies from which the long-term frequency scaling ratio was extracted. When the first correlation excersises were

attempted for XPD data taken at two frequencies, the antenna imperfections completely masked any correlations that might have been present. This was particularly true of the early measurements undertaken at 6 and 4 GHz [37]. The need to introduce static depolarisation cancellation was recognised and the first experiment at 6/4 GHz to demonstrate conclusively the excellent statistical correlation between the data at the two frequencies was conducted in Indonesia using a static depolarisation cancellation network [55]. Fig. 5.45 illustrates the frequency correlation trend of the XPD data which include both rain and ice crystal depolarisation measurements.

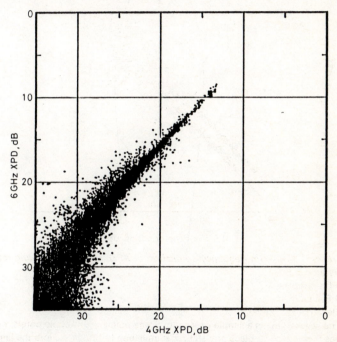

Fig. 5.45 *Short-term correlation of 6 and 4 GHz XPD (from Fig. 2 of Reference 55)*
The experiment used a static depolarisation cancellation network to remove the clear-sky residual depolarisation introduced by the satellite and earth station antennas
(Copyright © 1982 IEE, reproduced with permission)

The generally good instantaneous correlation between the XPD measured at two frequencies along the same path, usually the uplink and the downlink signals to the same satellite, offers hope that the depolarisation on the uplink can be compensated for (i.e. reduced) by measuring the depolarisation on the downlink and applying an appropriate scaling ratio to a dynamic depolarisation compensation network on the uplink path of the antenna feed. This is discussed in Chapter 7. The efficacy of such a (pre)compensation system relies on the accurate measurement of the downlink depolarisation. If the antenna has poor depolarisation characteristics and no static depolarisation cancellation is used, or if the antenna does not track accurately, erroneous XPD measurements will result. The importance of accurate tracking can be illustrated by reference to Figs. 5.46*a* and *b*.

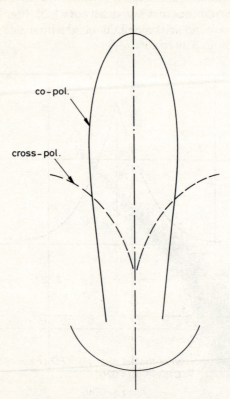

co-pol.

cross-pol.

Fig. 5.46a *Schematic presentation of the main lobe of the co-polar characteristic and the corresponding cross-polar characteristic of an antenna*

A typical earth station antenna will exhibit a main-lobe co-polar peak and a cross-polar null coincident with the axis of peak co-polar gain as shown in Fig. 5.46*a*. When the two characteristics are subtracted (in decibels) to arrive at the net isolation, or clear-sky XPD, Fig. 5.46*b* shows the typical result schematically. An antenna with poor depolarisation characteristics will give XPD values in the mid-to-low twenties; an on-axis XPD of 30 dB is excellent for a large earth station antenna. If the antenna does not track accurately, the achievable XPD falls rapidly, reaching 20 dB in some cases at the point where the co-polar gain has dropped by 3 dB from its peak value.

Most antennas will track to well within the 1 dB-down point on the co-polar characteristic unless conditions along the path deteriorate to such an extent that the satellite signal, upon which the antennas track, drops below the detection threshold. In the 6/4 GHz bands, such conditions can happen when ionospheric scintillation and rain depolarisation occur simultaneously. The simultaneous occurrence of these two phenomena has been reported [56]. In general, however, ionospheric scintillation of such severity to cause tracking problems will only occur rarely, the simultaneous and long-term correlation between the measured depolarisation at two frequencies along the same path usually being very good.

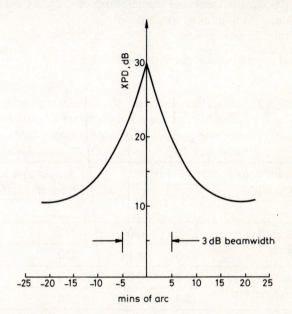

Fig. 5.46b *Schematic presentation of the net clear-sky XPD characteristic of a typical large antenna operating at C-band (6/4 GHz)*

More recent calculations by Fukuchi *et al.* [57], that have attempted to predict the scaling ratio between depolarisation occurring at two frequencies along the same path more accurately, introduce a number of parameters.

Fukuchi *et al.* show that [57]

$$XPD_{(f_1)} = XPD_{(f_2)} - W \text{ decibels} \tag{5.11}$$

where

$$W = u_2 \log (f_1/f_2) + 20 \log \left(\frac{l_1 \cos^2 \varepsilon_1 \sin |2(\phi_1 - \tau_1)|}{l_2 \cos^2 \varepsilon_2 \sin |2(\phi_2 - \tau_2)|} \right)$$

$$- 0 \cdot 0053 (\sigma_1^2 - \sigma_2^2) - (\Delta XPD_{(f_1)} - \Delta XPD_{(f_2)})$$

$$- u_4 (f_1 - f_2) \log (R) \text{ decibels} \tag{5.12}$$

given

$$(f_1 < f < f_2)$$

where l = path length through the rain, km
ε = elevation angle
ϕ = effective raindrop canting angle
τ = polarisation tilt angle of the incident wave with respect to the local horizontal
σ = standard deviation of the raindrop canting angle distribution
ΔXPD = differential attenuation, dB
R = rainfall rate, mm/h

The parameters u_2 and u_4 are given in Table 5.2

Table 5.2 *Parameters u_2 and u_4 associated with the Fukuchi et al. frequency scaling of polarisation over the ranges indicated (from Table 2 of Reference 57)*

Raindrop-size distribution	Parameters		Upper and lower frequencies (GHz)	
	u_2	u_4	f_{lower}	f_{upper}
Marshall–Palmer	21·0	0·00	3	20
	22·0	0·14	20	40
Joss thunderstorm	20·0	0·00	3	20
	18·3	0·15	20	40
Joss drizzle	21·0	0·00	3	24
	24·7	0·11	24	40

(Copyright © 1985 IEE, reproduced with permission)

In eqn. 5.12, if the two frequencies are transmitted along the same path with the same polarisation sense, the second and the third terms disappear. At frequencies below 10 GHz, and probably as high as 15 GHz, the differential attenuation term can be neglected in comparison with the other terms, yielding the modified equation

$$W = u_2 \log (f_1/f_2) - u_4(f_1 - f_2) \log (R) \text{ decibels} \tag{5.13}$$

Note that u_2 is close to 20 and so, as was noted earlier, a frequency-squared relationship yields a good approximation for the 6/4 and 14/11, 14/12 GHz bands. The effect of differential attenuation should not be ignored at higher frequencies if an exact ratio is required.

5.5.3 Correlation of attenuation and depolarisation

Depolarisation has been shown to occur in both rain events and ice crystal events. The path attenuation effects will not be the same in these two types of events which can produce similar levels of depolarisation. There would not therefore appear to be a strong likelihood that there would be a good correlation between attenuation and XPD and, at first sight, this appeared to be so (see Fig. 5.47 [58]).

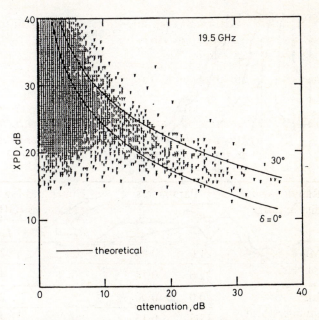

Fig. 5.47 *Long-term correlation of attenuation and depolarisation: scatter plot of simultaneous measurements (from Fig. 1 of Reference 58)*
(Copyright © 1985 IEE, reproduced with permision)

The parameter δ in Fig. 5.47 is the effective mean canting angle assumed to predict the attenuation XPD curves shown. The large scatter of the data points giving the simultaneous XPD versus attenuation measurements would seem to indicate little statistical correlation. If density contours are applied to the data, however, a clear trend emerges as is shown in Fig. 5.48 [58].

A different way of illustrating the density of the simultaneous data samples is shown in Fig. 5.49 [38]. Here the size of the point gives the number of samples at that particular joint impairment level. Probably of most significance, however, are the 10% and 90% contours which are also shown in Fig. 5.49. It is immediately obvious from these contours where the joint data samples are significant or not. From the contours of Figs. 5.48 and 5.49 a definite statistical correlation seems to exist between attenuation and depolarisation measured along the same path.

The correlation, or otherwise, of path attenuation and depolarisation is an important factor to know since the two impairments could individually, or acting together, cause an outage on a particular link. Simple equi-probable curves will tend to show the long-term average, Fig. 5.50 illustrating such curves for a three-year experiment [34], but it is the joint distribution of attenuation

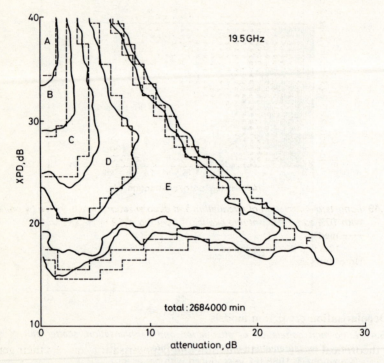

Fig. 5.48 *Long-term correlation of attenuation and depolarisation: contours illustrating the density of the simultaneous data points (from Fig. 2 of Reference 58)*
Solid lines show measured distribution and broken lines show approximated distribution n = density, in min/dB^2

A: $n > 2^{7 \cdot 5} \times 10$ B: $2^{7 \cdot 5} \times 10 > n > 2^{5 \cdot 5} \times 10$
C: $2^{5 \cdot 5} \times 10 > n > 2^{3 \cdot 5} \times 10$ D: $2^{3 \cdot 5} \times 10 > n > 2^{1 \cdot 5} \times 10$
E: $2^{1 \cdot 5} \times 10 > n > 2^{-1 \cdot 5} \times 10$ F: $2^{-1 \cdot 5} \times 10 > n > 2^{-2 \cdot 5} \times 10$

This figure should be compared with Fig. 5.47 where the same data are given in a different presentation
(Copyright © 1985 IEE, reproduced with permission)

versus XPD that is more descriptive of the link impairments. Fig. 5.51 gives the same data shown in Fig. 5.50 but in the form of a joint distribution with percentage time as a parameter [34].

In Fig. 5.50, taking the '1981' data curve, an XPD of 5 dB appears to occur when a path attenuation of 15 dB was observed at attenuation values of from 2·5 to 14·5 dB at a percentage time of 0·004%. The significance of the differences is discussed in Section 5.6.4

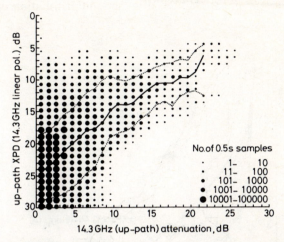

Fig. 5.49 *Long-term correlation of attenuation and depolarisation: density given by point size with 10% and 90% contours overlaid (from Fig. 8.19 of Reference 38)*
—— median
···· 10% and 90% Percentiles
No of 0·5 s samples

5.6 Depolarisation prediction models

Most theoretical models of attenuation and depolarisation have as their genesis the original Mie formulations [59]. The basic asymmetry of large raindrops led to the incorporation of both differential phase and differential attenuation into the theoretical prediction of depolarisation. For linear polarisation, the canting angle was also recognised as a significant parameter and a meteorological model was proposed to take account of this effect [60]. Successive improvements to the scattering theories of Morrison *et al.* [21, 61] and Oguchi [62–64] have led to an excellent understanding of the basic mechanism of depolarisation but, to apply the theories to practical situations, several assumptions need to be made with regard to the satellite-to-ground path. This led to a number of semi-empirical models being proposed that all had as their basis the measured (or predicted) attenuation along the same path. Since ice crystals do not cause significant attenuation, the empirical methods based on attenuation were immediately suspect when this phenomenon was discovered and two types of depolarisation models were developed for a while: rain depolarisation models and ice depolarisation models.

5.6.1 Rain depolarisation models

The general form of the semi-empirical models developed to relate XPD to path attenuation was

$$XPD = a - b \log(A) \text{ decibels} \qquad (5.14)$$

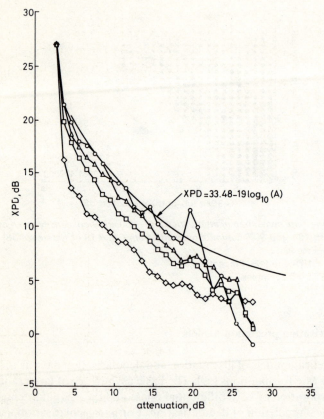

Fig. 5.50 *Long-term correlation of attenuation and depolarisation: equi-probable values for three consecutive years of measurements (from Fig. 12 of Reference 34)*
□ Combined
○ 1979
△ 1980
◇ 1981
The solid curve giving the XPD prediction is from a model developed by the experimenters
(Copyright © 1986 IEEE, reproduced with permission)

where a and b were constants and A was the path attenuation in decibels. At frequencies below 10 GHz, the path attenuation is quite low and a relationship was proposed [65] that replaced attenuation with effective path length and

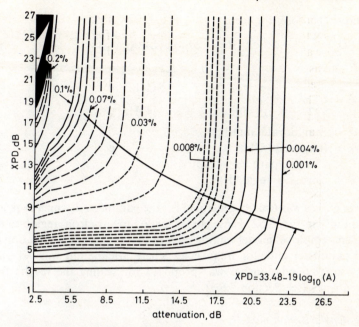

Fig. 5.51 *Long-term correlation of attenuation and depolarisation: joint probability distri-bution (from Fig. 11 of Reference 34)*

These data should be compared with those in fig. 5.50, which give a different presentation for the same measured data

(Copyright © 1986 IEEE, reproduced with permission)

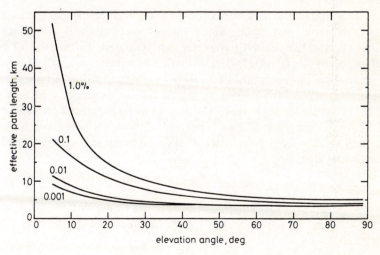

Fig. 5.52 *Typical family of effective path length curves with annual percentage times as parameter*

rainfall rate. This took the form

$$XPD = U - V \log(R) - 20 \log(L) \text{ decibels} \qquad (5.15)$$

where $U = 90 - 20 \log(f) - 40 \log(\cos \theta)$

$$V = \begin{cases} 25 & \text{for } 1 \leqslant f \leqslant 15 \,\text{GHz} \\ 27 - 0\cdot13f & \text{for } 15 < f \leqslant 35 \,\text{GHz} \end{cases}$$

$R =$ rainfall rate, mm/h
$f =$ frequency, GHz
$L =$ effective path length, km
$\theta =$ elevation angle, deg

A family of curves was usually generated for the parameter L which took a form similar to those in Fig. 5.52. These curves tended to vary from region to region and so a semi-empirical model was introduced that took path attenuation as the principal parameter and, for frequencies where attenuation was very low, a frequency-squared relationship was proposed for scaling the results from a higher frequency (see Section 5.6.3).

5.6.2 Ice depolarisation models
For frequencies below about 30 GHz, Rayleigh scattering theory can be applied to calculate the depolarisation induced by ice crystals [31]. There are two fundamental problems, however, in developing prediction models for ice depolarisation: finding a correlating parameter and being able to isolate the parameters.

(a) Correlating Parameter
A 'correlating' parameter is a meteorological or engineering parameter that is easily obtainable and upon which, using semi-empirical formulations, predictions for the the desired parameter can be based. In the case of path attenuation, point rainfall rate is the correlating parameter and, as will be seen, for depolarisation in rain, path attenuation is the correlating parameter. No such parameter exists for ice crystals depolarisation, principally because the bulk of the constituents never reach the ground to be measured. In addtion, if they do approach the ground, their fundamental characteristics have probably altered considerably from their original state. Attempting to measure them *in situ* is equally daunting.

(b) Isolation of Parameters
Ice crystals come in two fundamental shapes: plates or needles [66, 67]. Usually both are present at the same time in two separate layers along the path. Other parameters that vary are [67]:

● Size of the particles
● Number of particles per cubic metre

- Canting angle of each crystal
- Canting angle distribution
- Depth/height of the ice layer(s)

Unfortunately, even a dual-polarised, dual-frequency radar lacks sufficient independent measurable variables to enable all the parameters in an ice cloud to be isolated and so no general predictive model has been possible. Some attempt has been made, however, to include ice effects in the general CCIR predictive model.

5.6.3 General CCIR depolarisation model
The CCIR depolarisation prediction model [68] has evolved through successive refinements to the original semi-empirical model of Olsen and Nowland [65], replacing the rainfall rate and path length dependence with a path attenuation dependence. The CCIR model from Section 4.1 of Reference 68 is reproduced below with only the equation numbers changed for consistency with the text in this Chapter.

"To calculate long-term statistics of depolarisation from rain attenuation statistics the following parameters are needed:

A_p = rain attenuation (dB) exceeded for the required percentage time p for the path in question, commonly called co-polar attenuation (CPA)

τ = tilt angle of the linearly polarised electric field vector with respect to the horizontal (for circular polarisation use $\tau = 45°$)

f = frequency, GHz

θ = path elevation angle

The method described below to calculate XPD (cross-polarisation discrimination) statistics from rain attenuation statistics has been derived from Section 2 of Report 722 where a more detailed discussion of the background can also be found.

The method is valid for $8 \leqslant f \leqslant 35\,\text{GHz}$ and $\theta \leqslant 60°$ and consists of the following steps 1 to 8 to compute the respective contributions due to separate factors (for $f < 8\,\text{GHz}$ see below):

Step 1: Calculate the frequency-dependent term:

$$C_f = 30 \log f \quad \text{for } 8 \leqslant f \leqslant 35\,\text{GHz} \tag{5.16}$$

Step 2: Calculate the rain attenuation dependent term:

$$C_A = V \log (A_p) \tag{5.17}$$

where: $V = 20$ for $8 \leqslant f \leqslant 15\,\text{GHz}$
$\quad\quad\quad V = 23$ for $15 < f \leqslant 35\,\text{GHz}$

Some measurements seem to indicate that the above-mentioned values of V will result in too low XPD values for high rain attenuation (e.g. $A_p > 10\,dB$ at 12 GHz).

When studying differences in XPD statistics as a function of polarisation type and tilt angle, the value of A_p to be used should be the measured or predicted value for circular polarisation. This establishes a common reference, thus providing greater accuracy in the study of XPD differences.

Step 3: Calculate the polarisation improvement factor:

$$C_\tau = -10 \log [1 - 0.484(1 + \cos (4\tau))] \tag{5.18}$$

Note that the improvement factor $C_\tau = 0$ for $\tau = 45°$ and reaches a maximum value of 15 dB for $\tau = 0$ or 90°.

The above relation is approximately equivalent to eqn. 6 in Report 722, with σ_m (the standard deviation of the storm-to-storm variation of the effective canting angle) being assumed to be 5°. For regions where reliable values of σ_m have been measured, eqn. 6 of Report 722 should be used in preference to the above relation.

Step 4: Calculate the elevation angle-dependent term:

$$C_\theta = -40 \log (\cos \theta) \quad \text{for } \theta \leq 60° \tag{5.19}$$

Step 5: Calculate the canting angle dependent term:

$$C_\sigma = 0.0052\sigma^2 \tag{5.20}$$

σ is the effective standard deviation of the raindrop canting angle distribution, expressed in degrees. σ takes the value 0°, 5°, 10° and 15° for 1%, 0.1%, 0.01%, and 0.001% of the time, respectively. C_σ is the canting angle dependent term used in Report 722 where $C_\sigma = K^2$.

Step 6: Calculate rain XPD not exceeded for $p\%$ of the time:

$$XPD_{rain} = C_f - C_A + C_\tau + C_\theta + C_\sigma \text{ decibels} \tag{5.21}$$

Step 7: Calculate the ice crystal dependent term:

$$C_{ice} = XPD_{rain} \times (0.3 + 0.1 \log p)/2 \text{ decibels} \tag{5.22}$$

Step 8: Calculate the XPD not exceeded for $p\%$ of the time and including the effects of ice:

$$XPD_p = XPD_{rain} - C_{ice} \text{ decibels} \tag{5.23}$$

As noted in Section 2 of report 722, a design approach based on joint probability cumulative distributions of XPD and A_p is preferable, particularly for Earth–space paths where the variability in XPD for a given A_p is large for the low attenuation margins normally employed. It should be noted, however, that the use of an equi-probability relation between XPD and A_p for outage calculations may give the same results as the use of the joint probabilities if it is applied to the calculations of fading margins in systems using dual polarisation. Administrations are urged to give particular attention to this problem in future analyses."

The same sub-Section of the CCIR Report gives a procedure for scaling the XPD results from one frequency to another and one polarisation tilt angle to another along the same path. This is [68]

$$XPD_2 = XPD_1 - 20 \log \left[\frac{f_2 \sqrt{1 - 0.484(1 + \cos 4\tau_2)}}{f_1 \sqrt{1 - 0.484(1 + \cos 4\tau_1)}} \right] \text{ decibels}$$

(5.24)

for

$$4 \leqslant f_1, f_2 \leqslant 30 \text{ GHz}$$

where XPD_1 and XPD_2 are the XPD values not exceeded for the same percentage of time at frequencies f_1 and f_2 and polarisation tilt angles τ_1 and τ_2, respectively.

The factor C_{ice}, introduced in the CCIR model to account for the ice depolarisation, was developed using data that were obtained at elevation angles generally above 10°. For this reason, the C_{ice} factor becomes progesssively larger as the percentage time increases; i.e. ice crystal depolarisation is assumed to be negligible at low percentage times. Recent results for low elevation angles paths at frequencies above 10 GHz have tended to show a more constant ice crystal depolarisation impact over the whole range of percentage times and so the C_{ice} factor may need to reflect this in later models. Overall, however, the statistical impact of ice crystal depolarisation is low and so the CCIR prediction model gives good results.

Other approaches to the development of semi-empirical models have concentrated on either a more rigourous modelling technique than that adopted by the CCIR, followed by more curve fitting to obtain the best fit prediction of XPD versus attenuation [69], or an extension of the original Olsen and Nowland model with new coefficients for the effective path length and frequency dependence [70]. The former approach, called the SIM model [69], gives good results, but since it does not afford significantly superior results to the general CCIR model, the CCIR general prediction model is recommended. The latter approach can also give good results but requires an accurate knowledge of the path length through the rain medium.

5.6.4 Joint attenuation versus XPD prediction models
The semi-empirical procedure used in the general CCIR depolarisation model makes use of the supposition that, for a given rain attenuation, the XPD along the same path will be the same irrespective of climate; the particular rain

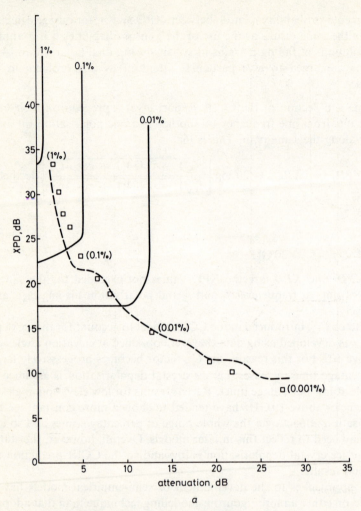

Fig. 5.53 *Reference distribution of joint attenuation/XPD curves in climate K at a frequency of 14 GHz (from Figs 11(b) and 10(b) of Reference 73)*
    a Elevation angle of 30°
    b Elevation angle of 10°
          Frequency: 14 GHz
          Polarisation tilt angle: 45°
    —— Climate K joint distribution
    - - - Scaled measured data from NJ
    □□□ Equiprobability values for climate K
(Copyright © 1986 John Wiley & Sons Ltd., reproduced with permission)

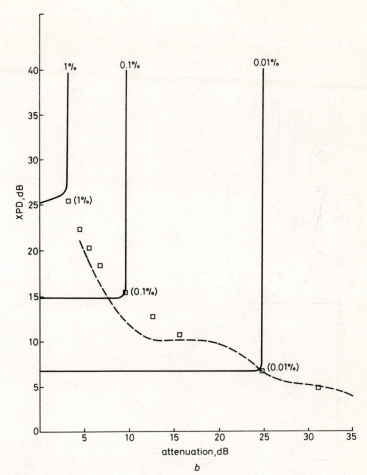

Fig. 5.53 *Continued*

parameters that are inducing the measured attenuation will induce an XPD that is generally independent of location, path geometry and rain zone. The approach assumes an equi-probability distribution is applicable to the correlation between attenuation and XPD. A growing body of measurements tends to support this supposition.

In most cases, however, an equi-probable distribution will imply a pessimistic XPD value for a given attenuation; i.e. the XPD is generally better than the predicted value when joint probability statistics are used. A simple joint probability model has been proposed [71] that uses the method proposed by Thirlwell and Howell for displaying joint XPD/attenution statistics [72] and connects together the 'knees' of the joint distribution. The curve through the joint distribution in Fig. 5.51 illustrates the procedure.

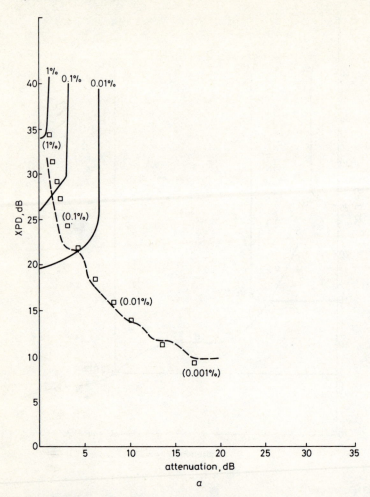

Fig. 5.54 *Reference distribution of joint attenuation/XPD curves in climate K at a frequency of 11 GHz (from Figs. 9(b) and 8(b) of Reference 73)*
  a Elevation angle of 30°
  b Elevation angle of 10°
     Frequency: 11 GHz
     Polarisation tilt angle: 45°
  —— Climate K joint distribution
  - - - Scaled measured data from NJ
  □□□ Equiprobability values for climate K

(Copyright © 1986 John Wiley and Sons Ltd., reproduced with permission)

A semi-empirical procedure for establishing the joint XPD/attenuation distributions, as opposed to just the locus, was proposed [73] that made use of a reference joint distribution. It was assumed that the scaling laws for frequency, elevation angle and polarisation tilt angle evolved for median and equi-probable distributions were also valid for joint distributions and, using the original joint

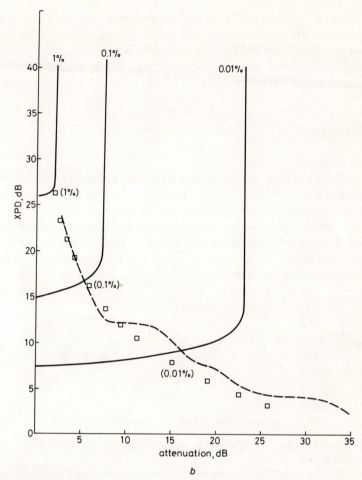

Fig. 5.54 *Continued*

distribution obtained for New Jersey [74], a set of reference curves was evolved for 11 and 14 GHz links in CCIR rain climate zones E, K and M [73]. Four such distributions are shown in Figs. 5.53a and b and 5.54a and b for 14 and 11 GHz links at elevation angles of 30° and 10° in climate K.

To scale the reference distributions to other elevation angles, frequencies and tilt angles, the following scaling equations were used [73]:

$$A(f_2, \theta_2) = A(f_1, \theta_1)(f_2/f_1)^2(\sin \theta_1/\sin \theta_2) \text{ decibels} \qquad (5.25)$$

$$XPD(f_2, \theta_2, \tau_2) = XPD(f_1, \theta_1, \tau_1)$$

$$- 20 \log \left\{ \frac{f_2}{f_1} \frac{\cos^2 \theta_2}{\cos^2 \theta_1} \frac{\sin \theta_1}{\sin \theta_2} \right\}$$

$$\times \sqrt{\left(\frac{1 - 0.484(1 + \cos 4\tau_2)}{1 - 0.484(1 + \cos 4\tau_1)} \right)} \text{ decibels} \qquad (5.26)$$

In this approach [73], a simplified attenuation scaling procedure is used compared with the CCIR procedure (see eqn. 4.46) and an elevation angle factor is introduced into the CCIR XPD frequency scaling procedure given in eqn. 5.24. The former was used since the applicable range of the prediction model is approximately 10–35 GHz and the latter to allow elevation angle scaling.

5.7 System impact

5.7.1 Co-channel interference

Unlike path attenuation, depolarisation does not cause any increase in the perceived system noise since it is not an absorptive phenomenon. Depolarisation, however, will cause energy from one polarisation to be transferred to the polarisation of the opposite sense, thus lowering the isolation between the two co-channel (i.e. same frequency) polarisations. A reduction in signal-to-noise C/N is therefore principally an attenuation effect while a reduction in signal-to-interference signal C/I is a depolarisation effect.

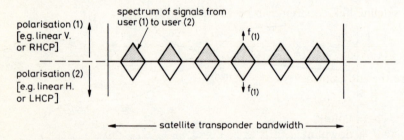

Fig. 5.55 *Schematic presentation of full co-channel frequency re-use by means of dual-polarisation operation*

The triangles represent the specturm of the FM carriers, a number of which can be contained within a typical satellite transponder. The carriers are assumed to be of equal size (i.e. the same number of channels) with full co-channel operation, that is the centre-frequencies of the carriers in polarisation (1) are exactly the same as those in polarisation (2).

In Fig. 5.55, a number of small FM carriers are shown within a transponder. These could equally well be small digital carriers. A severe reduction in C/I will, in the limit, cause the signal in polarisation 1 at frequency 1 to interfere with the co-channel signal in the opposite polarisation sense, polarisation 2, also operating with a carrier frequency of frequency 1.

It is rare that full co-channel operation is used, however, the usual procedure being to offset the centre frequencies of the carrier signals. This is shown schematically in Fig. 5.56.

With large FM carriers or large digital carriers, frequency interleaving is not possible and so a means must be found to reduce the interference if it causes any operational channels to decrease to below the specified performance threshold for longer than is permitted. Before this is attempted, however, it should be assessed whether the reduction in C/I by itself causes the link to experience an outage.

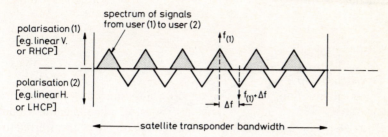

Fig. 5.56 *Schematic presentation of frequency-interleaved frequency re-use by means of dual-polarised operation*

In this representation, the carriers are offset by Δf in their centre frequencies to reduce any co-channel interference. For simplicity, equal-size carriers are shown.

A satellite link equation can be written as [75]

$$\frac{1}{(C/N)_t} = \frac{1}{(C/N)_u} + \frac{1}{(C/N)_d} + \frac{1}{(C/N)_{im}} + \frac{1}{(C/I)} \tag{5.27}$$

where the subscripts refer to:

t = total
u = uplink
d = downlink
im = intermodulation products

Note that the terms are numerical power ratios and have not been converted to logarithmic (i.e. dB) values.

The $(C/N)_t$ in eqn. 5.27 can drop below the required performance margin if any of the four terms becomes severely degraded. A link that is noise-limited or intermodulation-limited will tend to be relatively insensitive to C/I unless the C/I is exceptionally severe.

A practical system will try and balance the impairing phenomena in such a way that the cost of the overall system is minimised. There is no point in striving for a C/I of 45 dB if the $(C/I)_{im}$ is 20 dB. Typically, a clear-sky C/I of from 27 to 30 dB is about optimum, giving a good balance between cost-effective technology and co-channel interference. Techniques for improving the C/I under impaired conditions are discussed in Chapter 7.

5.7.2 Scintillation/depolarisation impact

Tropospheric and ionospheric scintillation, since they are essentially on-axis phenomena that effect the co- and cross-polarised channels equally, will not cause any perceptible depolarisation. The effect of scintillation, however, can have secondary impacts on the isolation of the link that can be equally severe if compensation mechanisms are installed that depend on an accurate correlation between attenuation and depolarisation or downlink depolarisation and uplink depolarisation.

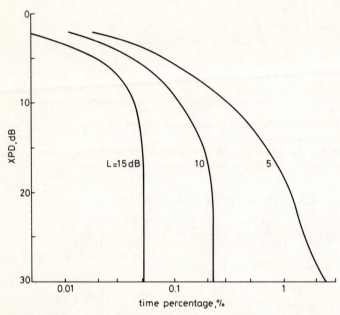

Fig. 5.57 *Joint attenuation/XPD distributions with attenuation as a parameter (from Fig. 6 of Reference 48)*
Ohita
11·452 GHz
Elevation angle = 6·6°
Jan.–Dec. 1983
Note the departure of the 5 dB curve at high percentage times from the trend of the 10 dB and 15 dB curves due, it is thought, to the effect of tropospheric scintillation
(Copyright © 1985 IECE, reproduced with permission)

(a) Tropospheric scintillation: Impact on depolarisation
The long-term correlation and joint distributions of path attenuation and depolarisation have been shown in earlier Sections to be generally well bounded by semi-empirical modelling. At low elevation angles, severe tropospheric scintillation (which causes negligible depolarisation) can cause a significant shift in the equi-probable and joint distribution of path attenuation versus XPD. An example of the latter is shown in Fig. 5.57 [48].

In Fig. 5.57, the 'tail' of the curve corresponding to 5 dB attenuation does not follow the trend of the curve for higher attenuation values. An uplink power control system, with a matching uplink depolarisation compensation system, would probably not function satisfactorily in this region.

(b) Ionospheric scintillation: Impact on depolarisation
Fig. 5.58 illustrates schematically the simultaneous occurrence of ionospheric scintillation and rain (or ice crystal) depolarisation.

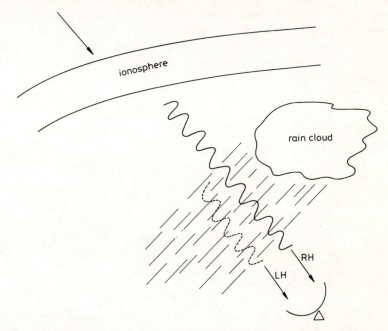

Fig. 5.58 *Schematic presentation of the joint occurrence of ionospheric scintillation and rain depolarisation*
The scintillations will appear in both the co- and cross-polarised channels, but, if the antenna tracks accurately, the XPD will remain relatively constant. Once the antenna loses track, however, depolarisation will occur on the uplink and downlink that departs from the theoretical values.

The severe ionospheric scintillation and simultaneous rain depolarisation observed in one experiment [56] would have led to a total depolarisation-induced outage (due to scintillation-induced antenna tracking problems) amounting to 0·06% on an annual basis if full co-channel operation had been in operation. These would have been in addition to the 'normal' rain depolarisation induced outages. Poor antenna tracking has been pointed out earlier as one of the potentially major causes of depolarisation observed on a link. Severe scintillation (from whatever origin) can be sufficient to cause an antenna tracking mechanism to fail.

Ionospheric scintillation is a cyclic phenomenon, peaking twice each year at the equinoxes with an underlying periodicity of approximately 11 years. Ionospheric scintillation activity tracks the sunspot number, the variation of which is shown in Fig. 2.15. A hypothetical presentation of those periods when joint ionospheric scintillation and rain depolarisation could cause unacceptable performance is shown in Fig. 5.59. Note, however, that such joint scintillation/ depolarisation impairments are only likely to occur at equatorial latitudes where ionospheric scintillation is generally the most severe; in the 6/4 GHz bands which is the only satellite communications band where the two phenomena are both significant; at the equi-noctal periods; and only at sunspot maxima periods. The average impact (over an 11 year period) is therefore negligible but, if outage criteria are defined in terms of any year or any worst month, they could be significant.

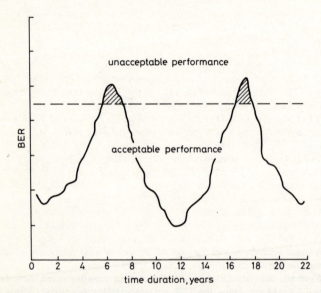

Fig. 5.59 *Hypothetical presentation of the periods when the performance of a 6/4 GHz link could be degraded to an unacceptable level by simultaneous rain depolarisation and ionospheric scintillation*

5.8 References

1 IPPOLITO, L. J., KAUL, R. D., and WALLACE, R. G.: 'Propagation effects handbook for satellite systems design', NASA Reference Publication 1982(03), 1983
2 WATSON,, P. A. and GHOBRIAL, S. I.: 'Off-axis polarisation characteristics of cassegranian and front-fed paraboloidal antennas', *IEEE Trans.*, 1973, **AP-20**, pp. 891-898
3 PRATT, T., and BOSTIAN, C. W.: 'Satellite communications (John Wiley, 1986)
4 MORRISON, J. A., and CHU, T. .S.: 'Perturbation calculations of rain-induced differential attenuation and differential phase shift at microwave frequencies', *Bell Syst. Tech. J.*, 1973, **52**, pp. 1907-1913.

5 ALLNUTT, J. E., and ROGERS, D. V.: 'System Implications of 14/11 GHz path depolarisation. Part II: Reducing the impairements', *Int. J. Satellite Commun.*, 1986, **4**, pp. 13–17

6 BRUSSAARD, G.: 'A meteorological model for rain-induced cross-polarisation', *IEEE Trans*, 1976, **AP-24**, pp. 5–11

7 UZUNOGLU, N. K., EVANS, B. G., and HOLT, A. R.: 'Scattering of electromagnetic radiation by precipitation particles and propagation characteristics of terrestrial and space paths', *Proc. IEE*, 1977, **124**, pp. 417–424

8 WATSON, P. A., and ARBABI, M.: 'Cross-polarisation isolation and discrimination', *Electron. Lett.*, 1973, **9**, pp. 516–517

9 COX, D. C., and ARNOLD, H. W.: 'Comparison of measured cross-polarisation isolation and discrimination for rain and ice on a 19 GHz space-earth path', *Radio Science*, 1984, **19**, pp. 617–628

10 BRUSSAARD, G.: Private communication, Sept. 1987

11 STUTZMAN, W. L.: 'Mathematical formulations and definitions for dual-polarized reception of a wave passing through a depolarizing medium (A polarisation primer)', Supplemental Report 1 on A Depolarization and attenuation experiment using the CTS and COMSTAR satellites, June 1977 (revised Jan. 1980), prepared for NASA Goddard Space Flight Center, Greenbelt, Maryland 20771, USA

12 MIYA, K. (Ed.): 'Satellite communications technology', (KDD Engineering & Consultancy, Inc., Tokyo, 1985, 2nd Edn. (English Language Edition)).

13 BRYANT, D. L.: 'Uplink depolarisation pre-compensation experiment (INTEL-156)' Final Report on INTELSAT contract INTEL-156, (British Telecom International, Landsec House, 23 New Fetter Lane, London, EC4A 1AE, 1984)

14 ALLNUTT, J. E., and GOODYER, J. E.: 'Design of receiving stations for satellite-to-ground propagation research at frequencies above 10 GHz', *IEE J. Microwaves, Optics, & Acoustics*, 1977, **1**, pp. 157–164

15 THIRLWELL, J.: 'Private communication, March 1984

16 CHANDRA, M., and McEWAN, N. J.: 'Use of dual-polarisation radar for XPD prediction due to rain along a slant path having polarisation tilt', URSI Commission F/IEE Open symposium, Bournemouth, UK, 1982, pp. 195-200

17 HOLT, A. R.: 'Some factors affecting the remote sensing of rain by polarization radar in the 3 to 35 GHz frequency range', *Radio Science*, 1984, **19**, pp. 1399–1412

18 POIARES BAPTISTA, J. P. V., and McEWAN, N. J.: 'Prediction of slant-path rain cross-polarisation radar data', *Electron. Lett.*, 1985, **21**, pp. 460–461

19 'Special issue on DRS-8 Digital Radio System', *Telesis*, Dec. 1977

20 WATSON, P. A., and ARBABI, M.: 'Rainfall cross-polarisation at microwave frequencies', *Proc. IEE*, 1973, **120**, pp. 413–418

21 MORRISON, J. A., CROSS, M. J., and CHU, T. S.: 'Rain-induced differential attenuation and differential phase shift at microwave frequencies', *Bell Syst. Tech. J.*, 1973, **52**, pp. 599–604

22 PRUPPACHER, H. R., and PITTER, R. L.: 'A semi-empirical determination of the shape of cloud and rain drops', *J. Atmos. Science*, 1971, **28**, pp. 86–94

23 OGUCHI, T.: 'Scattering from hydrometeors: A survey', *Radio Science*, 1981, **16**, pp. 691–730

24 TAUR, R. R.: 'Rain depolarisation: theory and experiment', *COMSAT Tech. Rev.*, 1974, **4**, pp. 187–190 (More fully reported in Technical Memoranda Cl-14-73 and CL-40-73. COMSAT Labs., 22300 Comsat Drive, Clarksburg, Md 20871, USA)

25 BOSTIAN, C. W., STUTZMAN, W. L., MANUS, E. A., WILEY, P. H., and MARSHALL, R. E.: 'Depolarisation measurements on the ATS-6 20 GHz downlink: A description of the VPI & SU experiment and some initial results', *IEEE Trans.* 1975, **MTT-23**, pp. 1049–1053

26 SHUTIE, P. F., ALLNUTT, J. E., and MacKENZIE, E. C.: 'Depolarisation results at 30 GHz using transmissions from the ATS-6 satellite', URSI Commission F Open Symposium, 1977, La Baule, France, pp. 367–369

27 SHUTIE, P. F., MacKENZIE, E. C., and ALLNUTT, J. E.: 'Depolarisation measurements

at 30 GHz using transmissions form ATS-6', Proc. of ATS-6 meeting, 1977, ESTEC, Noordwijk, ESA SP-131, pp. 127–134

28   COX, D. P., and ARNOLD, H. W.: 'Preliminary results from the Crawford Hill 19 GHz COMSTAR beacon propagation experiment', US National Committee of the International Union of Radio Science (USNC/URSI) Meeting, Oct. 1976, Amherst, Ma,. USA

29   SHUTIE, P. F., ALLNUTT, J. E., and MacKENZIE, E. C.: 'Satellite–Earth signal depolarisation at 30 GHz in the absence of significant fading', *Electron. Lett.*, 1977, **13**, pp. 1–2

30   McEWAN, N. J., WATSON, P. A., DISSANAYAKE, A. W., HAWORTH, D. P., and VAKILI, V. T.: 'Cross polarisation from high-altitude hydrometeors on a 20 GHz satellite radio path', *Electron. Lett.*, 1977, **13**, pp. 13–14

31   BOSTIAN, C. W., and ALLNUTT, J. E.: 'Ice-crystal depolarisation on satellite-earth microwave radio paths', *Proc. IEE*, 1979, **126**, pp. 951–960

32   MOYER, V., HORVATH, N., and THOMPSON, A. H.: 'The College Station, Texas, Halo Complex of 22 March 1979', *Bull. Amer. Meteorological Soc.*, 1980, **61**, pp. 570–572

33   HAWORTH, D. P., McEWAN, N. J., and WATSON, P. A.: 'Relationship between atmospheric electricity and microwave radio propagation', *Nature*, 1977, **266**, pp. 703–704

34   BOSTIAN, C. W., PRATT, T., and STUTZMAN, W. L.: 'Results of a three-year 11·6 GHz low-angle propagation experiment using the SIRIO satellite', *IEEE Trans.*, 1986, **AP-34**, pp. 58–65

35   OGAWA, A.: '6/4 GHz depolarisation correlation measurements', Tech. Memo., IOD-P-83-01, INTELSAT, 3400 International Drive, NW, Washington, DC 20008-3098, USA, 1983

36   STRUHARIK, S. J.: 'Rain and ice depolarisation measurements at 4 GHz in Sitka, Alaska', *COMSAT Tech. Rev.*, 13, pp. 403–436

37   ALLNUTT, J. E.: 'The system implications of 6/4 GHz satellite-to-ground signal depolarisation results from the INTELSAT propagation measurements programme', *Int. J. Satellite Commun.* 1984, **2**, pp. 73–80

38   Final Report on Phase 3 of the INTELSAT contract INTEL-159/238 'INTELSAT V low angle propagation measurements carried out at Martelsham Heath, England'. British Telecom Research Labs.

39   COX, D. C.: 'Depolarization of radio waves by atmospheric hydrometeors in earth-space paths: A review', *Radio Science*, 1981, **16**, pp. 781–812

40   STUTZMAN, W. L., BOSTIAN, C. W., TSOLAKIS, A., and PRATT, T.: 'The impact of ice along satellite-to-earth paths on 11 GHz depolarization statistics', *Radio Science*, 1983, **18**, pp. 720–724

41   BOSTIAN, C. W., PRATT, T., and STUTZMAN, W. L.: 'Operation of the SIRIO transportable receiving terminal'. Final Report VPI & SU/EE/SATCM-83/5, Electrical Engineering Dept., VPI & SU, Blacksburg, Va 24061, USA, 1983

42   ARNOLD, H. W., COX, D. C., HOFFMAN, H. H., and LECK, R. P.: 'Ice depolarisation statistics for 19 GHz satellite-to-Earth propagation', *IEEE Trans.*, 1980, **AP-28**, pp. 546–550

43   Recommendations and Reports of the CCIR, XVIth Plenary Assembly, Dubrovnik, 1986 Volume V (Propagation in non-ionized media); Report 723-2: 'Worst month statistics'.

44   YON, K. M., STUTZMAN, W. L., and BOSTIAN, C. W.: 'Worst-month rain attenuation and XPD statistics for satellite paths at 12 GHz', *Electron. Lett.*, 1984, **20**, pp. 646–647

45   FUKUCHI, H., KOZU, T., and TSUCHIYA, S.: 'Worst month statistics of attenuation and XPD on Earth-space paths', *IEEE Trans.*, 1985, **AP-33**, pp. 390–396

46   THIRWELL, J.: 'Depolarisation measurements at 11 and 14 GHz with OTS', IEE Conf. on Results of tests and experiments with the European OTS Satellite, London, 1981

47   HEWITT, M. T., EMERSON, D. J., RABONE, D. C., and HOWELL, R. G.: 'Depolarisation rate-of-change and event duration statistics for the OTS slant-path propagation experiments', British Telecom Research Labs., Research and Technology Executive Internal Research Memo R6.2.3 No 5005/85 Issue 1, 1985

48   YAMADA, M., YASUKAWA, K., FURUTA, O., KARASAWA, Y., and BABA, N.: 'A

propagation experiment on Earth-space paths of low elevation angles in the 14 and 11 GHz bands using the INTELSAT V satellite', ISAP 85, Vol. 1 1985, pp. 309–312

49   EVANS, B. G., and UPTON, S. A. J.: 'Analysis of 4 GHz signal attenuation and depolarisation data and concurrent 12 GHz radiometer data', Final Report on INTELSAT contract INTEL-118, Dept. of Electrical Engineering, University of Essex, Wivenhoe Park, Colchester, 1982

50   FANG, D. J.: 'IS-898 magnetic tape data analysis', Final Report on INTELSAT Contract INTEL-222, Task RAD-003, COMSAT Labs., 22300 Comsat Drive, Clarksburg, Maryland 20871, USA, 1982

51   LAU, P.-K., and ALLNUTT, J. E.: 'Attenuation and depolarisation data obtained on 12 GHz satellite-to-Earth paths at four Canadian locations', *Electron. Lett.*, 1979, **15**, pp. 565–567

52   LARSON, J. R.: 'Results of XPD site-diversity measurements at 11·8 GHz', *Electron. Lett.*, 1982, **18**, pp. 81–82

53   MAWIRA, A., and NEESEN, J. T. A.: 'Propagation data for the design of 11/14 GHz satellite communications systems', AIAA 10th Communications Satellite Systems Conference, 1984, Orlando, Florida, USA, pp. 629–639

54   HOWELL, R. G., and THIRLWELL, J.: 'Cross-polarisation measurements at Martlesham Heath using OTS', URSI Commission F International Symposium on Effects of the lower atmosphere on radio propagation at frequencies above 1 GHz, 1980, Lennoxville, Canada, pp. 26–30

55   OGAWA, A., and ALLNUTT, J. E.: 'Correlation of 6 and 4 GHz depolarisation on slant paths', *Electron. Lett.*, 1982, **18**, pp. 230–232

56   FANG, D. J., and ALLNUTT, J. E.: 'Satellite signal degradation due to simultaneous occurrence of rain fading and ionospheric scintillation at equatorial earth stations', IEE International Conference on Antennas and Propagation (ICAP 87), IEE Conf. Publ. 274, 1987, pp. 281–284

57   FUKUCHI, H., AWAKA, J., and OGUCHI, T.: 'Frequency scaling of depolarisation at centimetre and millimetre waves', *Electron. Lett.*, 1985, **21**, pp. 10–11

58   FUKUCHI, H.: 'Two dimensional probability distribution of attenuation and depolarisation of Earth-space paths', *Electron. Lett.*, 1985, **21**, pp. 445–447

59   MIE, G.: 'Beitrage zur optik truber medien speziell kolloidaler metalosungen', *Ann. Phys.*, 1908, **25**, pp. 377–445

60   BRUSSAARD, G.: 'A meteorological model for rain-induced cross polarisations', *IEEE Trans.*, 1976, **AP-24**, pp. 5–11

61   MORRISON, J. A., and CROSS, M. J.: 'Scattering of a plane electromagnetic wave by axi-symmetric raindrops', *Bell syst. Tech. J.*, 1974, **53**, pp. 955–1019

62   OGUCHI, T.: 'Attenuation and phase rotation of radiowaves due to rain: calculations at 19·3 and 34·8 GHz', *Radio Science*, 1973, **8**, pp. 31–38

63   OGUCHI, T., and HOSOYA, A.: 'Scattering properties of oblate raindrops and cross-polarisation of radio waves due to rain (Part II): Calculations at microwave and millimeter wave regions', *J. Radio Res. Labs. (Japan)*, 1974, **21**, pp. 191–259

64   OGUCHI, T.: 'Scattering properties of Prupacher-and-Pitter form raindrops and cross-polarisation due to rain: calculations at 11, 13, 19·3, and 34·8 GHz', *Radio Science*, 1977, **12**, pp. 41–51

65   OLSEN, R. L., and NOWLAND, W. L.: 'Semi-empirical relations for the prediction of rain depolarisation statistics: their theoretical and experimental basis', Proceedings of the International Symposium on Antennas and Propagation (ISAP-78), 1978, Sendai, Japan

66   STUTZMAN, W. L., OVERSTREET, W. P., BOSTIAN, C. W., TSOLAKIS, A., and MANUS, A. E.: 'Ice depolarisation on satellite radio paths', Final Report on INTELSAT contract INTEL-123, Dept. of Elec. Engg., VPI & SU, Blacksburg, Va 24061, USA, April, 1981

67 TSOLAKIS, A., and STUTZMAN, W. L.: 'Calculation of ice depolarisation on satellite radio paths', *Radio Science*, 1983, **18**, pp. 1287–1293

68 Recommendations and Reports of the CCIR, XVIth Plenary Assembly, Dubrovnik, 1986 (Propagation in non-ionized media); Report 564-3: 'Propagation data and prediction methods required for Earth-space telecommunications systems'

69 STUTZMAN, W. L., and RUNYON, D. L.: 'The relationship of rain-induced cross-polarisation discrimination to attenuation for 10 to 30 GHz Earth-space radio links', *IEEE Trans.*, 1984, **AP-32**, pp. 705–710

70 FUKUCHI, H., AWAKA, J., and OGUCHI, T.: 'Improved theoretical formula for the relationship between rain attenuation and depolarisation', *Electron. Lett.*, 1984, **20**, pp. 859–860

71 GAINES, J. M., and BOSTIAN, C. W.: 'Modeling the joint statistics of satellite path XPD and attenuation', *IEEE Trans.*, 1982, **AP-30**, pp. 815–817

72 THIRLWELL, J., and HOWELL, R. G.: '20 and 30 GHz slant-path propagation measurements at Martlesham Heath, UK', Proceedings AGARD 26th Symposium on Electromagnetic wave propagation, London, 1980, pp. 1–9

73 ROGERS, D. V., and ALLNUTT, J. E.: 'System implications of 14/11 GHz path depolarization. Part I: Predicting the impairment', *Int. J. Satellite Commun.*, 1986, **4**, pp. 1–11

74 ARNOLD, H. W., COX, D. C., HOFFMAN, H. H., and LEAK, R. P.: 'Characteristics of rain and ice depolarisation for a 19 and 28 GHz propagation path from a COMSTAR satellite', *IEEE Trans.*, 1980, **AP-28**, pp. 22–28

75 FLAVIN, R. K.: 'Rain attenuation considerations for satellite paths in Australia', *ATR*, 1982, **16**, pp. 11–24

76 MAEKAWA, Y., CHANG, N. S., and MIYAZAKI, A.: 'Seasonal variations of cross-polarisation statistics observed at CS-2 experimental earth station', *Electron. Lett.*, 1988, **24**, pp. 703–704

Chapter 6

Terrain, multipath and other particulate effects

The overriding importance of rain attenuation and depolarisation within the lower atmosphere and ionospheric scintillation in the upper atmosphere on satellite-to-ground signals tend to overshadow a number of other propagation phenomena that can be of significance in Earth-space communications systems. These phenomena relate to links that, by virtue of their elevation angle, can cause an interaction with surface obstacles to occur or, if located in regions of unusual concentrations of non-aequeous precipitation (e.g. dust), can exhibit apparent impairments in excess of free-space propagation conditions. The former will be discussed under two categories, terrain and multipath, while the latter will be under the general classification of other particulates.

6.1 Terrain effects

The general effects of terrain irregularities, and the interaction between the types of terrain cover (e.g forests, snow-capped ridges) and the radio signal, are dealt with conceptually in Reference 1. These effects are significant mainly for line-of-sight, terrestrial radio-relay systems. In these systems, great care is usually taken to avoid not only obstacles in the path that might reduce the free passage of the radiowave signal but also potentially reflective surfaces that might cause severe multipath effects.

With the increase in the number of small earth stations located close to or within large urban areas has come the enhanced potential for interference. In some cases, natural terrain effects can be taken advantage of to site earth stations so that there is no direct path between the earth station and the potential source(s) of interference. This is the concept of site shielding (see Fig. 6.1 and also Section 1.4.2).

An obstacle placed between the earth station and a potentially interfering source will not provide complete blockage. Some portion of the interfering signal will be received owing to diffraction around the object. An idealised obstacle, with a very sharp (compared to the wavelength) edge will give rise to knife-edge diffraction .

6.1.1 Knife-edge diffraction

Knife-edge diffraction formulae make use of the dimensionless parameter v which can be calculated in a number of ways [2]:

$$v = h\sqrt{\frac{2}{\lambda}\left(\frac{1}{d_1} + \frac{1}{d_2}\right)} \tag{6.1}$$

$$v = \theta\sqrt{\left\{\frac{2}{\lambda\left(\frac{1}{d_1} + \frac{1}{d_2}\right)}\right\}} \tag{6.2}$$

$$v = \sqrt{\left(\frac{2h\theta}{\lambda}\right)} \tag{6.3}$$

$$v = \sqrt{\left\{\frac{2d}{\lambda}\alpha_1\alpha_2\right\}} \tag{6.4}$$

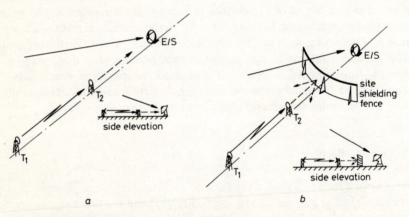

Fig. 6.1 *Concept of site shielding using a metallic fence*
a Without site shielding
b With site shielding
In (*a*), a terrestrial link (T_1, T_2) is causing unacceptable interference into an earth station (E/S), or vice versa. One technique to remove the mutual interference is to erect a metallic fence between the two systems, as shown in (*b*). A site shielding fence is not often used for single-source interference, other available interference reduction techniques such as signal cancellation being cheaper, but a site shielding fence provides general protection and is usually used in a 'cluttered' urban environment.

The parameters h, d_1, d_2, θ, α_1 and α_2 are shown in Fig. 6.2 (from Fig. 6 of Reference 2) with the angles in radians and the distances and the wavelength λ in the same units (e.g. metres).

Eqns. 6.2–6.4 generally assume that the angles θ and α are small, i.e. less than or equal to 0·2 rad, or 12° [2], which is generally true for shielding obstacles remote from the earth station (e.g. a mountain ridge). For artificial site shielding

obstacles placed close to the earth station, this condition is not normally true
and eqn. 6.1 is to be preferred to derive values of v. Given v, the transmission
loss, $J(v)$, can be calculated and is shown in Fig. 6.3 (Fig. 7 of Reference 2). An
approximate formula for the transmission loss $J(v)$ is given in Reference 3.

$$J(v) = 6{\cdot}9 + 20 \log \left[\sqrt{\{(v - 0{\cdot}1)^2 + 1\}} + v - 0{\cdot}1\right] \text{ decibels} \quad (6.5)$$

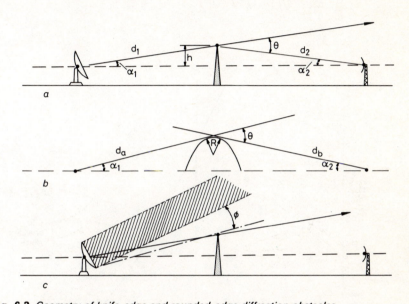

Fig. 6.2 *Geometry of knife-edge and rounded-edge diffraction obstacles*
In (*a*), the critical dimensions are the height *h* above the beam centres, and the
respective distances d_1, d_2 and angles α_1, α_2 the antennas make with the knife-edge.
The angle between the beams, θ, is also important.
    In (*b*), the diffraction obstacle is a rounded-edge and the curvature radius, *R*, is
important.
    In general, when the diffraction obstacle is in the near field of an antenna, two
factors are important. All obstructions should be clear of the cylinder described by a
projection of the antenna area along the propagation path (shown cross-hatched in
(*c*)). Secondly, there should preferably be the equivalent of two 3 dB beamwidth
separations from the edge of the projected cylinder and the minimum clearance from
the bottom of the antenna. This is the angle ϕ in (*c*).

If the obstacle is rounded, rather than being a knife edge, the diffraction loss
tends to be greater [2]. On the other hand, if the obstacle is such that the
path angle between the earth station and the interfering site at the obstacle
(angle θ) is very small, diffraction effects over the obstacle can cause an enhancement of the received signal/interfering signal to be experienced over that
expected due to only free space loss alone. This is referred to as obstacle gain
[1]. In general, however, if the angle θ is small (less than 0·2 rad), but not close

to a grazing angle, the additional diffraction loss ΔA, due to the obstacle being rounded rather than a knife-edge, is [3]

$$\Delta A = 7(1 + 2v)\varrho \text{ decibels} \tag{6.6}$$

for $v > 0$ and $\varrho < 1$. The value of ϱ can be obtained from [2]

$$\varrho^2 = \left(\frac{d_a + d_b}{d_a d_b}\right) \bigg/ \left[\left(\frac{\pi R}{\lambda}\right)^{1/3} \frac{1}{R}\right] \tag{6.7}$$

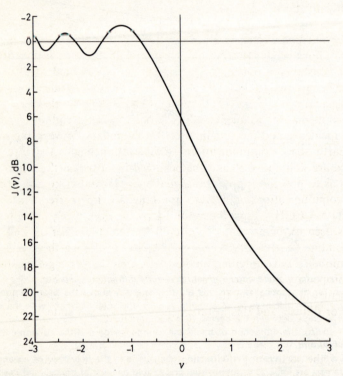

Fig. 6.3 *Transmission loss relative to free space caused by knife-edge diffraction (from Fig. 7 of Reference 2)*
(Copyright © 1986 ITU, reproduced with permission)

with d_a, d_b, and R as shown in Fig. 6.2 and λ the wavelength of the radiowave signal. The total transmission loss A over a rounded obstacle is therefore

$$A = J(v) + \Delta A \text{ decibels} \tag{6.8}$$

Despite the increasing accuracy of the site shielding predictions, some of the old 'rules of thumb' still apply [4] with regard to the clearance angles and the minimum height of the site shielding fence, or natural obstacle. To prevent detectable 'obstacle gain' from occurring, the angle between the earth station's beam direction and the top of the obstacle must exceed 2° for a frequency of 4 GHz using a 30 m diameter antenna and the top of the obstacle must be 1·5D above the centre of the antenna (where D is the diameter of the antenna). For higher frequencies or smaller antennas, the 2° clearance angle can be scaled accordingly, always ensuring the same number of beamwidth clearances as in the example given. When more than one diffraction edge is involved, the procedure is more complicated [2, 5].

6.2 Multipath effects

6.2.1 Maritime mobile communications

When a radiowave signal encounters a boundary between media of different dielectric constants, a portion of the signal will be reflected with the angle of reflection being equal to the angle of incidence. The magnitude and phase of the reflected signal will depend on many factors, including the electrical properties of the two media, the frequency and polarisation of the radiowave signal, the angle of incidence and the roughness of the boundary between the media.

For Earth–space communications, the elevation angles that are normally employed are well above 5° and the beamwidths of the earth station antennas are such as to prevent appreciable signal energy from striking the ground. In mobile communications, however, not only are lower frequencies used at present (1·5–1·6 GHz) than those for Fixed Services using Satellites but the antennas used on board ships or aircraft are much smaller as well. The two effects combine to give larger beamwidths for the ship earth station antennas and the mobile service antennas on board the aircraft, thus significantly increasing the propensity for incurring multipath impairments. The type of multipath impairment will depend on the nature of the reflections, i.e. whether they are specular or diffuse, and this in turn depends on the sea state.

(a) Effect of the sea state

In Fig. 6.4, the occurrence of significant specular reflection is determined by the sea state. The smoother the sea, the smaller the reflective area and the greater the degree of coherent (specular) reflection.

The sea state has been defined [6] and the terminology (from [7]) is given in Table 6.1. Quite often a sea roughness factor u is invoked, where u is given by the following definition [6]:

$$u = \frac{4\pi}{\lambda} h_0 \sin \theta_0 \text{ radians} \tag{6.9}$$

where λ is the wavelength in metres, θ_0 is the elevation angle in degrees, and h_0 is the RMS profile height of the sea surface in metres. Assuming the sea surface height distribution to be gaussian, the parameter h_0 is related to the significant wave height H by [8]

$$H = 4h_0 \text{ metres} \tag{6.10}$$

The significant wave height H is defined as the average value of the peak-to-peak trough heights of the highest one-third of all the waves [7].

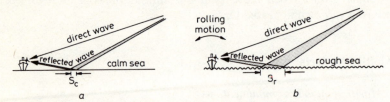

Fig. 6.4 *Illustration of multipath effects for maritime mobile communications in calm and rough seas*
a Calm sea conditions
b Rough sea conditions
In (*a*), the sea surface is smooth and the area of the sea surface S_c that will have facets that can reflect the satellite signal into the ship's antenna is very small. This reflected wave, since it will be only from a very small region giving only slightly different pathlengths for the collection of reflected signals, will be formed of generally coherent components and is termed as specular reflection. In (*b*), since the sea is rough, the probability of there being wave facets that will reflect the satellite signal into the ship's antenna is fairly high for a large surface area S_r sometimes called the 'glistening surface' from optical terminology. The large variations in path length of the reflected signals will lead to a generally incoherent multipath wave which is termed 'diffuse reflection'.

Table 6.1 *Description of sea states (from Table III of Reference 7 after Reference 9)*

Sea state number	State of the sea	Significant wave height H (m)	β_0 (deg)
0 and 1	Calm	< 0·15	1
2	Smooth	0·15–0.8	1–5
3	Slight	0·8–2·0	5–12·5
4	Moderate	2·0–4·0	12·5–23·5
5 and 6	Rough	4·0–6·5	23·5–35
7	Very rough	6·5–9·5	35–46·5
8	High	> 9·5	> 46·5

The parameter β_0 is the maximum value of the angle formed by a facet of the wave in question with the horizontal plane. Sometimes an RMS value of β_0 is used for all, or a portion, of the waves as opposed to a maximum value
(Copyright © 1986 ITU, reproduced with permission)

The magnitude of the coherent reflection relative to the direct signal is a function of both the significant wave height and the elevation angle to the satellite. Some examples are given in Fig. 6.5 (from Fig. 1 of Reference 7).

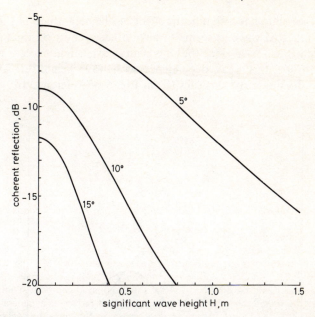

Fig. 6.5 *Magnitude of the coherent reflection relative to the direct signal for elevation angles of 5°, 10° and 15° at a frequency of 1·5 GHz for circular polarisation (Fig. 1 of Reference 7)*
(Copyright © 1988 ITU, reproduced with permission)

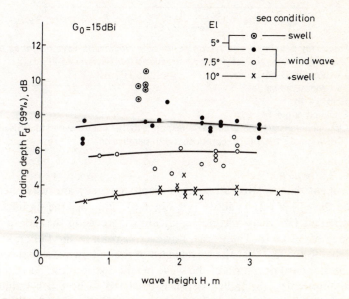

Fig. 6.6 *Fading depth as a function of wave height (Fig. 5 of Reference 10)*
(Copyright © 1986 IECE, reproduced with permission)

While the sea state has a significant effect on the magnitude of the coherent (specularly) reflected wave, it seems to have a lesser effect on the total fading depth on a statistical basis once the significant wave height exceeds about 1 m. Fig. 6.6 illustrates this weak dependence with wave height; elevation angle appearing to dominate the statistics (from Fig. 5 of Reference 10).

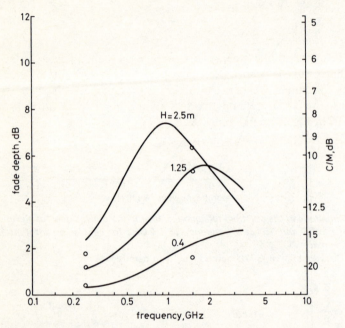

Fig. 6.7 *Frequency dependence of fading depth with significant wave height as parameter (Fig. 5 of Reference 7 abstracted and revised from Reference 11)*
        —— Predicted values
        ● Measured values
Gain: 14 dB
Elevation angle: 5°
Circular polarisation; fade depth corresponds to that not exceeded for 99% of the time
(Copyright © 1986 ITU, reproduced with permission)

(b) Effect of frequency

An experiment that isolated all effects except frequency and wave height [11] confirmed the earlier results above (see Fig. 6.6) that, once the significant wave height exceeded 1 m, there was no significant difference in the fading depth at a frequency of 1·5 GHz. There was, however, a peak in the fading at around 1·5 GHz due to the fact that the wavelength of the radiowave signal is comparable to the standard deviation of the wave height distribution. At much higher or lower frequencies, the fading depth is expected to be much reduced. The experimental and theoretical results of Reference 11 are illustrated in Fig. 6.7 (from Fig. 5 of Reference 7).

(c) *Effect of polarisation*

At a frequency of about 1·5 GHz, there is a marked difference between the reflection coefficients for horizontally and vertically polarised signals, particularly at low elevation angles (see Fig. 3.10). The maritime mobile signals are circularly polarised and so, at low angles become horizontally oblate, elliptically polarised signals. This can be used to advantage to reduce the multipath effects [10] and is discussed in more detail in Chapter 7, section 7.4.3.

(d) *Effect of antenna gain*

As is to be expected, the higher the antenna gain the narrower the beamwidth, and the narrower the beamwidth the smaller the susceptibility to multipath. In rough sea conditions, where nearly all of the multipath effects are due to diffuse reflections, there is a definite trend of fading depth with antenna gain (see Fig. 6.8 from Fig. 4 of Reference 7).

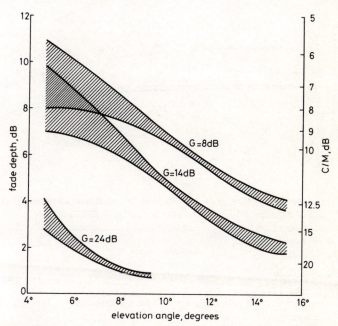

Fig. 6.8 *Fading depth versus elevation angle with antenna gain as parameter* (*Fig. 4 of Reference 7*)
The fade depth corresponds to that not exceeded for 99% of the time using circular polarisation and a frequency of 1·5 GHz. The data refer to diffuse multipath fading in rough sea conditions
(Copyright © 1986 ITU, reproduced with permission)

6.2.2 *Aeronautical mobile communications*

All the maritime mobile communications effects apply to aeronautical mobile communications over the sea. There are, however, two main difference between the two mobile systems, and one secondary difference. The main differences are the antenna height above the sea and the relative antenna (aircraft) speed. The secondary difference applies to an aircraft while it is ascending or descending.

(a) Effect of antenna height

Data taken in rough sea conditions ($u > 2$; significant wave height about 1·4 m) at elevation angles of 5° and 10° from an aircraft showed the same trend as with maritime mobile experiments [12]. Fig. 6.9 gives some examples (from Fig. 1 of Reference 2)

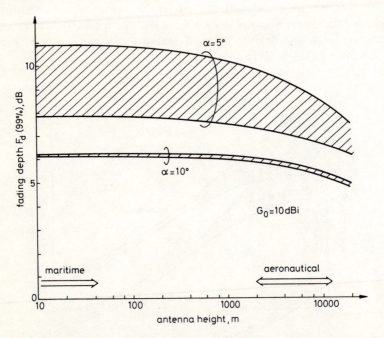

Fig. 6.9 *Fading depth versus antenna height with elevation angle as parameter (Fig. 1 of Reference 12)*
A frequency of 1·5 GHz was used with circular polarisation. The data refer to rough sea conditions
(Copyright © 1986 IECE, reproduced with permission)

Fig. 6.9 seems to indicate that, for the same frequency, elevation angle and antenna, the multipath experienced on aeronautical mobile systems will be less than that on maritime mobile systems but that the difference amounts to only 1 or 2 dB at most. As the elevation angle increases, the difference decreases to 1 dB or less [12].

An additional effect of the large height of the antenna above the sea is to reduce the 'correlation frequency'. If the correlation frequency is defined as the minimum frequency difference for which the correlation coefficient becomes $1/e$ (that is 0·37), there is a linear reduction in correlation frequency with height [12]. Fig. 6.10 illustrates the effect (from Fig. 6 of Reference 12).

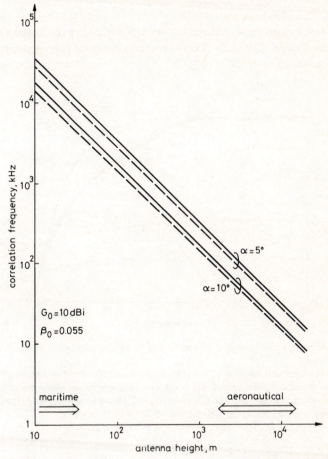

Fig. 6.10 *Correlation frequency versus antenna height with elevation angle as parameter (Fig. 6 of Reference 12)*
——— coherent component
– – – incoherent component
A frequency of 1·5 GHz was used with circular polarisation. The definition of β_0 is given in Table 6.1 and it relates to the sea surface state. In Table 6.1, β_0 is the maximum value of the sea surface facet angle while, in this figure, β_0 is the RMS slope of the sea surface
(Copyright © 1986 IECE, reproduced with permission

(b) Effect of speed
The major impact of the aircraft speed is on the 'corner frequency' of the fading spectrum. Maritime multipath spectra (see Fig. 3.26) tend to have higher corner

frequencies than tropospheric scintillation, indicating the more rapid fluctuations that occur in multipath phenomena. The corner frequency for maritime mobile multipath fading illustrated in Fig. 3.26 is around 1–2 Hz, the higher corner frequency being for larger wave heights (3 m) and higher elevation angles (10°). With aeronautical mobile, the corner frequency can be much higher. If the aircraft climbs or dives, the corner frequency is even higher, exceeding 10 Hz in most cases. Fig. 6.11 gives some experimental results (from Fig. 3 of Reference 12).

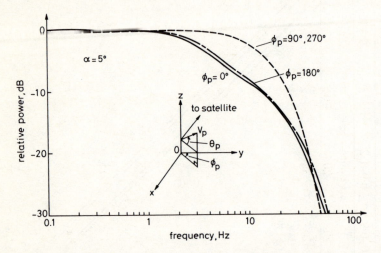

Fig. 6.11 *Frequency spectrum of multipath fading for aeronautical mobile (Fig. 3 of Reference 12)*

The flight altitude was 10 km with a speed of 1000 km/h. A frequency of 1·5 GHz with circular polarisation was used. The data are for an elevation angle α of 5° in rough sea conditions. The solid line ($\phi_p = 0°$) is for level flight while the even broken line is for climbing/diving attitudes of 5° to the horizontal

(Copyright © 1986 IECE, reproduced with permission)

6.2.3 Variability in space and time of mobile multipath effects

Little data exist for estimating the variability in space and time of aeronautical mobile multipath effects but, since the fading characteristics of aeronautical and maritime mobile systems using almost the same frequencies are very similar, it is expected that the variability in space and time of multipath phenomena on the two systems will be generally the same.

With the exception of the ionosphere (the effects of which are considered in Chapter 2), the atmosphere *per se* has little effect on 1·5/1·6 GHz (L-band) transmissions. The variability in space and time of L-band multipath will therefore have more to do with the secondary effects of the atmosphere, e.g. what are the statistics of the sea state conditions induced by the wind forces?

Table 6.2 *Significant wave height statistics for various regions of the world (Table IV of Reference 7 after Reference 13)*

Region	Height of waves (m)					
	0–0·9	0·9–1·2	1·2–2·1	2·1–3·6	3·6–6	> 6
North Atlantic, between Newfoundland and England	20	20	20	15	10	15
Mid-equatorial Atlantic	20	30	25	15	5	5
South Atlantic, latitude of southern Argentina	10	20	20	20	15	10
North Pacific, latitude of Oregon and south of Alaskan peninsula	25	20	20	15	10	10
East-equatorial Pacific	25	35	25	10	5	5
West wind belt of South Pacific, latitude of southern Chile	5	20	20	20	15	15
North Indian Ocean, Northeast monsoon season	55	25	10	5	0	0
North Indian Ocean, Southwest monsoon season	15	15	25	20	15	10
Southern Indian Ocean, between Madagascar and northern Australia	35	25	20	15	5	5
West wind belt of southern Indian Ocean on route between Cape of Good Hope and southern Australia	10	20	20	20	15	15
Averages over all regions	22	23	20·5	15·5	9·5	9·0

(a) Sea state statistics

Some data exist for estimating the likely occurrence of certain significant wave heights on a world wide basis. Table 6.2 gives some of these data (Table IV of Reference 7 after Reference 13)

At L-band, the fading depth does not seem to vary much with wave height above a significant wave height of about 1 m and so it is probably simplest to divide the time into the two periods: the total time when the significant wave height is below 1 m and the total time when the significant wave height is above 1 m. On this basis, there does not seem to be much spatial variability from ocean region to ocean region.

(b) Fade duration statistics

With fade duration time T_D and fade occurrence interval T_I defined as shown in Fig. 6.12, some assumptions were made with regard to the amplitude distribution in order to calculate the power spectrum [14].

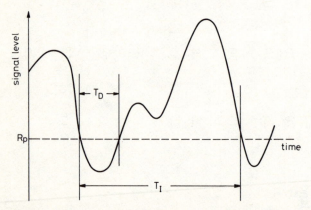

Fig. 6.12 *Definition of fade duration and fade occurrence interval (Fig. 1 of Reference 14)*
R_p = threshold signal level for a given percentage of the time
The fade duration T_D is the time required for the signal amplitude to return to a certain level after it has dropped below that level. The fade occurrence interval T_I is the time period from the point where the signal intensity drops below the set threshold level to the next time it drops below the same threshold level
(Copyright © 1987 IEEE, reproduced with permission)

The amplitude distribution of maritime multipath fading is best described by a Nakagami–Rice distribution but in Reference 14 the assumption was made that for the percentage times (50–99%) and amplitude levels (less than 10 dB) involved, a gaussian distribution could be invoked with only a small error. From the power spectra (see Fig. 3.26), distributions of T_D and T_I were predicted which fitted the measured data quite well. These predictions are shown in Fig. 6.13 (from Fig. 5 of Reference 14).

The above data and analyses assume that the incoherent component is dominant. In some cases, particularly in fairly calm sea conditions where one parameter is in addition changing slowly (e.g. elevation angle because either the

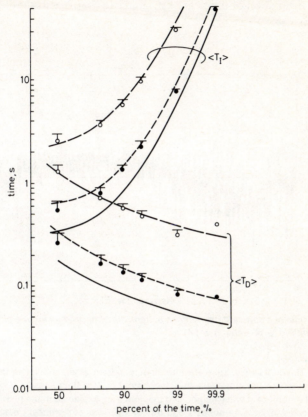

Fig. 6.13 *Mean fade duration* $\langle T_D \rangle$ *and mean fade occurrence interval* $\langle T_I \rangle$ *as a function of time for percentages from 50 to 99·9% (from Fig. 5 of Reference 14)*
The dots and circles are measured data with the horizontal line above some of the points representing the limits when a 'hysteresis' of 0·2 dB is employed in deciding on a cross-over point. The solid curves are estimates using the power spectra for the three cases (see Fig. 3.26)

Case	Elevation angle	Wave height	Ship velocity	Ship rolling
─ ─·─ *a*	5°	0·5 m	11 knots	1°
─ ─ ─ ─ *b*	10°	3 m	11 knots	5°
───── *c*	10°	5 m	20 knots	30°

ship is moving or the satellite is in a highly inclined orbit), a coherent interference effect will be superimposed on the incoherent phenomena. An example of this is shown in Fig. 6.14 (from Fig. 9 of Reference 14).

From Fig. 6.14 it can be seen that a coherent effect can cause very large fade durations and fade duration intervals to occur. It can also be seen that these coherent phenomena can significantly bias the fading amplitude statistics since the apparent mean level of the clear-sky signal can change substantially for long periods.

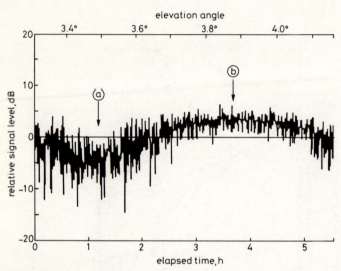

Fig. 6.14 *Example of a coherent interference pattern superimposed on incoherent interference*
(from Fig. 9 of Reference 14)
$H \simeq 0.4 - 0.6 \, m$
$(u \simeq 0.5)$
The slowly varying, coherent interference pattern essentially alters the mean level about which the rapidly varying, incoherent pattern fluctuates.
H, the significant wave height, and the term *u* are explained in the text. The sampling interval was 5 s. The points *a* and *b* depict periods of significant destructive interference and constructive interference, respectively
(Copyright © 1987 IEEE, reproduced with permission)

6.2.4 Prediction models

Diffuse multipath effects can be characterised by a dominant (direct-path) signal plus a large number of smaller (reflected) signals that have random phases with respect to the dominant signal. The reflected signal follows a Rayleigth distribution and the overall process, the interaction of the Rayleigh-distributed reflected signal with the dominant direct signal, yields a distribution that follows Ricean statistics [6].

The key parameter in a Ricean distribution [15] is the ratio of the (dominant) direct component power C to the power in the (reflected) multipath component power M. This is defined as the Rice factor C/M, and Fig. 6.15 gives the probability that the composite signal is above a certain level, with respect to the total signal power, for the Rice factor values of interest (Fig. 1 of Reference 15 after Reference 16)

Using the Ricean probability distribution and measured data, an empirical model was evolved [15] to give C/M for various antenna gains and elevation angles. This is shown in Fig. 6.16 for L-band maritime mobile communications (Fig. 2 of Reference 15).

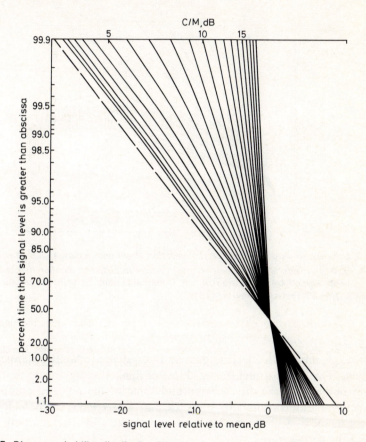

Fig. 6.15 *Ricean probability distribution function for various values of the Rice factor, C/M (Fig. 1 of Reference 15 from Reference 16)*
(Reproduced by permission of the Communications Satellite Corporation from COMSAT Technical Review [15])

Like all empirical prediction models, particularly those evolved from limited measured data, the accuracy of the C/M predictions given in Fig. 6.16 can vary for different paths and sea state situations. Implicit in the model is that the sea is rough. Where the sea is smooth, coherent reflection effects can be significant [15] and the model described in Fig. 6.16 breaks down. Similarly, the model predictions in Fig. 6.16 are those to be used in system design considerations, i.e.

they pertain to 99% of the time the C/M does not exceed the values given in Fig. 6.16. For worst case situations (similar to worst month data in the case of Fixed Services using Satellites), the Rice factor should be lowered between 1 or 2 dB [15], the former value applying to elevation angles increasingly below 10° and the latter to elevation angles increasing above 10°.

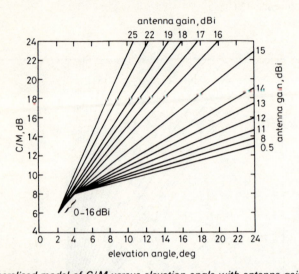

Fig. 6.16 *Generalised model of C/M versus elevation angle with antenna gain as parameter* (*Fig. 2 of Reference 15*)
(Reproduced by permission of the Communications Satellite Corporation from COMSAT Technical Review [15])

6.2.5 System effects

Two principle multipath effects impact on the system availability and perform-ance levels. The first is straightforward loss of signal (fading) due to destructive interference between the direct signal and some of the reflected components which determines the availability statistics, and the second is a general lowering of performance due to the background 'noise' effect of the multipath power lowering the C/M ratio.

The depth of the expected fade for a given percentage time can be estimated and the signal power, antenna gain etc. increased if the availability predicted falls below the desired value. Simply increasing the raw power, however, will have no effect on the C/M ratio since this ratio is independent of the transmitted power. It is therefore important to be able to estimate the Bit Error Ratio (BER) of a digital signal for a given C/M, so that, if need be, the C/M can be improved to the desired level.

As has been noted earlier, the amplitude distribution can be described for various portions of the statistics by gaussian, Nakagami–Rice, Ricean, and m distributions (see Section 2.6.1 on ionospheric scintillation for more discussions on those distributions). The connection between these distributions is noted in Fig. 6.17.

The relation between the direct signal-to-multipath signal power (C/M) and the equivalent degradation value of carrier-to-noise power ratio (C/N) is shown in Fig. 6.18 for two types of digital modulation [17]. As noted in Reference 17, the equivalent degradation value is the difference in the E/N_0 values required to

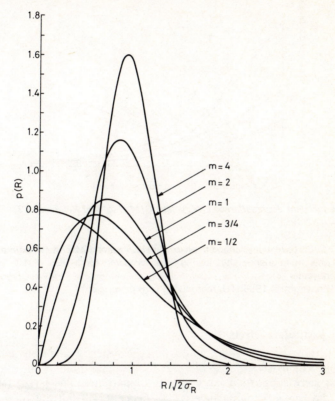

Fig. 6.17 *Examples of the m-distribution with m as parameter (Fig. 1 of Reference 17)*
R = amplitude of the reflected signal
σ_R = RMS value of the reflected signal
The figure represents one-sided gaussian fading when m = 0·5 and Rayleigh fading when m = 1, where m is the inverse of the normalised variance of the reflected signal amplitude squared
(Copyright © 1986 ITU, reproduced with permission)

obtain a BER of 10^{-5} in the presence of thermal noise, between the case when a sea-reflected signal exists and the case when it does not. That is, the equivalent degradation in E/N_0 is the additional margin required to combat multipath fading in a maritime mobile system.

Figs. 6.16 and 6.18 should be utilised for estimating the antenna gain needed to provide a required *C/M* under given conditions (e.g. elevation angle) to obtain the necessary BER. The process, naturally, is iterative and seems to be borne out by measurements [18, 19].

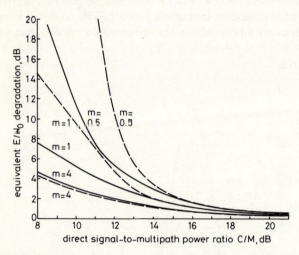

Fig. 6.18 *Equivalent degradation of E/N_0 versus C/M (Fig. 5 of Reference 17)*
--- 2 CPSK
—— 2 NFSK
2 CPSK (two-phase coherent phase shift keying) and 2 NFSK (two-frequency non-coherent frequency shift keying) are compared for a BER of 10^{-5} with *m* as parameter
(Copyright © 1986 ITU, reproduced with permission)

6.3 Other particulate effects

6.3.1 Range of particles

The atmosphere supports a large variety of particles. The larger (and hence generally the heavier) the particle is, the faster it will fall out of suspension. Fig. 6.19 shows the range of particle sizes and their fall velocities in still air (from Fig. 1.1 of Reference 20).

The effect of water in its various phases has been considered in earlier Chapters, as have gaseous absorption and ionospheric scintillation effects. It is uncertain whether viruses and bacteria will prove to have a detectable effect on Earth–space propagation which leaves smoke–dust(fine)–dust(coarse) as the only particles whose effects have not been considered so far.

Smoke falls into that class of particles that are referred to as aerosols. Aerosols have a size range up to about 50 μm [21] and are characterised by their negligible or extremely small fall velocities. In any sort of updraft, aerosols will remain airborne for days and sometimes years.

Smoke *per se* will have a negligible effect in the microwave and millimetre wave bands; it is only as the optical frequencies are approached that the extinction cross-section becomes appreciable. Tests through a hydrocarbon flame [22] showed that, although little attenuation was observed, the level of scintillation increased significantly. In essence, the refractive effects of the turbulent air in the heated up-current were more significant than the combustion products in terms of their effect on the microwave and millimetre wave transmissions. This was true for all five of the measurement frequencies (37, 57, 97, 137 and 210 GHz). The effects of fine dust and coarse dust (usually referred to as sand) are more noticeable.

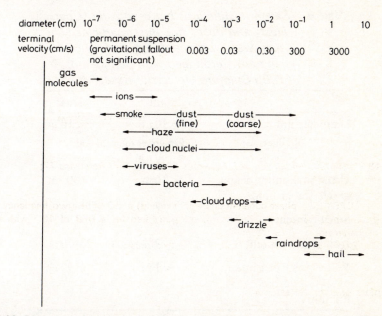

Fig. 6.19 *Diameter and terminal velocities of particles in the atmosphere (Fig. 1.1 of Reference 20)*

Cloud drops, drizzle and raindrops refer to different stages of rain production, the last being the heavier shower and thunder-shower variety. Cloud nuclei are usually ice particles at some stage although cloud nuclei can form around any particle initially as a focal point for growth. Smoke, dust and haze refer to those particles/phenomena that restrict visibility to various degrees. While all three are due to non-aequeous particles, haze can also be due to moisture in the air

(Copyright © 1981 Merrill Publishing Co., reproduced with permission)

6.3.2 Sand and dust effects

An intensive investigation of the effects of sand and dust particles on 6/4 and 14/11 GHz signals on satellite-to-Earth paths [23] collected together all the available data on sand and dust effects that were of relevance to Earth–space communications systems at that time (1985).

A clear distinction was made in Reference 23 between dust and sand (fine dust and coarse dust, respectively, as defined in Fig. 6.19). The distinction can be made on several grounds and these are summarised in Table 6.3.

Sand particles, because they are so much larger than dust particles, tend to stay close to the ground as noted in Table 6.3 and, in general, move by a process called saltation [24]. Since sand particles are rarely observed above a height of 10 m, their significance in satellite-to-ground links from a propagation impairment viewpoint is very small. For all practical purposes, therefore, sand has no noticeable effect on microwave Earth–space links at elevation angles above 5°.

Table 6.3 *Salient distinctions between sand (Coarse Dust) and dust (fine dust) particles (The data were abstracted from Reference 23 and are reproduced with permission, © 1985 Bradford University Research Ltd.)*

	Sand	Dust
Diameter limits	> 10 μm, generally > 100 μm	< 10 μm
Usual maximum height above the ground	< 10 m	> 1000 m
Silica content	> 80%	< 55%
Iron oxide content	~ 7%	20–30%
Water absorbed in air of 91% relative humidity	< 1%	6–9%

Dust particles, which have been observed at heights in excess of 1 km, can exist along a significant portion of an Earth–space path and it is worthwhile to identify the regions where dust storms can occur and to predict the likely impairments on an Earth–space communications link.

(*a*) *Variability in space and time of dust storms*
Fig. 6.20 [23] shows the major desert regions of the Earth and the locations and directions of motion of the dust storms.

In reviewing the literature [23], eight dust storm types were identified that had clearly definable features, and these are depicted in Tables 6.4–6.6. These Tables also give the main classifications of the eight dust storm types [23].

Of the eight dust storm types, only three were found to give rise to dust extents and suspended dust densities that could cause propagation impairments, namely [23]:

Cyclogenic

Frontal

Haboob

Table 6.4 *Dust storm types indicating their typical wind speeds (from Table 2.4.78 of Reference 23, copyright © 1985 Bradford University Research Ltd., reproduced with permission)*

Type of storm	Average wind speed (m/s)	Max. gust speed (m/s)
I Planetary winds	6–17	23
II Cyclogenic	7–18	27–50+
III Frontal	9–17	38
IV Katabatic winds	12–21	36–50+
V Haboob	11–21·5	41
VI Constriction	14	18
VII Dust devils	5–10	15
VIII Diurnal winds	8–13	15

Table 6.5 *Dust storm types indicating their typical structures (from Table 2.4.77 of Reference 23, copyright © 1985 Bradford University Research Ltd., reproduced with permission)*

Type of storm	Width (km)	Length (km)	Height (km)
I Planetary winds	0·3–250	40–8000	0·4–3
II Cylogenic			
A Low level jet	500–1000	500–2000+	3–5
B Upper level jet	500–1000	500–2000+	3–5
C Surface storm circulation	50–150	50–150	0·4–0·8
III Frontal	500–1000	50–2000+ (The length of the cold front)	1–5
IV Mountain katabatic winds	15–150	100–450	1–5
V Haboob	3–75	3–300	0·5–12
VI Constriction	0·5–10	Length of the valley	Height of the valley
VII Dust devils	0·01–0·50	Localised	0·5–3
VIII Diurnal wind cycle	0·1–50	1–40	< 1

The Haboob, since it is very local in nature and is the direct result of a thunderstorm, which itself is an intense meteorological event, tends to give rise to very high suspended dust densities. Fig. 6.21 gives a simplified schematic presentation of a vertical cut through a Haboob (from Fig. 2.3.8*a* Reference 23)

Fig. 6.20 World map giving major dust sources, dust transport and areas affected by dust storms (from Fig. 2.2.1 of Reference 23 as modified later by the authors of Reference 23)

Area likley to be affected by Haboobs are specifically shown

▨ Deserts of the Earth

■ Main source areas for dust

::: Areas where haze may occur at sea on >10 days in any season

→ Major directions and distances of dust transport

C Areas affected by Haboobs (Type V dust storms)

D Areas which are not desert but where dust storms also occur

(Copyright © 1985 Bradford University Research Ltd, reproduced with permission)

Since suspended dust particles tend to reduce (optical) visibility in direct proportion to the suspended dust density, visibility statistics have been used to estimate the variability in time of dust storms.

Table 6.6 *Dust storm types indicating the typical visibilities and expected durations of dust storm events (from Table 2.6.2.1 of Reference 23, copyright © 1985 Bradford University Research Ltd., reproduced with permission)*

Type	Visibility expected	Duration
I Planetary wind flow patterns	10 m– < 11 km	< 24 h–2 weeks
II Cyclogenic	0– < 1000 m (0–50 m–severe storms)	6–24 h
III Frontal	0– < 1000 m (0–50 m–severe storms)	1–8 h (Up to 7 days)
IV Mountain katabatic winds	3– < 1000 m	0·5–18 h
V Haboob	(0–50 m–severe storms) 200–400 m	0·5–6 h
VI Constriction	(3 m–severe storms) 800–1000 m	0·5 h
VII Dust devil	< 1000 m	0·1–0·5 h
VIII Diurnal wind cycle	< 1000 m Near zero (severe)	< 1 h

The principal criterion applied to defining the occurrence of a dust storm is when the visibility drops below 1 km. On that basis, the number of dust storm days per year was found to be between 0·1 and 174 days per year for the various regions of the world where such events occur [23]. To derive statistics, a lognormal characteristic of the visibility exceedence statistics was assumed with the number of dust storm days as one point on the curve and the severe dust storm events reported for the region in question used as the second point on the curve [23]. With this simplistic, but reasonable, model, visibility exceedance statistics were estimated. Table 6.7 (from Table 2.6.3.1 of Reference 23) gives a few examples of the 'worst' regions of the world as far as visibility statistics are concerned, based upon the somewhat sparse data that are available today and some educated guesses by the authors of Reference 23.

(b) Propagation impairment prediction models for dust effects
Sand and dust particles attenuate electromagnetic waves by the same mechanisms as do rain drops [25]. As with rain drops, estimating the number of dust particles per unit volume and their size distribution, is fundamental in predicting

the propagation impairments due to dust storms. To obtain these, it is usual to infer a dust concentration from the visibility statistics.

The most commonly used expression for relating the suspended dust mass density ϱ to the visibility V (in kilometres) is an empirical formula due to Chepil and Woodruff [26]:

$$\varrho = \frac{56 \times 10^{-9}}{V^{1 \cdot 25}} \text{ grams/cm}^3 \tag{6.11}$$

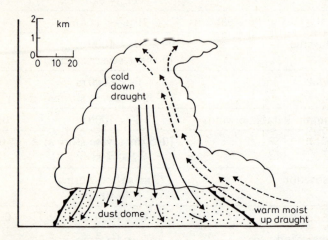

Fig. 6.21 *Cross-section through a thunderstorm that is producing a Haboob (from Fig. 2.3.8a of Reference 23 as modifed later by the authors of Reference 23)*
Since the strong gusts producing the dust storm result from the cold down draught within the thunderstorm, the Haboobs are sometimes referred to as cold dust storms. The air is usually fairly humid, leading to a significant uptake of water by the dust which will increase the propagation effects
(Copyright © 1985 Bradford University Research Ltd., reproduced with permission)

It should be noted that the suspended particle density and the visibility are referred to that which exists at a height of 2 m above the ground. This density decays with height and, in order to calculate the integrated effects along a slant path to a satellite, an effective path length needs to be determined in much the same way as rain attenuation is calculated. Typically, a power law decay with height of particle density is assumed with a ceiling of 2 km [27] for the dust particles. The power law index varies with the visibility reported at a height of 2 m with, in one case, values of 0·77, 0·57 and 0·33 selected for visibility distances of 2 m, 10 m and 1 km, respectively [27]. The same model also assumes that the dust is contained within a cylinder 10 km in diameter with the dust horizontally uniform inside the cylinder and negligible outside. Fig. 6.22 illustrates the concept.

Table 6.7 *Estimated yearly statistics of dust storm visibility for various locations around the world (Data abstracted from Table 2.6.3.1 of Reference 23, copyright © 1985 Bradford University Research Ltd., reproduced with permission)*

Location	Country	Humid storm (time as a percentage)	Percentage of the year that the visibility is less than the stated value (m)				Visibility at the stated percentage	
			1000 m	100 m	10 m	1 m	0·17%	0·01%
Mexico City	Mexico	70%	0·090	0·019	0·0015	0·0001	>1000	50
Arizon (Yuma)	USA	50%	0·034	0·005	0·0003	–	>1000	200
Great Plains (Lubbock)	USA	34%	0·130	0·031	0·0039	0·0003	>1000	27
Alberta, Man., Sask.	Canada	50%	0·011	0·00029	–	–	>1000	800
Yukon	Canada	0%	0·170	0·0045	–	–	>1000	150
Tunisia, Algeria, Morocco	N. Africa	12%	0·210	0·016	0·0003	–	700	70
Libya	N. Africa	12%	0·220	0·0015	–	–	800	200
Faya Largeau/Bodele Depression	Central and W. Africa	25%	0·180	0·050	0·0076	0·0007	950	14
Kano	Nigeria	30%	0·110	0·025	0·003	0·0001	>1000	40
Khartown, Atbara, Dongola	Sudan	50%	0·820	0·150	0·010	0·0002	120	10
Djibouti	Ethiopia	15%	0·068	0·0037	–	–	>1000	190
Nasiriyah	M. East	38%	2·200	0·200	0·0044	–	80	16
Beersheba	Israel	20%	0·210	0·0079	–	–	800	120
Kerman	Iran	40%	0·300	0·038	0·0012	–	440	38
Karachi	Pakistan	25%	0·410	0·060	0·0029	–	310	24
Ganganagar	N.W. India	50%	0·870	0·004	–	–	400	140
Jinkiang	China	50%	1·800	0·100	0·0009	–	140	32
S.E. Uzbekistan	USSR	40%	1·400	0·110	0·0019	–	140	25
Alice Springs	Australia	50%	0·150	0·022	0·0012	–	>1000	50

Note: These statistical estimates were compiled using somewhat sparse data and are, in the authors' of [23] opinion, the "best guesses" available at present. More data are required to verify the above estimates.

The number of dust particles per cubic metre, N can be related to the visibility V (in km) by Reference 33

$$N = \frac{5 \cdot 51 \times 10^{-4}}{Va^2} \qquad (6.12)$$

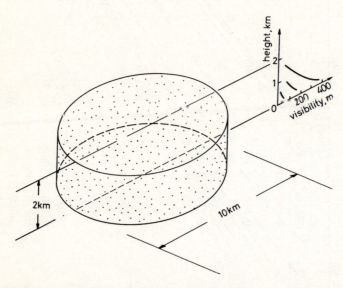

Fig. 6.22 *Schematic of the dust cell model proposed by McEwan et al (Reference 27)*
The dust is assumed to be of uniform density horizontally with all of the significant dust contained within a symmetrical right cylinder of diameter 10 km. The dust density falls off with height, following a power law decay, the index of which is determined by the initial visibility at a height of 2 m.

where a is the radius of the particles in metres. If k_r and k_i are the real and imaginary parts of the complex relative dielectric constant of the dust particles, the specific attenuation α_p (in dB/km) can be given by [33]

$$\alpha_p = \frac{189}{V} \frac{a}{\lambda} \left[\frac{3k_i}{(k_r + 2)^2 + k_i^2} \right] \text{decibels/km} \qquad (6.13)$$

where λ is the wavelength in metres

Eqns. 6.12 and 6.13 invoke a uniform radius a for the dust particles. In practice, not only the size but the shape and orientation of the dust particles will vary.

The shape and alignment of the dust particles will contribute to the attenuation and depolarisation predictions, particularly the latter. Measurements of dust particles [28, 29] have given an average value of 0·71 to the ratio of the horizontal minor axis to the major axis, and values of 0·57 [28] and 0·53 [29] to the ratio of the vertical minor axis to the major axis. The average eccentricity of the dust particles, plus their anticipated alignment by aerodynamic [28] and

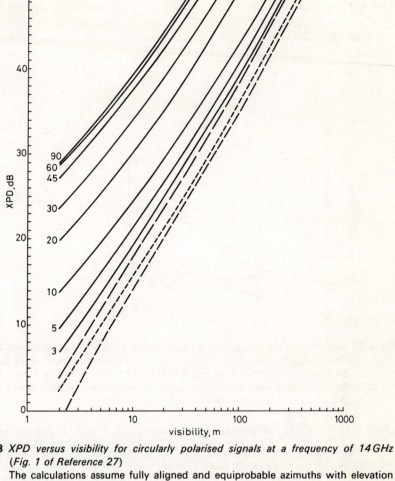

Fig. 6.23 *XPD versus visibility for circularly polarised signals at a frequency of 14 GHz*
(*Fig. 1 of Reference 27*)

The calculations assume fully aligned and equiprobable azimuths with elevation
angle shown as a parameter. The air humidity is 0% and the antenna height 20 m
unless otherwise indicated. The parameters on the curves are the elevation angles.

—— 3° Elevation   0% humidity
                  2 m antenna height
– – – – 3° Elevation 60% humidity
                  20 m antenna height
– — – 3° Elevation 60% humidity
                  2 m antenna height

(Copyright © URSI 1986 reproduced with permission)

electrical forces [23], leads to potentially significant depolarisation effects, particularly at low elevation angles and frequencies well above 10 GHz. Fig. 6.23
gives some predicted XPD values for a frequency of 14 GHz (from Fig. 1 of
Reference 27).

It is interesting to note the effect of water uptake on the calculated XPD values in Fig. 6.23. With a relative humidity of 60%, the water uptake by dust particles can be almost 10% by mass. This can lead to significant variations in predicted attenuation. Fig. 6.24 gives the predicted specific attenuation with frequency for a visibility of 10 m under low (0·3%) and high (10%) water uptake percentages.

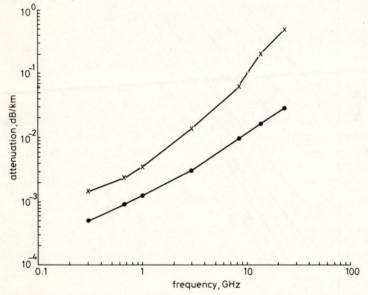

Fig. 6.24 *Attenuation as a function of frequency for falling dust particles (Fig. 16 of Reference 30)*

●—● *(0·3% (g H$_2$O/g soil)*
×—× *(10% (g H$_2$O/g soil)*
A visibility of 10 m is assumed. It is further assumed that all the dust particles are spherical and uniformly distributed along the path. While correction factors that take into account the decay of particle density with height and the non-sphericity of the dust particles will need to be used to arrive at a total path attenuation, the relative shape of the two curves will be unchanged

Based on the slant path attenuation and depolarisation levels that are likely to occur under visibility conditions that are prevalent for significant time percentages, it is unlikely that dust storm induced propagation impairments will prove limiting to satellite-to-ground links until either frequencies well above 30 GHz and/or elevation angles well below 10° are used. Some effects, however, have been observed on both terrestrial and Earth–space paths at frequencies below 10 GHz.

(c) System impact of dust effects
Despite theoretical predictions to the contrary, noticeable effects have been

observed on both terrestrial and Earth–space paths during dust storms [23]. After eliminating effects of a mere 'mechanical' nature, such as dust contamination in the waveguides and mispointing of the antenna due to high gust loadings, some system effects remained that needed explaining. These included the monopulse tracking of the INTELSAT earth station at Umm Al–Aish (Kuwait) appearing to 'hunt' for periods between 1 and 30 s and terrestrial links fading up to 30 dB in Bagdad, Oman and New South Wales during dust storms [23].

The tentative explanation for these phenomena is that the bulk refractivity of the air mass, not the dust contained in the air mass, changes relatively rapidly over a short distance [Reference 32 reported in Reference 31]. Rafuse [32] noted that the refractivity at the boundary of a Haboob can change by as much as 26 N units. Even when the visibility is as low as 10 m, the refractivity change due to the dust particles alone is only 7 N units [27]. Fig. 6.25 illustrates a specific Earth–space geometry that, coupled with the rapid refractivity changes across the Haboob boundary, could possibly explain the system effects noted on earth satellite links in dust storms [27]. The same rapid change of the refractivity across the Haboob boundary could lead to super-, or sub-, refractive effects on terrestrial systems that could cause severe multipath problems.

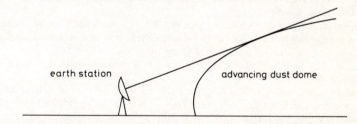

earth station   advancing dust dome

Fig. 6.25 *Schematic of a Haboob approaching on earth station (Fig. 2 of Reference 27)*
This simplified schematic of a Haboob approaching an earth station indicates how the look angle of the earth station can cause the link to be at a grazing incidence with the advancing dust dome (see Fig. 6.21). The rapid change of N units across the dust dome boundary (20 N units is a typical figure) can cause ray bending and even multipath effects to be observed
(Copyright © 1986 URSI, reproduced with permission)

6.4 References

1 Recommendations and Reports of the CCIR, XVIth Plenary Assembly, Dubrovnik, 1986, Volume V (Propagation in non-ionized media); Report 236-6: 'Influence of terrain irregularities and vegetation on tropospheric propagation'. (Note, however, that Report 236-6 has now been replaced by a Draft New Report, Document 5/168 of the Interim Meeting of Study Group 5: 'Transmission Loss Over Irregular Terrain With And Without Vegetation', 20 April 1988)

2 Ibidem; Report 715-2: 'Propagation by diffraction'.

3   Input Document 5/101: 'Use of approximate formulae'. Submitted to the Interim Meeting of CCIR Study Group 5 [1986–1990], 25 Feb. 1988

4   DALGLEISH, D. I.: 'The Influence of interference on the siting of earth stations'. International Conference on Satellite Communications, 1975, IEE Conf. Publ. 126, pp. 21–27 (and see COST Project 210: 'The influence of the atmosphere on interference between radio communications systems at frequencies above 1 GHz'. 3rd Annual Report, 1986–1987, XIII/043/87, EUCO/TELE/210/AR1/87)

5   Conclusiosn of the Interim Meeting of Study Group 5 (Propagation in non-ionized media), Geneva, 11–26 April 1988, Document 5/204; Report 884-1 (MOD-I) 'Propagation data for maritime mobile-satellite systems for frequencies above 100 MHz'

6   BECKMAN, P., and SPIZZICHINO, A.: 'The scattering of electromagnetic waves from rough surfaces' (Pergamon Press, NY, 1963)

7   Recommendations and Reports of the CCIR, XVIth. Plenary Assembly, Dubrovnik, 1986, Volume V (Propagation in non-ionized media); Report 884-1: 'Propagation data for maritime mobile-satellite systems for frequencies above 100 MHz'.

8   KINSMAN, B.: 'Wind waves: their generation and propagation on the ocean surface' (Prentice-Hall, USA, 1965)

9   HOGBEN, N., and LUMB, F. E.: 'Ocean wave statistics' (HM Stationery Office, London, 1967)

10 KARASAWA, Y., YASUNAGA, M., NOMOTO, S., and SHIOKAWA, T.: 'On-board experiments on L-band multipath fading and its reduction by use of the polarization shaping method', *Trans. IECE Japan*, 1986, **E69**, pp. 124–131

11 OHMORI, S., IRIMATA, A., MORIKAWA, H., LONDO, K., HASE, Y., and MIURA, S.: 'Characteristics of sea reflection fading in maritime satellite communications'. *IEEE Trans.*, 1985, **AP–33**, pp. 838–845

12 YASUNAGA, M., KARASAWA, Y., SHIOKAWA, T., and YAMADA, M.: 'Characteristics of L-Band multipath fading due to sea surface reflection in aeronautical satellite communications', *Trans. IECE Japan*, 1986, **E69**, pp. 1060–1063

13 LONG, M. W.: 'Radar reflectivity of land and sea' (Lexington Books, Lexington, Mass., USA, 1975)

14 KARASAWA, Y., and SHIOKAWA, T.: 'Fade duration statistics of L-Band multipath fading due to sea surface reflection', *IEEE Trans.*, 1987, **AP–35**, pp. 956–961

15 SANDRIN, W. A., and FANG, D. J.: 'Multipath fading characteristics of L-Band maritime mobile satellite links', *COMSAT Tech. Rev.*, 1986, **16**, pp. 319–337

16 DFVLR: 'Technical assistance study of lightweight shipborne terminals'. Final Report for ESA /ESTEC Contract 4786/81/NL/MD, 1982

17 Recommendations and Reports of the CCIR, XVIth, Plenary Assembly, Dubrovnik, 1986, Volume VIII-3 (Mobile Satellite Services [Aeronautical, Land, Maritime, Mobile and Radio-determination]-Aeronautical Mobile Service); Report 762-1: 'Effects of multipath on digital transmissions over links in the maritime mobile-satellite service'.

18 DISSANAYAKE, A. W., JONGEJANS, A. W., and DAVIES, P. E.: 'Preliminary results of PROSAT martime-mobile propagation measurements', International Conference on Antennas and Propagation ICAP 85, IEE Conf. Publ. 248, 1985, pp. 338–342

19 EDBAUER, F.: 'Influence of multipath propagation of the maritime satellite channel on PSK-modulated systems'. *Ibid*, 1985, pp. 333–337

20 MILLER, A., and ANTHES, R. A.: 'Meteorology', (Charles E. Merrill Publishing Co., 1981, 4th Edn.)

21 Recommendations and Reports of the CCIR, XVIth. Plenary Assembly, Dubrovnik, 1986, Volume V (Propagation in non-ionized media); Report 883-1: 'Attenuation of visible and infra-red radiation'.

22 GIBBINS, C. J., and PIKE, M. G.: 'Millimetre, Infra-red, and optical propagation studies on a 500 m range'. International Conference on Antennas and Propagation ICAP 87, IEE Conf. Publ. 274, Part 2, 1987, pp. 50–53

23 McEWAN, N. J., BASHIR, S. O., CONNOLLY, C., and EXCELL, D.: 'The effects of sand and dust particles on 6/4 and 14/11 GHz signals on satellite to Earth paths'. Final Report to INTELSAT under contract INTEL-349, School of Electrical and Electronic Engineering, University of Bradford, Bradford, West Yorkshire, BD7 1DP, England, 1985

24 BAGNOLD, R. A.: 'The physics of blown sand and desert dunes' (Chapman & Hall, 1941 and 1973)

25 Recommendations and Reports of the CCIR, XVIth. Plenary Assembly, Dubrovnik, 1986, Volume V (Propagation in non-ionized media); Report 721–2: 'Attenuation by hydrometeors, in particular precipitation, and other atmospheric particles'

26 CHEPIL, W. S., and WOODRUFF, N. P.: 'Sedimentary characteristics of dust storms: 2. visibility and dust concentration', *Am. J. Science*, 1957, **255**, pp. 104–114

27 McEWAN, N. J., CONNOLLY, C., EXCELL, D., and BASHIR, S. O.: 'Attenuation, cross polarisation and refraction in dust storms'. Paper VIII-3, URSI Commission F Open Symposium, 1986, University of New Hampshire, Durham NH, USA

28 McEWAN, N. J., and BASHIRE, S. O.: 'Microwave propagation in sand and dust storms: the theoretical basis of particle alignment'. International Conference of Antennas and Propagation, ICAP 83, IEE Conf. Publ. 219, 1983, pp. 40–44

29 GHOBRIAL, S. I., and SHARIEF, S. M.: 'Microwave attenuation and cross polarisation in dust storms', *IEEE Trans.*, 1987, **AP–35**, pp. 418–425

30 ABDULLA, S. A. A., AL-RIZZO, H. M., and CYRIL, M. M.: 'Particle-size distribution of Iraqi sand and dust storms and their influence on microwave communication systems', *IEEE Trans.*, 1988, **AP–36**, pp. 114–126

31 BASHIR, S. O., and McEWAN, N. J.: 'Microwave propagation in dust storms − Part 1: A review,' *IEE Proc. H*, 1986, **133**, pp. 241–247

32 RAFUSE, R. P.: 'Effects of sand storms and explosion-generated atmospheric dust on radio propagation'. Project Report DCA-16, ESD-TR-81-290, MIT Lincoln Lab., Lexington, Mass., USA, 1981

33 FLOCK, W. L.: 'Propagation effects on satellite systems at frequencies below 10 GHz: A handbook for satellite system design'. NASA Reference Publication 1108(2), 1987, 2nd Edn.

Chapter 7

Restoration of performance during signal impairments

7.1 Introduction

The previous Chapters have detailed the impact of various propagation impairment phenomena on the radiowave signals between earth stations (both fixed and mobile) and satellites. In each Chapter, the system effects of the impairment in question were discussed and some means of reducing the impact of the impairment alluded to. This Chapter will present in some detail the various schemes suggested and, in some cases, already implemented to restore performance during a signal impairment.

The restoration of performance can be exactly what it implies, i.e. the continuous application of a resource to improve the system performance under all conditions, or it can be the dynamic allocation of a resource to improve the system performance only under 'degraded' conditions. The latter approach does not improve the system performance *per se*, it only improves the availability of the system by increasing the margin of the system only when it is necessary. Ideally, such an approach would provide a constant performance level in clear sky and in degraded conditions by dynamically allocating sufficient resources to maintain the C/N, C/M, E_b/N_0, or whatever other criterion of performance is being used, constant. In this Chapter, although the restoration schemes are essentially aimed at improving the availability of the system, by doing so the performance of the system is improved in degraded conditions. In that sense, restoring performance and increasing availability are synonymous.

Where the propagation impairment essentially causes a decrease in amplitude (e.g. scintillation and rain attenuation), a restoration technique that will reduce the impairments caused by one impairment (e.g. rain attenuation) may also be useful in reducing the impairments of the other (in this case, scintillation). Some restoration techniques, however, are unique to the phenomenon they are applied against. For this reason, the restoration discussion will be broken down into the five principal impairment phenomena: ionospheric effects, tropospheric scintillation effects, multipath effects, (rain) attenuation effects, and depolarisation affects.

7.2 Ionospheric scintillation effects

Ionospheric scintillation is essentially an 'on-axis' phenomenon as far as an earth station antenna is concerned; that is, the fluctuations in the signal parameters occur without any apparent directional deviation that could cause angle-of-arrival impairments. Severe amplitude fluctuations caused by ionospheric scintillation can lead to a breakdown in the automatic tracking capability of an earth station, however, and, if this results in a significant depointing of the antenna, other impairments not strictly related to the propagation effects present on the path can occur [1]. These apparent impairments are a reduction in the mean signal level and significant depolarisation due to the signal arriving away from the main-beam axis of the antenna when it has been depointed from the correct look angle. The only way of reducing impairments due to antenna depointing is to incorporate some programme tracking capability whereby the auto tracking commands are verified before being executed. Such programme tracking techniques are utilised in propagation experiments and have proved to be very successful.

The ionosphere will cause a rotation of the electric vector, called Faraday rotation, the magnitude of the rotation being a function of the Total Electron Content (TEC) and the frequency. For circularly polarised signals, the Faraday rotation is not significant, being largely undetectable in most systems. Indeed, for circularly polarised signals, ionospheric scintillations do not cause any appreciable depolarisation that can be attributable solely to the ionosphere [2, 3]. Of the several variations in the radiowave signal that can be attributed to the ionosphere (amplitude scintillations, phase scintillations, Faraday rotation etc.) only two are of significance for satellite-to-ground links at frequencies above 1 GHz: amplitude scintillations and Faraday rotation.

7.2.1 Ionospheric amplitude scintillation amelioration
In most cases, the impact of ionospheric amplitude scintillations will not be system limiting on an annual average basis [4] but, for some services, the worst month criteria may not be met, particularly for those earth stations close to the equator in periods of high sunspot activity.

Ionospheric amplitude scintillations cannot normally be reduced by antenna diversity schemes since the correlation distances are quite large; much larger separations would be required to achieve appreciable decorellation than site and height diversity schemes for rain attenuation and tropospheric-scintillation/multipath-phenomena, respectively. The only effective way of overcoming the ionospheric amplitude scintillation impairment is to add something to the signal. For digital systems this means coding.

In examining the best way of coding a signal to overcome ionospheric amplitude scintillations, it is useful to construct a general representation of the signal fluctuations with time. Fig. 7.1 gives a schematic of such a representation.

The scintillations can be generally classified as those that appear to give rise to a relatively constant degradation, similar to Additive White Gaussian Noise (AWGN), during relatively 'shallow fades', and those that are sporadic in nature and give rise to burst errors during relatively 'deep fades' (see S and D in Fig. 7.1). The extreme fluctuations usually last less than a second and the 'burst period' of the sequences of extreme fluctuations is usually less than 30 s. A constant Forward Error Correction (FEC) code will counteract the AWGN component but another coding scheme will be required to correct the burst errors. Two basic schemes can be employed, FEC coding with interleaving and FEC coding with concatenated outer code [5].

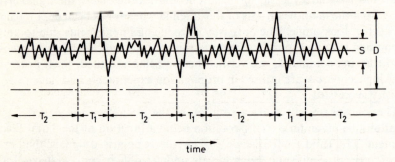

time

Fig. 7.1 *Schematic representation of ionospheric scintillation*
In general, the fluctuations can be divided into those that give an almost constant 'shallow fade', denoted in level by S, and those that give sporadic 'deep fades', denoted in level by D.
During time intervals T_1, deep fades giving rise to burst errors in a digital system occur; during time intervals T_2, shallow fades giving rise to an AWGN component occur.

(a) FEC coding with interleaving
The basic objective of interleaving is to distribute bits randomly in time so that errored bits are jumbled up with non-errored bits, thus permitting the coding/de-coding scheme an additional measure of testing for errors. Thus, in Fig. 7.1, bits from time periods T_1 are randomly dispersed with bits from time periods T_2. Fig. 7.2 illustrates this schematically.

If there are n_1 bits in the deep fade periods, T_1, and n_2 bits in the shallow fade periods, T_2, and assuming that these periods are relatively constant and periodic in nature, a total of $n_1 + n_2$ bits should be interleaved so that the bits transmitted in the period of deep fades are totally scrambled before entering the coder [5]. The approximate size of the buffer needed to store the bits prior to coding and transmission (and prior to de-interleaving and de-coding on reception) is given by $n_1 + n_2$. A block of bits is usually a lot smaller than $n_1 + n_2$ (as can be inferred from Fig. 7.2) and the blocks, having been stacked in the buffer, are read in an orthogonal manner to the input sequence prior to transmission. In this way, errored bits are spread between blocks on de-interleaving on reception.

(b) FEC coding with concatenated outer code

When the deep fading persists for periods in excess of T_1 (or n_1 bits), it may still be feasible to recover from the errors by nesting an inner code (for the AWGN component) inside a powerful outer code [5]. The length of the outer code has to be at least $n_1 + n_2$ in order to handle the really deep fades and this has implications with regard to buffer length. The delay in coding and de-coding such concatenated codes will also add appreciably to the normal delay of a satellite circuit and so such coding techniques are not normally recommended for commercial, two-way voice traffic.

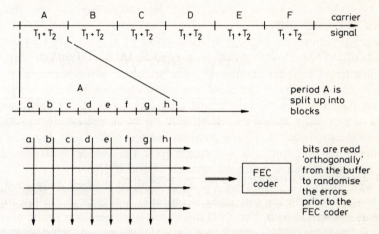

Fig. 7.2 *Schematic of interleaving followed by FEC coding prior to transmission*
The signal is assumed to have approximately periodic errors with the period given by $T_1 + T_2$. Some of the blocks *a, b, c . . . h* will have no errors while some will have many. Interleaving them prior to transmission will ensure that, in most cases, errored bits have 'good' bits on both sides of them.

(c) FM transmissions

For FM signals, unless the scintillations are extremely severe, the modem will be able to 'flywheel' through the negative spikes since they are of relatively short duration. FM communications links degrade in a 'soft' manner unlike digital links which, although having generally superior performance for most of the time, fail 'hard' when the carrier drops below the threshold detection level.

7.2.2 Faraday rotation amelioration

The Total Electron Content (TEC) of the ionosphere will vary directly as the sunspot cycle intensity varies, and the magnitude of the Faraday rotation of the electric vector of a linearly polarised signal will vary directly with the TEC (see Figs. 2.11 and 2.15). Within the larger variation, which has a period of about 11 years, there are seasonal and diurnal variations. The larger and faster these variations are, the greater the system impact.

For linearly polarised systems at frequencies around 4 GHz, the maximum variation in the rotation is about 10°. This amounts to less than 0·1 dB in signal loss if the linearly polarised receiving antenna is mismatched by 10° from the incoming polarisation orientation. The situation is more significant if dual-polarised operation is required.

For a rotation of θ degrees in the electric vector, the XPD is given by

$$XPD = 20 \log \cot \theta \text{ decibels} \qquad (7.1)$$

From eqn. 7.1 it can be seen that the XPD has reduced to 13·5 dB for a rotation of 10°. Typical rotation values are closer to 5° [6] (and see Fig. 2.11) but this still yields an XPD of 21·2 dB.

An examination of the Faraday rotation effect as it applied to Earth–space 6/4 GHz links in the USA established a procedure for minimising the depolarisation due to the Faraday rotation [7]. The procedure involves measuring the minimum depolarisation in the pre-dawn hours when the Faraday rotation is least and then rotating the uplink and downlink feeds of the earth station antenna to a position somewhere in between the measured minimum rotation value and the predicted maximum rotation value expected that day. The predicted maximum rotation value was based upon a regression analysis and the typical rotations required averaged between 1° and 2° [7].

The same analyses [7] evaluated the need to adjust the feed daily, monthly, annually and never, the last case being a constant dis-alignment to allow for the peak years of solar activity. The XPD due to mis-alignment was found never to fall below 26 dB for 0·01% of any year under all adjustment schemes and, on balance, it would appear that a simple one-time mis-alignment of between 2° and 2·5° at 4 GHz and of between 0·88 and 1·1° at 6 GHz would suffice for the links examined. Note that the downlink offset rotational direction will be opposite to that used on the uplink.

7.3 Tropospheric scintillation effects

Tropospheric scintillation effects increase with frequency and with decreasing antenna size and elevation angle. The anticipated rapid introduction of satellite telecommunications services that utilise small earth stations, now commonly known as VSATS (Very Small Aperture Terminals), in the 14/11 and 30/20 GHz bands has stimulated investigations into tropospheric scintillation effects. The effects are essentially due to refractive phenomena but, at very low elevation angles (approximately 1°), multipath effects begin to dominate. Between 10° and 1° there is a 'grey area' where there is a gradual shift in importance from purely refractive phenomena to purely multipath phenomena. Counteracting the two effects requires different techniques.

7.3.1 Ameliorating the refractive effects

Tropospheric scintillation is very similar to the shallow fades discussed in the amelioration of ionospheric scintillation, and can be considered generally as AWGN. A straightforward FEC code will reduce the impairment; increasing the transmitted power, within the power flux density limitations, will have the identical effect, but typically at more overall system expense.

7.3.2 Ameliorating the multipath effects

There is only one general solution to counteracting multipath effects and that is to be able to select a different path. Changing the frequency will effectively provide a different path since the propagation effects producing the loss of signal, due either to destructive interference or ray bending, will usually not be the same at two different frequencies provided there is a large frequency difference. The more normal technique is to use antenna diversity to obtain the required path separation. For very low elevation angles, orienting the diversity antennas such that there is an apparent height difference between them as viewed by the satellite will provide a substantial reduction in the signal fading and scintillation amplitude observed. This is discussed in more detail later in this Chapter when space and site diversity techniques are reviewed for rain attenuation amelioration.

7.4 Maritime multipath effects

Ionospheric scintillation is much more severe at the present maritime (and aeronautical and land) mobile frequencies at L-band than at 6/4 GHz but, on an annual average basis, multipath effects are much more significant and are the system limiting impairment for link budget calculations in these mobile services. As with the general class of randomly fluctuating signal impairments that are known generically as scintillations, the shallow-fade portions of the maritime multipath effects can be treated as AWGN and standard FEC codes used to improve the margin. The deeper fades induced by severe multipath activity are much longer than most interleaved or concatenated codes can handle and so a number of other approaches have been suggested for reducing the multipath effects, including:

● Frequency diversity
● Height/space diversity
● Polarisation-shaping antennas
● Beam-shaping antennas

7.4.1 Frequency diversity

In this method, the carrier frequency is changed when deep fading occurs in order to establish a 'second' path that is not subject to multipath effects. To do this effectively, a relatively large difference must exist between the two frequencies. Unfortunately, the maritime mobile communications frequency assignment at present is less that 10 MHz wide at approximately 1·5 GHz, i.e. much less than 1% bandwidth. With the small percentage bandwidth available, multipath effects reduction is unlikely with frequency diversity and this has proved to be the case [9].

7.4.2 Height/space diversity

In terrestrial systems, multipath effects due to refractive and reflective phenomena are counteracted by using two antennas simultaneously at different heights [10]. This is referred to generically as space diversity but, by virtue of the physical disposition of the antennas, is also called height diversity. Since the wavelength at L-band is about 20 cm, just separating the antennas by a few tens of centimetres will completely decorrelate the multipath effects. It has been found [9] that a vertical separation of as little as 40 cm gives quite a good diversity effect. The problems associated with buffering and combining the communications signals on the downlink and selecting an appropriate transmission switching diversity technique on the uplink, particularly when the ship is rolling, have yet to be solved economically, however.

7.4.3 Polarisation-shaping antennas

As noted in Chapter 6, the different reflection characteristics of horizontal and vertical polarisations on the sea surface have led to the proposal to use a polarisation-shaping antenna to reduce the effect of multipath phenomena [11]. At elevation angles below about 20°, horizontally polarised signals will be reflected amost perfectly from the sea surface while those that are vertically polarised will be significantly attenuated on reflection. An incident, circularly polarised wave will therefore be reflected as an elliptically polarised wave. In addition, since the horizontally polarised wave goes through a phase reversal on reflection, the elliptically polarised wave is of the opposite sense to the direct wave. KDD of Japan have utilised these facts to produce a polarisation-shaping antenna called a Modified Short Backfire (MSBF) antenna [11].

The MSBF antenna utilises crossed dipoles to generate the circularly polarised signals. By adjusting the relative phase and amplitude of the signals going to the two dipoles, an elliptically polarised beam can be generated that is orthogonal to the reflected (multipath) signal. The polarisation mismatch between the direct wave and the antenna will be more than offset by the reduction in the multipath fading, even in very rough seas [12]. Fig. 7.3 illustrates the enhanced performance obtainable with this restoration technique (from Fig. 13 of Reference 12). The technique can be passive, i.e. set periodically and left at this setting, or active, i.e. dynamically adjusted for optimum performance.

7.4.4 Beam-shaping antenna

This technique is similar to polarisation-shaping except that it is the gain that is changed, not the polarisation. Instead of a symmetrical beam pattern about the boresight axis of the antenna, the beam is shaped to reduce the gain in the vertical direction towards, and below, the horizon. The reduced gain of the antenna in the anticipated direction of the incoming multipath-generated signal will reduce the impact of that phenomen.

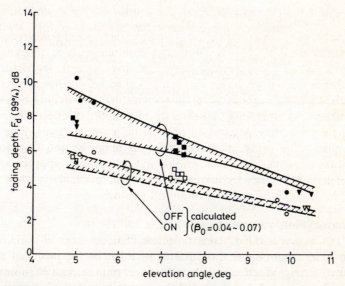

Fig. 7.3 *Illustration of the reduction in multipath effects using a Modified Short Backfire (MSBF) antenna (from Fig. 13 of Reference 12)*

Measured Data

Wave height	OFF	ON
1–2 m	●	○
2–3 m	■	□
3–4 m	▼	▽

OFF: circular polarisation
ON: optimally (elliptically) shaped polarisation
The antenna diameter was 40 cm
(Copyright © 1986 IECE, reproduced with permission)

7.5 Rain attenuation effects

In determining the link budget for a satellite-to-ground communications link, only two conditions need to be established in most cases: the 'clear sky' conditions and the 'degraded sky' condition. As has been noted, the former establishes the performance level of the system while the latter fixes the availability. The difference between the two link budget levels is termed the margin of the link.

For digital satellite systems, the measurement variable for circuit performance is the Bit Error Rate (BER). The BER criteria for satellite links in the fixed satellite service has been established by the CCIR [13] and are summarised in Table 7.1 [14].

Table 7.1 *Allowable BER at the output of the hypothetical reference digital path for systems in the fixed-satellite service using pulse-code modulation for telephony (from Table I of Reference 14 as summarised in Reference 13, copyright © 1986 ITU, reproduced with permission.)*

BER	Integrating period	To be achieved for:
10^{-6}	10 min	At least 80% of any month
10^{-4}	1 min	At least 99.7% of any month
10^{-3}	1 s	At least 99.99% of any year

Most digital satellite systems use 4-phase shift keying (4-phase PSK or simply QPSK) modulation with coherent detection and the theoretical graph of BER versus E_b/N_0 is given in Fig. 7.4 for 4-phase PSK.

As can be seen from Fig. 7.4, the desired BER can be obtained by establishing the required E_b/N_0. When rain intercepts the transmission path, however, both the received signal level will drop and the perceived system noise will increase. Both will cause a reduction in signal-to-noise and so, for the initial satellite systems, an adequate margin was designed in to provide sufficient degraded-sky capabilities. The need to reduce the size of the earth stations in order to reduce the cost of ownership, coupled with an increasing use of the bands above 10 GHz, has caused a re-assessment of the amount of margin that it is economical to allocate on a fixed basis.

In even the wettest climates it only rains in measureable amounts for about 10% of an average year. For most temperature climates, this percentage is nearer 3–5%. Allocating a fixed margin of, say, 7 dB (the original margin of a typical standard C earth station operating at 14/11 GHz in the INTELSAT system) that is not utilised for 95% or more of the time, is a generally poor allocation of resources, particularly if the 7 dB margin is obtained by over-sizing the antenna or the high power transmitting amplifier. A large number of techniques have been proposed that can be used to allocate additional margin [30], sometimes in a dynamic manner, so that the available resources can be used in an optimal, and economical, manner.

There is no simple way to catalogue the various techniques for restoring the signal during a fade. Some techniques apply a constant resource (e.g. FEC coding) while others are dynamic (e.g. shared-resource TDMA); some apply a fixed margin increase (e.g. site diversity), others a variable margin increase (e.g. uplink power control); some are only utilisable if on-board processing is incorporated in the satellite (e.g. signal regeneration) while others can be applied only if the satellite acts as a simple transponder (e.g. frequency addressable antennas). The classification that has been adopted in the following discussion makes a broad division between fixed resources and dynamic resources and, for the latter, further subdivides into earth-based or satellite-based resource allocation and into fixed-level and variable-level application. Table 7.2. illustrates the classification.

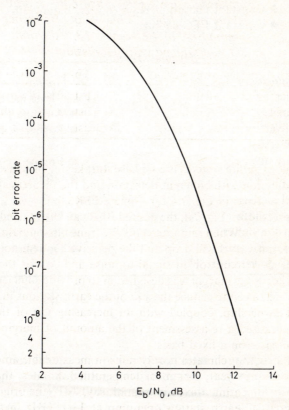

Fig. 7.4 *Theoretical curve of E_b/N_0 versus BER for ideal 4-phase PSK (QPSK) with coherent detection*

7.5.1 Fixed resource allocation to counteract signal attenuation

(a) Constant margin increase

This was the traditional approach to the allocation of propagation margins in a link budget. If the margin was insufficient, the transmitted power was increased either by increasing the power amplifier output or the antenna gain. At C-band (6/4 GHz) and below, signal attenuation was very small and typical margins were 2–3 dB. In the bands above 10 GHz, attenuation is much higher than at C-band for a given rainfall rate and, following trade-off studies, it was determined that the continuous application of an error correcting code would permit economical margins to be achieved in these bands.

Table 7.2 *Classification of restoration techniques for overcoming signal attenuation*

Fixed resources
- Constant margin increase (power, bandwidth)
- Constant FEC coding

Dynamic resources

Earth-based allocation	*Satellite-based allocation*
Fixed-level increase:	Fixed-level increase:
• Site diversity	• Antenna gain
• Height diversity	• Regeneration
• Angle diversity	
• Frequency diversity	• Frequency diversity
• Orbital diversity	• Orbital diversity
Dynamic-level increase:	Dynamic-level increase:
• Signal bandwidth	• SSFEC coding
• Transmission buffering	• Shared resource TDMA
• Uplink power control	• Downlink power control
• Downlink power control*	

---

* Only possible with a linear satellite transponder under single-carrier, single-destination conditions. Fading on the downlink is compensated for at the transmitting end of the (up)link

(b) Constant FEC code
The choice of a coding scheme is complex. For rain attenuation, the periods of signal fading are appreciably longer than in scintillation events and interleaving is of no real benefit. The number of check bits available in a block for correcting possible errors must be a balance between the number of errors expected and the increase in 'non-communication' bandwith. A typical block length could be 255 bits with 40 bits used as check bits. That is, about 15% of the bits are 'non-communication' bits. Some of the trade-offs in selecting codes can be found in Chapter 6 of Reference 15.

One of the factors in selecting an FEC code is the original uncoded transmission rate. INTELSAT has selected Rate 3/4 FEC for information rates of 10 Mbit/s or less and Rate 7/8 FEC for higher information rates. Fig. 7.5 illustrates the performance of a Rate 3/4 FEC code [16].

From Fig. 7.5 it can be seen that there is a reduction of about 2 dB in the E_b/N_0 required to meet a given BER. This has exactly the same effect as increasing the power amplifier output by 2 dB or the antenna gain by 2 dB. For this reason, the difference between the 'No FEC' curve and the 'Typical Performance with Rate 3/4 FEC' curve at a given BER is referred to as coding gain.

In the application of additional resources to reduce the impact of propagation impairments, two measures are used. One is of the effective increase in the margin due to the application of the resource. This is nearly always measured in decibels and is referred to generically as 'gain', e.g coding gain, diversity gain. The other measure is in terms of the outage time reduction. This is usually measured as a ratio of the percentage outage times, with and without the application of the resource, and is termed generically as 'advantage', e.g. diversity advantage.

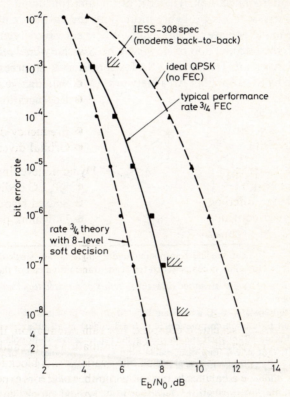

Fig. 7.5 *Effect of channel coding on the BER versus E_b/N_0 performance of a digital channel*
    (Fig. 11 of Reference 16)
    (Reproduced with permission from INTELSAT)

7.5.2 Dynamic resource allocation to counteract signal attenuation
7.5.2.1 Earth-based allocation
A Fixed-level increase
(a) Site diversity
The meteorological background as to why site diverstiy is an effective technique to combat severe rain attenuation due to thunderstorm activity is contained in

Section 1.3.3(*b*) and a detailed description of site diversity application and modelling is given in Section 4.5.4(*b*).

While the amount of gain achievable from a site diversity scheme is variable, the main application of the technique is to combat severe rain events. The site diversity gain characteristics are fairly linear in this high attenuation region (see Fig. 4.4) and so the technique is considered to be a fixed level technique.

Dual-site diversity, with a few rare exceptions, has not been applied to commercial satellite communications operations owing to the considerable cost of setting up not only two sites with complex earth stations but also of putting in a diversity interconnect link. The cost of the link (including the cost of the land, right-of-ways, etc.) is probably more than the cost of implementing an oversized antenna to achieve the same gain. Where site diversity may come into its own is with the introduction of slow-switched diversity between a multitude of Very Small Aperture Terminals (VSATs) spread over an urban area that is well connected by Metropolitan Area Networks (MANs). Fig. 7.6 illustrates the concept.

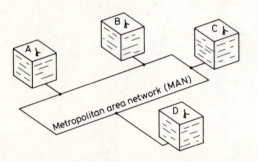

Fig. 7.6 *Illustration of a slow-switched diversity network connected together by a MAN in an urban area*

VSAT terminals A, B, C and D are located on the premises of buildings owned or operated by a large corporation (e.g. hotel chain) in an urban area. Typically, the buildings are 10 km apart and are connected together in a Metropolitan Area Network (MAN), probably a fibre-optic cable loop system. Should one of the terminals be obscured by a rain storm, any of the other terminals may be selected. No fast switching occurs during a fade. Should the link go down during a transmission/reception session, the link is simply re-established using another terminal that is not carrying traffic.

VSAT operators usually do not require outage times of less than 2% and the activity ratio, the ratio of the time the terminal is in active use to the overall time, is typically low. A figure of 10% is a reasonable activity ratio for a VSAT. With these outage and activity ratio requirements, a slow-switching diversity scheme (similar in many respects to a normal telephone switching system) that can select an 'open' VSAT from among many connected to the MAN is entirely feasible. Such a scheme will be almost a mandatory requirement when 30/20 GHz VSAT operations are introduced. The availability of a relatively wide-band (1 to

2 MHz, or more) MAN will be necessary for such a wide-area, slow-switching diversity scheme.

(b) Height diversity

Height diversity makes use of the fact that meteorological structures in the atmosphere that give rise to refractive and reflective effects tend to have larger correlation distances horizontally than vertically. To a satellite system designer, this means that the separation of the two antennas to achieve the same decorrelation would be smaller if the antennas were placed vertically above each other than if they were spaced horizontally apart. (Note that this is exactly the opposite for site diversity to counteract rain attenuation when lateral separations are better than vertical separations). Fig. 7.7 illustrates the technique of height diversity.

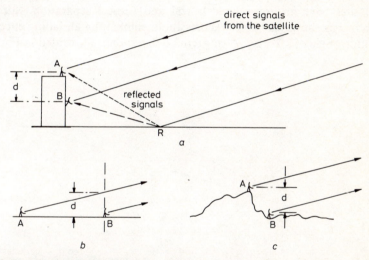

Fig. 7.7 *Illustration of height diversity*

Terminals A and B have a height separation such that reflected signals received from a region R are not correlated in phase on arrival. In addition, the separate direct paths are not correlated well on passing through a region causing tropospheric scintillation. This lack of correlation assists greatly in countering scintillation and multipath effects. In illustration (*a*) above, the terminals are located on the same office building, one on the top and the other two or three floors lower down. The terminals could equally well have been separated horizontally on level ground to give an apparent height separation *d* as in illustration (*b*) or made use of terrain irregularities to enhance the height separation as in illustration (*c*).

Case (*a*) is utilised in maritime height diversity with the superstructure or mast giving the required height difference.

At very low elevation angles (below 5°), multipath effects begin to be observed although it is difficult to differentiate their characteristics from those of low elevation angle tropospheric scintillation effects which can themselves be severe. A litmus test is the correlation in the amplitude fluctuations observed on the uplink and the downlink. If they are correlated, the effect is due to scintillation (or simple attenuation); if they are decorrelated, the effect is due to multipath.

In an experiment in the Canadian High Arctic [17] at a frequency of 6/4 GHz and at an elevation angle of 1°, the variations in the level of the uplink and the downlink signals were essentially decorrelated. The scintillations were therefore due to ray bending and/or multipath. An experimental site diversity scheme, with an appreciable height difference between the two terminals (see Fig. 7.7 case (c)) significantly reduced the impact of the scintillations [18] and was quickly followed by the introduction of the first active site diversity system at 6/4 GHz [19].

At elevation angles between about 1° and 5°, it may well be that a staggered orientation of a dual-site diversity system is optimum. A staggered orientation is one in which the remote site is placed so that, when viewed from the satellite, the two sites have an appreciable lateral and vertical separation. Such an orientation was shown to be very successful in combatting all fading effects at an elevation angle of 3·2° [20]. A staggered orientation is illustrated in Fig. 7.8.

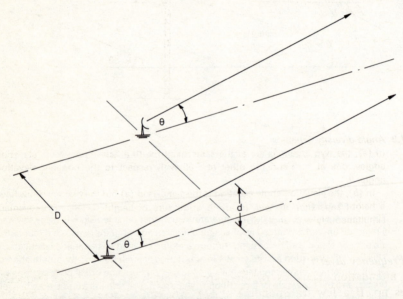

Fig. 7.8 *Illustration of staggered dual-site diversity*
    In this scheme, the two earth stations are offset so that a lateral separation D and a
    vertical separtion d combine to produce impairment reduction due to both rain
    attenuation and scintillation/multipath effects.

(c) *Angle diversity*
Angle diversity is the displacement in elevation angle of two co-located earth station antennas such that both are slightly depointed in elevation angle either side of the nominal satellite look angle (see Fig. 7.9a).

For very low elevation angle paths, the elevation angle can vary due to changes in the bulk refractive index of the atmosphere (see angle-of-arrival effects in Section 3.2.5). By pointing the two antennas to either side of the nominal satellite position, the appropriate antenna can be selected depending on whether the incoming signal is super-refracted or sub-refracted away from the nominal path. The restoration technique does not combat the main clear sky impairment at very low elevation angles − multipath effects − unless, as well as being mutually depointed in elevation angle, the two antennas are also separated in height to give a height diversity gain (see Fig. 7.9*b*).

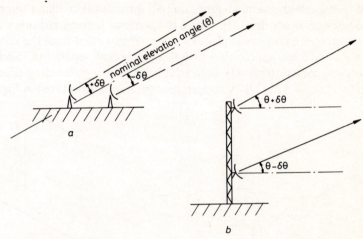

Fig. 7.9 *Angle diversity schemes*

In (*a*), the two antennas are at the same height with a $2\delta\theta$ difference in elevation angles, one at $+\delta\theta$ and the other at $-\delta\theta$ with respect to the nominal elevation angle, θ.

In (*b*), the elevation angle pointing is the same as in (*a*) but the two antennas have a height separation as well to afford a measure of height diversity improvement simultaneously with angle diversity.

(*d*) *Frequency diversity*

Path attenuation increases rapidly, approximately as f^2, as the frequency f goes up. If a few transponders at 6/4 GHz can be reserved in a satellite then, when the links at 14/11 GHz or 30/20 GHz are severely attenuated, switching the impaired channel to the 6/4 GHz band can afford significant attenuation reduction [21].

Historically, earth station complexes carrying significant traffic have tended to utilise the 6/4 GHz bands before the 14/11 GHz bands are used. In general, it is only when traffic saturation occurs at 6/4 GHz that the 14/11 GHz bands are used; the relatively benign propagation environment at 6/4 GHz can reduce the cost of an earth station significantly from that of a comparable 14/11 GHz earth station. Since it is likely that 6/4 GHz facilities will be available at earth station complexes with major 14/11 GHz traffic requirements, the capability of diverting traffic from a 14/11 GHz link (or a 30/20 GHz link) to one at 6/4 GHz during severe rain storms exists. The capacity of the 14/11 GHz earth station would be quite large, however, and a significant amount of 6/4 GHz capacity would have to lie empty for a considerable portion of an average year in order to have the capacity available for quasi-instantaneous switching. Frequency diversity for large traffic streams does not therefore appear to be an economic proposition from the earth station viewpoint.

(e) Orbital diversity
Orbital diversity achieves a reduction in signal attenuation by diverting traffic from one satellite to another [22]. Fig. 7.10 illustrates the technique.

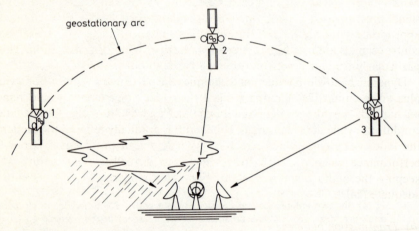

Fig. 7.10 *Illustration of orbital diversity*
The earth-station antenna communicating with satellite no. 1 suffers a rain fade and has to transfer operations to satellite no. 2 or 3. Depending on the size and location of the storm, a significant reduction in signal attenuation can be achieved.
In the above sketch, three antennas are shown separately for ease of understanding. In practice, only one antenna may be used.

Orbital diversity does not require the long diversity interconnect link of a dual-site diversity scheme nor does it, in theory, require more than one earth station. In general, however, orbital diversity gain is less than site diversity gain and, to obtain useful levels of orbital diversity gain, the angular separation of the satellites must be large. Significant orbital diversity gain will not usually be possible for transoceanic links since there is only a limited range of orbital arc accessible to a given earth station and, even if a large arc is available, to use a large angular separation will require the utilisation of low elevation angles at both ends of the arc which could reduce the available orbital diversity gain to very small values (1 or 2 dB). Intersatellite Links (ISLs) would improve the situation [22]. In fact, the extensive use of ISLs would enable most earth stations to operate at relatively high elevation angles and, while not a direct extension of orbital diversity, would certianly facilitate the introduction of 14/11 GHz services to high rainfall rate regions and 30/20 GHz services world wide.

B Dynamic-level increases
(a) Signal bandwidth variations
The noise power in a communications channel is directly proportional to the bandwidth. A 3 dB reduction in the channel bandwidth, therefore, will increase the C/N by 3 dB. Reducing the bandwidth by 3 dB will usually require the transmission rate to be decreased. This technique, if applied dynamically [24], however, can afford significant protection against fading while not reducing the signal throughput significantly on a system basis. To enable the technique to work efficiently without on-board processing at the satellite node, a control signal must be used between the co-responding earth stations to facilitate the appropriate bandwidth/signal-rate changes in synchronism. In general, such control signals are usually processed by a hub, or main control station. This is particularly true for a network of small earth stations.

Dynamic bandwidth reduction techniques are particularly suitable for a video plus voice circuit [24]. During a signal fade, the voice circuit would remain unchanged, providing a high quality channel under all but the deepest fading conditions, while the video circuit bandwidth is gradually reduced to counteract the fade. For digital video circuits, this would lead progressively from full performance video, through freeze frame, to virtually a still picture while keeping the audio quality constant. For video conferencing, this is entirely adequate [24].

(b) Transmission buffering
Transmission buffering is very similar to signal bandwidth variation except that, when the signal rate is reduced owing to a rain fade being present on the link, the bits that cannot be sent are stored in a buffer until the rain fade is no longer present and transmission at a higher rate than normal can take place to 'catch up' with the normal, unfaded rate. Such a scheme will require an inordinate amount of buffer capacity for high data systems and is therefore only economical for very low data rates. Paradoxically, it is the low data rate systems that can stand longer outage times and so, except for buffering that is necessary to take out Doppler shift and variations in transmission time due to satellite motion, transmission buffering is unlikely to be used as a signal restoration technique.

(c) Uplink power control
Uplink power control (ULPC) is the dynamic application of additional power to the transmitted signals at an earth station in an attempt to compensate for attenuation occurring on the uplink. In an ideal situation, the received power at

the satellite will be held virtually constant. This has many advantages from interference, capacity and availability aspects.

(i) *Interference*

In Fig. 7.11, six FM carriers of equal capacity are shown schematically in a satellite transponder. To minimise adjacent channel interference and mutual intermodulation effects, the carriers are held at similar received power levels at the satellite [25]. An uplink fade on carrier B will cause a relative increase in adjacent channel interference into carrier B. Not only will the carrier-to-noise C/N of carrier B be reduced but also the carrier-to-interference C/I. If the transmitted level of carrier B can be increased by U decibels, then the original C/N and C/I can be restored.

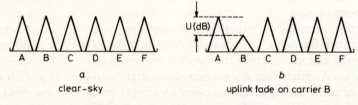

Fig. 7.11 *Illustration of an uplink fade in multicarrier operation*
(*a*) and (*b*) illustrate the loading of six equal FM carriers in one satellite transponder. In (*a*), all six carriers from six different earth stations are received in clear sky conditions. In (*b*), earth station B is in a rain storm and carrier B is attenuated on the uplink by an amount *U* decibels.

(ii) *Capacity*

The number of telephone channels, data rate etc. in a given bandwidth is directly proportional to the signal (carrier) power [25]. Exotic modulation schemes that use 16, 32 or even 64 information states, as opposed to the 4 of QPSK, can provide commensurately higher information rates, but the threshold detection level for the same BER is higher the more information states there are. An examination from the viewpoint of channel capacity [26] and number of earth stations accessing the transponder [27] showed that ULPC conferred significant advantages. In the latter case, up to 3·7 times more earth stations could access the satellite when ULPC was used compared to the case when ULPC was not employed.

(iii) *Availability*

In most satellite communications systems, it is usual for the outages on the uplink and the downlink to be equalised. If the propagation outage allowance is 0·04% in an average year, 0·02% is allocated to the uplink and 0·02% to the downlink. If uplink power control is utilised, the propagation outages on the uplink can be significantly reduced. The outage time thus 'saved' can be used to increase the overall total availability or transferred to the downlink budget. For example, if the use of ULPC reduces the outages on the uplink from 0·02% to 0·01% in an average year, the overall uplink/downlink system availability can be increased from 99·96% to 99·97% in an average year. On the other hand, if the system availability remains at 99·96%, the outage on the downlink can be increased from 0·02% to 0·03% by 'transferring' 0·01% of the uplink outage allocation to the downlink allocation.

The transferrence of the uplink outage 'saved' to the downlink is the usual procedure adopted since this will greatly reduce the propagation margin required on the downlink, thus reducing the earth station antenna diameter.

There are a number of ways of implementing uplink power control. These fall into the generic categories of Closed-Loop ULPC, Open-Loop ULPC, and Feed-Back ULPC [28].

I *Closed-loop ULPC*

If the transmitting and receiving earth stations are in the same satellite antenna coverage, then the two earth stations will have access to both their transmitted and received carriers. If earth station A is transmitting to earth station B and earth station A suffers rain attenuation on its satellite-to-ground path, it will be able to measure the signal attenuation by simply measuring the level of its transponded carrier.

In Fig. 7.12, the application of uplink power control at earth station A keeps the received level at the satellite constant and the received level at earth station B constant (assuming the satellite employs a linear transponder). A number of methods can be used to calculate the amount of uplink power to apply.

The simplest technique is to maintain the received levels of signals A and B at earth station A the same by adjusting the uplink power of signal A. This assumes, however, that signal B does not suffer a simultaneous uplink fade (at earth station B) and that it is held constant under all conditions.

The second method would be to measure the received level of signal A at earth station A with respect to its normal clear sky level. The attenuation measured would be the total, looped-back attenuation of the link, i.e. the attenuation on the uplink plus the attenuation on the downlink. If the attenuation ratio connecting the uplink attenuation to the downlink attenuation is known, the uplink attenuation can be extracted and that amount applied to the uplink power transmitted. The variation in the attenuation ratio can be considerable, however, leading in its turn to errors in the level of uplink power to be applied (see Figs. 4.43–4.46). In addition, changes in transponder loading or setting can cause significant changes to the transponded signal levels. These changes in signal level could be interpreted as signal attenuation due to propagation impairments on the path and an erroneous application of uplink power addition commanded. Additional errors can be caused if the earth station antenna tracking is not accurate. A change in the level of the downlink signal A could be interpreted as rain attenuation on the link instead of as antenna depointing, and an erroneous application of uplink power addition commanded. A cross-check between the levels of signals A and B however, would provide an additional assurance that the correct amount of uplink power has been added.

A third method would measure a downlink beacon provided by the satellite and, assuming the ratio between the uplink and downlink attenuation is known, apply an appropriate increase in uplink power to counteract the calculated uplink attenuation. A cross-check between signals A and B would enhance the accuracy of the technique as in the second method.

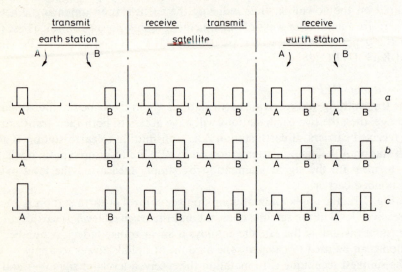

Fig. 7.12 *Illustration of close-loop ULPC*

    (*a*) The clear-sky situation is illustrated. On the transmit side, earthstations A and B transmit at different frequencies in the same transponder and are received at an equal power level at the satellite. The transponded signals leave the satellite at the same level and are received on the ground at both earth stations since both earth stations are within the same satellite coverage area. The signals are assumed to be digital carriers in both cases.

    (*b*) A rain fade over earth station A causes the received level of that signal at the satellite to drop. The transponded signal levels will differ by this amount in a linear transponder and at earth station A, both signals A and B are reduced owing to passing through the same rainstorm as signal A did on the uplink.

    (*c*) ULPC has restored the transponded signals to the same level but, at earth station A, both signals A and B still suffer downlink attenuation. At earth station B, however, both signals are back at their previous clear-sky level.

A fourth method would use a radiometer at a suitable frequency to estimate the uplink attenuation instead of using a satellite beacon or transponded carrier level detection on the downlink. While radiometers are inherently simpler than beacon or carrier receivers, the accuracy of a beacon receiver at attenuations above 5 or 6 dB is higher than that of a radiometer. The addition of a beacon receiver in the receiver chain of an already relatively complex earth station is also, perhaps, easier than installing a radiometer which will require a completely separate feed and receiver subsystem behind the feed horn.

All the four methods above, however, have the advantage of a cross-check with the transponded carrier so that the accuracy of the uplink power control system can be monitored at the earth station applying ULPC.

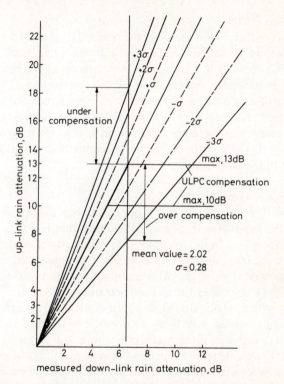

Fig. 7.13 *Illustration of the magnitude of errors that can occur with open-loop ULPC (Fig. 4.10 of Reference 29)*
The mean value (2·02) and the RMS deviation of this ratio value between 14 and 11 GHz (0.28) were derived from measurements at an elevation angle of about 6·5°. The errors are shown for the ranges of ULPC, 10 dB and 13 dB.

II *Open-loop ULPC*

Trans-oceanic satellite communications systems have widely spread coverages and it is rare that a transmitting earth station has access to its own carriers following retransmission by the satellite (transponding). The application of any uplink power control is therefore necessarily of an open-loop nature; i.e. there is no cross-check possible on the accuracy of the uplink power control system at the transmitting earth station. The only way to estimate the amount of uplink power to add when using an open-loop system is to measure the path attenuation on the downlink and calculate the attenuation that would occur at the uplink frequency. An example of the errors that can occur are illustrated in Fig. 7.13 (see also Figs. 4.43–4.46). An open-loop system will never, on average, attain the accuracy of a closed-loop system and so great care must be excercised in its operation so as not to overdrive the satellite excessively or greatly to underestimate the uplink attenuation.

III *Feedback-loop ULPC*

In this scheme, a control station monitors the received level of all the transmitted/ transponded carriers and signals back to the transmitting earth stations the required changes necessary in their uplink power to optimise the system. If the control station is generally remote from all the transmitting earth stations, uncorrelated Earth–space paths can be assumed from a meteorological aspect. This system is particularly suitable for Time Division Multiple Access (TDMA) networks that utilise one, or more, control earth stations to synchronise the TDMA signals [28]. The addition of carrier level monitoring equipment would be insignificant in the overall scheme of things.

(d) *Downlink power control*

While Downlink Power Control (DLPC) appears to be a satellite-based resource allocation, DLPC is possible from an earth station that is accessing a satellite with linear transponders. In this version of downlink power control, the downlink attenuation at earth station A, for example, is compensated for by increasing the uplink power at all of the other stations that are transmitting to earth station A. Fig. 7.14 illustrates the method schematically.

The difficulty with applying downlink power control on the uplink is that the satellite transponder would most likely be driven into the non-linear, saturated portion of its characteristics if too much power was applied on the uplink from those stations that were increasing their power in essentially clear sky conditions as far as their uplink path was concerned. In addition, adjacent channel interference and intermodulation effects will be significantly increased. For these reasons, downlink power control via uplink power control through a satellite with a linear transponder is not considered to be a practical solution for multi-carrier operations unless only small ULPC margins are used.

7.5.2.2 *Satellite-based allocation*
A *Fixed-level increase*
(a) *Antenna gain control*

The power flux density on the surface of the Earth generated by a satellite is a function of both the satellite amplifier and the gain of the satellite antenna. The gain of an antenna is determined mainly by the diameter of the antenna; changing the gain, therefore, usually implies changing the antenna since the diameter is a fixed quantity.

Fig. 7.15 depicts the north-west quadrant of the Earth as viewed from 307° E longitude in the geostationary arc. A single, circularly symmetric beam is shown covering most of the USA. The circular contour is the 3 dB-down locus of the antenna beam (i.e. the beamwidth). In this instance, the beamwidth is about 4°. If the antenna gain is to be increased by increasing the antenna diameter, the beamwidth will be correspondingly reduced. If the antenna diameter is doubled, giving a gain increase of 6 dB for the individual beam, the beamwidth is halved.

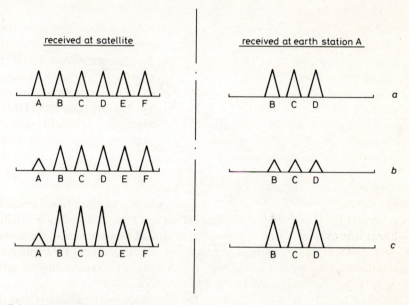

Fig. 7.14 *Illustration of DLPC applied on the uplink through a satellite with a linear transponder*
a Clear-sky
b Rain fade at earth station A: no DLPC
c Rain fade at Earth station A: DLPC applied on the uplink by earth stations B, C and D
Earth station A is only receiving the FM carriers transmitted from earth stations B, C and D. When a rain fade occurs over earth station A, earth stations B, C and D are requested to increase their uplink power by the same amount as the downlink fade at station A. (By the same token, earth stations receiving the FM carrier from earth station A may request an increase in the uplink power at earth station A to compensate for the uplink fade on carrier A. This is not shown in the above illustration)

Four of these beams are now needed to cover the USA (see Fig. 7.16a). If the four beams are used collectively, a composite contour is obtained as shown by the broken outer line. Progressively reducing the beamwidth of the individual spot beams will progressively increase the gain available in each beam. Going from the approximately 4° beamwidth in Fig. 7.15 to the 1·9°, 1·6° and 1° beamwidths of Figs. 7.16a, b and c will give gain increases of about 6, 8 and 12 dB, respectively.

A gain of 12 dB is significant, much more than is required in most 11 GHz links to overcome downlink attenuation. If a single 50 W Travelling Wave Tube Amplifier (TWTA) is required for each beam, however, the 50 W required for the single beam in Fig. 7.15 will have risen to 800 W for the 16 beams of Fig. 7.16*c*. This is within the reach of current spacecraft technology but is uneconomic to implement.

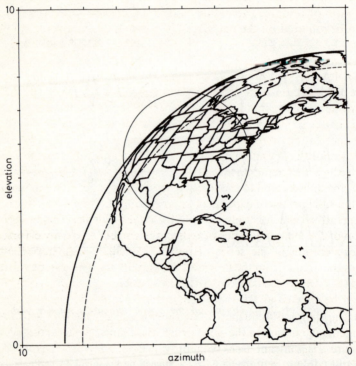

Fig. 7.15 *NW quadrant of the Earth with a single spot-beam coverage over the USA as viewed from 307°E*
From 307°E in the geostationary orbit, the USA is shown to be almost covered by a single spot beam, the 3 dB angular dimension being 4° in spacecraft co-ordinates. In general, an optimised elliptical coverage having the same forward gain as the circular beam would be used but a circular representation is shown for comparison with Figs. 7.16*a*, *b* and *c*.

A conceptual approach which reduces the amplifer requirements is to have a single spare TWTA that can be switched to the required small spot beam. Communications would be generally carried out with the wide beam (i.e. 4° beamwidth antenna) but, when a link needs an increase in downlink power, the 'spare' TWTA would be switched into the required beam. The approach is referred to as a 'hopping' beam since it hops to the requisite location in need of enhanced EIRP. A yet more advanced scheme would have all the communications carried over the multiplicity of small beams with each beam scanned in turn so that only one beam is actually 'on' at any one time. This is referred to as a scanning/hopping beam [31]. The dwell time over each spot beam coverage would be a function of the traffic being carried in that beam. Such techniques will eventually be used but they will require the development of adaptive phase array antennas (e.g. [32]) and a significant amount of on-board processing capabilities in the satellite. In addition, to be able to assign the correct ratio of resource capabilities to normal capabilities, more information will also be required on the correlation of rainfall over large areas [33, 34].

(b) Regeneration
The current transponders used in satellites are of the linear repeater type. While some of them are 'backed off' in power to be able to provide a more linear output response to the input received (generally for FM services) and others are operated closer to saturation (generally for wideband digital services), the fact remains that any errors/fades that occur on the uplink are translated directly on to the downlink. To try and reduce the overall transmission errors to a minimun, and to reduce the requirements on the satellite transmitted power as far as possible, the uplinks tend to provide substantially more EIRP than would be the case if the satellites were capable of regenerating the signal.

Regeneration, which is only feasible for digital signals, usually involves the detection of the incoming series of pulses, the correction of any errors, and the exact regeneration of the series of pulses [25]. Since regeneration effectively separates the uplink from the downlink, the uplink margins can be substantially reduced (5–7 dB) with a concomitant reduction in earth station costs. Regeneration can be effected at RF (i.e. at the received radio frequency), at an intermediate translational frequency in the satellite, or at baseband with the power requirements increasing in the same order. Baseband regeneration (processing) will require a significant technology advance to bring the overall spacecraft power (and cooling) requirements to within a feasible range.

(c) Orbital diversity
This is identical to that given in Section 7.5.2.1.*A*(*e*) and is included here in the satellite-based allocations for completeness. From the spacecraft viewpoint, orbital diversity requires the deployment of at least two satellites that can 'see' both ends of the link on the ground so that switching between satellites is feasible.

(d) Frequency diversity
This is identical to that given in Section 7.5.2.1.*A*(*d*) and is included here in the satellite-based allocations for completeness. From the spacecraft viewpoint, frequency diversity is far more efficient than earth station frequency diversity since a large number of earth stations can be served by the satellite using the same, relatively small, lower frequency resource to counteract rain attenuation.

a

b

Since it is the overall system costs, however, that dictate the utility of a technique, frequency diversity is not considered to be of prime significance owing to the inefficient use of the earth segment.

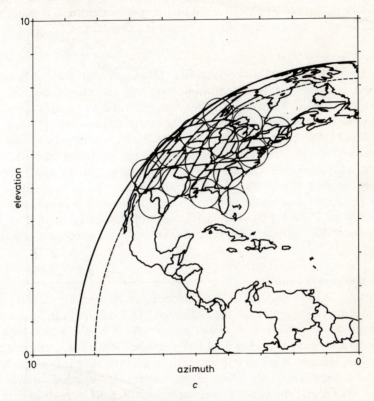

Fig. 7.16 *NW quadrant of the Earth with different numbers of spot beams over the USA as viewed from 307°E*
  (*a*) 1·9° spot beams
  (*b*) 1·6° spot beams
  (*c*) 1° spot beams

B Dynamic-level increase

(a) Satellite-switched forward error correction

The constant application of Forward Error Correction (FEC) can be relatively wasteful of spectrum if a very high coding gain is required. If the coding gain can be dynamically varied as required, the overall spectrum efficiency would be increased. By detecting a signal fade (or BER increase) at an earth station and signalling via a control station that the coding gain needs to be increased at the earth station, and the other earth stations communicating with the first earth station, only a relatively small additional bandwidth is required for the periods of high coding gain since the additional bandwidth can be shared between all the earth stations in the network. Such a scheme was found to have merit if there were a large number of earth stations carrying relatively equal amounts of traffic [35].

The controlling station need not be on the ground. Locating it at the satellite, giving satellite-switched FEC, will speed up the control element but require the implementation of advanced on-board processing capabilities.

(b) Shared resource TDMA

In any Time Division Multiple Access (TDMA) system there will be an established protocol for synchronisation, control, and communication. To synchronise a satellite TDMA system requires control stations to transmit the synchronising pulses onto which all the other earth stations lock [25].

With each TDMA burst there are the bits that carry the information. In a rain fade, there will be a higher chance of errors occurring. To reduce the probability of errors, additional bits can be added to the frame for coding purposes or simply to slow down the effective transmission rate. These additional bits can be shared amongst a pool of earth stations in a scheme known as shared-resource TDMA [36]. The logical control station for such a scheme is the satellite, but again, a significant level of on-board processing would be necessary.

(c) Downlink power control

In an exactly similar fashion to that described in Section 7.5.2.2.A(a) with respect to antenna gain increase, the additional EIRP required can be obtained from a high power TWTA. This higher power TWTA would be selectable for the appropriate beam requiring an EIRP increase. While more feasible than antenna gain increase from an implementation point of view (i.e. no on-board processing is necessary), only a limited gain increase is available without running into TWTA and spacecraft power limitations. A 6 dB increase in gain will require a 50 W TWTA to be replaced with a 200 W TWTA.

7.5.2.3 On-board processing

In many of the examples given for satellite-based resource allocation, the necessity for some form of active control by the satellite was highlighted. In many aspects of future satellite communications, not just the implementation of signal restoration techniques, the ready availability of on-board processing is critical. The technologies that will require the availability of reliable, low-power, on-board processing capabilities include:

- Optical inter-satellite links
- Scanning/hopping beams
- Resource sharing schemes
- Variable capacity traffic switching
- Autonomous spacecraft control

7.6 Depolarisation effects

Depolarisation effects are only significant if frequency re-use techniques are being employed. By re-using the same frequency twice, but with opposite sense polarisations, co-channel operation can effectively double the available spectrum. Depolarisation will cause the energy from one polarisation sense to be coupled into the other, thereby causing interference.

7.6.1 Techniques below 10 GHz

At frequencies of 6 and 4 GHz, rain attenuation effects are very low and so the limiting propagation impairment is depolarisation. Following early fundamental theoretical work [37], considerable effort went into devising means of restoring the orthogonality of the depolarised signals [38, 39] and in measuring the effects [40]. Since the propagation medium was relatively lossless at 6 and 4 GHz, the restoration schemes were all along the lines of incorporating networks of polarisers in the antenna feed systems (see Fig. 5.14)

The actions of the quarter-wavelength and half-wavelength polarisers are captured in Fig. 5.15. A convenient means of representing the change between polarisation senses is to use a Poincaré sphere (Reference 42 from Reference 43). Fig. 7.17 depicts a Poincaré sphere. The 'latitude' and 'longitude' of the Poincaré sphere are given in terms of the angles α and β, which are defined in Fig. 7.18.

The polarisation ellipse in Fig. 7.18 can be defined either in terms of the axial ratio r, and the inclination of the major axis to a reference plane, β, or in terms of the angles α and β. If $\alpha = 45°$, the polarisation is perfectly circular (and hence β is undefined). It is normal, therefore, to restrict α to $\pm 45°$. With this limitation $r = \tan \alpha$.

In Fig. 7.17, the equator of the Poincaré sphere always represents linear polarisation while the poles represent perfectly circular polarisation. At points in between on the surface of the sphere, an elliptical polarisation can be defined with 2α and 2β the latitude and longitude of the Poincaré sphere, respectively. Some interesting pictorial representations of the action of the half-wavelength and quarter-wavelength polarisers on the Poincaré sphere have been constructed [45] to illustrate that perfect restoration of a depolarised signal can be achieved through the use of polarisers.

As with signal restoration techniques for attenuation effects, downlink (i.e. closed loop) depolarisation compensation is relatively straightforward since a rapid feedback loop exists between the measured depolarisation and the sensing device rotating the polarisers. The difficulty is to effect pre-compensation on the uplink in an open-loop situation.

One solution adopted [44] was to measure the rotational movement, from their normal position, of the polarisers on the downlink that achieved exact cancellation/compensation of the incoming propagation medium-induced depolarisation and to use these measurements to derive the required rotational movements of the uplink polarisers to achieve precompensation on the uplink. If

$\Delta\alpha$ = a rotational movement from the initial setting of a polariser

qd and *hd* = suffices for the downlink quarter wavelength and half wavelength polarisers, respectively

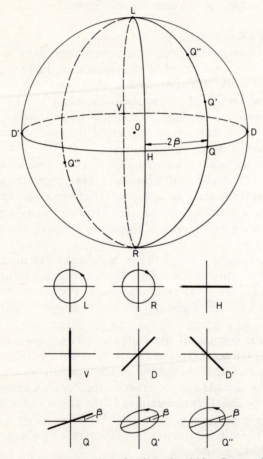

Fig. 7.17 *Poincaré sphere representation of polarisation (After Bryant, from Fig. 1 of Reference 44)*

At point Q on the 'equator', the polarisation is linear with a tilt angle of β. As the point moves from Q to Q'to Q" the polarisation becomes progressively circular but with the same tilt angle.

qu and *hu* = suffices for the uplink quarter wavelength and half wavelength polarisers, respectively

then [44]

$$\Delta\alpha_{qu} = \Delta\alpha_{qd} \tag{7.2}$$

and

$$\Delta\alpha_{hu} = \frac{\Delta\alpha_{qu}}{2} + \frac{\Delta\beta_u}{4} \tag{7.3}$$

where $\Delta\beta_u$ is the differential phase shift on the uplink.

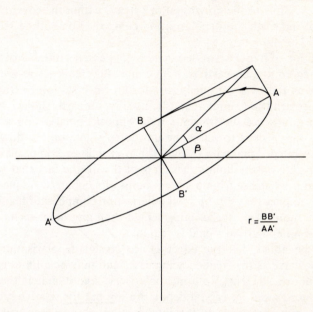

Fig. 7.18 *Definition of parameters α and β in a polarisation ellipse for application to a Poincaré sphere*
The axial ratio r, and the tilt angle β define the shape of the polarisation ellipse. Alternatively, the parameters α and β can be used to define the ellipse with $r = \tan \alpha$ (with $\alpha \leqslant 45°$)

But

$$\Delta\beta_u = r \times \Delta\beta_d \tag{7.4}$$

where r is the ratio of the uplink to downlink differential phase shift, and [44]

$$\Delta\beta_d = 4\Delta\alpha_{hd} - 2\Delta\alpha_{qd} \tag{7.5}$$

giving

$$\Delta\alpha_{hu} = \frac{\Delta\alpha_{qu}}{2} + \frac{r}{4}(4\Delta\alpha_{hd} - 2\Delta\alpha_{qd}) \tag{7.6}$$

The initial value for r in the 6/4 GHz experiment was 1.55 [44].

7.6.2 Techniques above 10 GHz
The increasing effects of attenuation as the frequency becomes greater than 6 GHz have led to a number of proposals to cancel both the differential attenuation and

differential phase effects separately to achieve exact cancellation/(pre)compensation of the depolarisation.

The necessity for exact cancellation has been questioned [46] and shown [47] to be unnecessary in most applications. For linearly polarised systems, if the orientation of the electric vector can be changed at the satellite so that the vector is as close as possible to the horizontal or vertical within the coverage area on the surface of the earth, no depolarisation compensation will be required for most systems [48, 49].

A classification of frequency re-use systems [50] noted a number of important factors with regard to most earth stations employing frequency re-use, the most significant being that it is extremely rare for an earth station to transmit and receive both polarisation simultaneously; usually, an earth station will receive in one polarisation sense and transmit in the opposite polarisation sense. There is, therefore, only a need to reject the unwanted signal on the downlink and transmit the wanted signal with as pure (or as exactly precompensated) polarisation as possible on the uplink. This, coupled with the fact that phase only compensation is adequate from a system point of view up to frequencies of at least 30 GHz, implies that sub-optimal polarisation compensation and precompensation are more likely to be implemented than any other technique if the restoration of a depolarised signal is required.

It should be noted that the effect of ice crystal depolarisation becomes progressively worse as the frequency increases and may be a limiting factor at frequencies above 30 GHz. Rotating polarisers cannot match the speed of lightning-induced rapid ice crystal depolarisation but the relative rarity of this phenomenon may not prevent the introduction of polariser, sub-optimum cancellation/precompensation systems. Depolarisation cancellation/precompensation at IF will be fast enough for ice crystal depolarisation but it is no longer considered to be a practical alternative to sub-optimal polariser systems for depolarisation cancellation/precompensation except on terrestrial paths.

7.7 Interference

Interference can take many forms. Intermodulation products can severely limit system capacity as can co-channel interference due, for instance, to depolarisation in a frequency re-use operation. Sometimes the interference is passive, as in most commerical satellite systems, and sometimes it is active, i.e. jamming interference. For the latter, frequency-agile communications allied to antenna nulling techniques are particularly effective counter-measures providing the jamming signal is not overwhelmingly powerful. Jamming can also take place in civilian operations [52] although here the jamming can take the form of slightly degrading the BER of the link to below the acceptable performance level without leading the user to suspect that a jamming device is being operated against his system. The following Section, however, will only deal with passive interference.

7.7.1 General representation

As has been discussed in earlier Chapters, the carrier-to-noise ratio (C/N) of a link can be calculated (using the 'dB' formulation rather than the arithmetic formulation) from [14]

$$C/N = -10 \log \{(C/N_{tu})^{-1} + (C/N_{td})^{-1} + (C/N_i)^{-1}$$
$$+ (C/I_u)^{-1} + (C/I_d)^{-1}\} - M \text{ decibels} \qquad (7.7)$$

where N_{tu} = uplink thermal noise per hertz
   N_{td} = downlink thermal noise per hertz
   N_i = equivalent intermodulation noise per hertz
   I_u = equivalent uplink interference noise per hertz
   I_d = equivalent downlink interference noise per hertz
   M = margin to allow for degradation due to high power amplifier non-linearities, dB.

Note that any additional margin required for rain attenuation etc. will need to be added to the C/N. The interference, both on the uplink and on the downlink, can arrive from a number of sources. Using the earth station as a reference point, a number of cases will be considered.

7.7.2 Sidelobe interference

(a) Direct interference

Earth stations normally operate at elevation angles above 5°. For large earth stations (15–30 m diameter antennas) this means that there is very little energy directed towards the horizon at grazing incidence along the azimuth angle of the main beam.

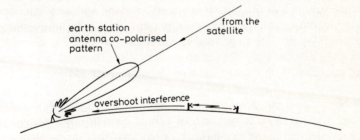

Fig. 7.19 *Schematic of terrestrial interference into an earth station*
There is normally a large separation (minimum of 100 km) between a terrestrial system that is aligned closely with the operating azimuth of an earth station using the same frequency band. In some meteorological situations, however, anomalous ducting can lead to enhanced levels of interference into the earth station.

In Fig. 7.19, interference from a terrestrial system is entering on a sidelobe of the earth station antenna. Careful co-ordination usually elliminates such interference paths for all but a minute fraction of the time, but occasionally [53] meteorological conditions that may persist for unusually long periods can render the affected earth station channels inoperative for an unacceptable percentage of the time. The technique used to counteract such single-path interference is to use an auxiliary antenna.

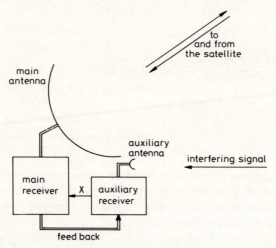

Fig. 7.20 *Use of an auxiliary antenna to cancel interference*
Any signals that are received via the auxiliary antenna that are above a certain threshold are adjusted in amplitude and phase so that, on being fed in at X to the main receiver, exact cancellation of any interference entering the main antenna from the same source occurs. A feedback loop is used to adjust the amplitude and phase of the auxiliary signal to ensure optimum cancellation occurs.

In Fig. 7.20, an auxiliary antenna is directed along the look angle of a known interference source and the output adjusted in phase and amplitude to cancel exactly the incoming interference, when present. Such a technique has proved to be successful in satellite-to-ground communications [53]. The technique of an auxiliary path has also been applied to terrestrial systems with some success for reducing the effect of multipath induced depolarisation [54 and e.g. 55].

(b) Differential path interference
This is the inverse, in some respects, of orbital diversity. If a second satellite, spaced a few degrees in the geostationary arc away from the wanted satellite, is transmitting at the same frequency towards the earth station, the possibility exists that rain along the wanted-satellite path will cause a higher attenuation than along the unwanted-satellite path, thereby enhancing any interfering signal that may have been present in clear sky conditions.

Measured results using radar data [56] have shown that the differential attenuation along the two paths can be significant. The absolute values, however, are not significant when compared to the suppression of the interfering signal by the earth station antenna characteristics. The sidelobes of an antenna are usually at least 20 dB below the main beam gain even a few degrees away from the main beam axis. As an example, at a frequency of 15 GHz, the differential attenuation was calculated to be 3 dB for a path separation of 6° at an elevation angle of 10° [56] at 0·01% of the time. A reduction in isolation from 20 dB to 17 dB for 0·01% of the time is not significant in most practical systems.

(c) Rain scatter coupling

An attempt was made [57] to measure the rain scatter coupling at a frequency of 19 and 28 GHz. Within the limits of the equipment (− 40 dB coupling at 19 GHz and − 45 dB coupling at 28 GHz in a 15 dB fade at either frequency), no significant coupling was observed at a displacement of 0·85° from the main beam axis. Since 15 dB is close to the practical limit for Earth–space communications links, rain scatter coupling along adjacent paths is unlikely to be a major factor in system interference protection.

Rain scatter coupling (forward scattering) is still more important than tropospheric scattering when it comes to earth station co-ordination [58], provided that the earth station elevation angle is above 5°. It should be noted, however, that rain scatter is essentially a 'common volume' phenomenon; a rainstorm is required to be exactly at the interaction of the wanted antenna beam and the interfering antenna beam. The likelihood of this conditional probability is quite low and, for this reason, superrefraction and ducting considerations tend to dominate earth station co-ordination [58] (see Chapter 1).

7.7.3 Mainlobe (mainbeam) interference

Earth stations with electrically small antennas will have correspondingly wide beamwidths. This is particularly true for mobile satellite system antennas (land, aeronautical and maritime) and the growing market of Very Small Aperture Terminal (VSAT) applications. The width of the mainlobes of these small antennas gives rise to an enhanced potential for mainlobe interference. Such mainlobe interference can include multipath effects (see Section 7.4); the discussion that follows, however, is restricted to mainlobe interference from the direct reception of signals from adjacent satellites.

VSAT antennas have large beamwidths, particularly those that operate in the 6/4 GHz bands. Typically, in these bands, up to four interfering signals can be received from satellites spaced 2° apart in the geostationary arc (see Fig. 7.21). VSAT operations usually have a low activity ratio, i.e. the 'on-time' divided by the total time, with typical values being well below 10%. The outage criteria are similarly much more relaxed than for the Public Switched Telephony Network (PSTN), a figure of 1–5% being typical as opposed to the 0·04% outage tolerated for the PSTN. The fade margins of such VSAT systems can therefore be relatively small, 1–5 dB, even for frequencies as high as 30 GHz. Interference can therefore be the limiting factor in many VSAT operations. Two distinct approaches have been taken to counteract interference in such services: spread-spectrum coding and frequency addressable antennas.

(a) Spread-spectrum coding

This involves spreading the power of a signal over a much larger bandwidth than normal, usually by convoluting the modulated carrier with a pseudo-random code that is running at a higher rate than the modulation [24]. This is referred to as direct sequence spread spectrum since no frequency hopping is employed. The information, although literally buried in the noise, is at a precisely known set of frequencies within the bandwidth. By a suitable

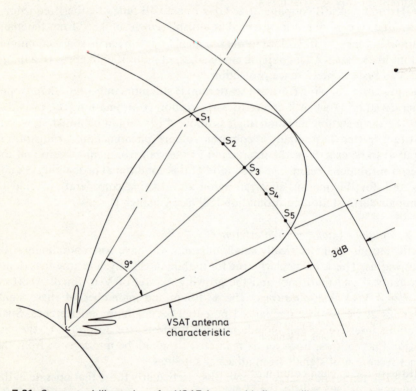

Fig. 7.21 *Conceptual illustration of a VSAT beam with five satellites contained within the* −3 dB *contour*

Signals from five satellites S_1 to S_5, spaced 2° apart in the geostationary arc, could be received by a 0.6 m diameter VSAT antenna operating at about 4 GHz (from Reference 59)

'de-spreading' of the signal, the information can be recovered at the known frequencies and decoded to provide the original information. The technique has been employed by the Global Positioning System (GPS) of satellites and the Tracking and Data Relay Satellite System (TDRSS) and is also proposed to be used for Telemetry, Tracking and Control (TT&C) functions [60].

(b) Frequency addressable antennas

This is a novel idea, believed to be attributable to the Hughes Aerospace Company, in which the forward directivity of an antenna (gain) is traded for frequency (bandwidth). The operation is similar to a prism splitting up white light to give a full spectrum of colours. For a frequency addressable antenna, the white light can be considered to be the input bandwidth to the antenna, the prism is a phased-array antenna, and the spectrum of colours is replicated by a series of adjacent frequency slots that together cover the required area. A schematic of this approach is shown in Fig. 7.22.

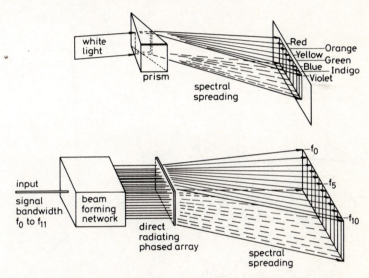

Fig. 7.22 *Prism analogy to the operation of a frequency addressable antenna*
The phase of the signals reaching the radiating elements of the antenna are adjusted by the beam-forming network so that the RF bandwidth is spread out spectrally just like the colours of white light are by a prism.

If the phase to each column of antenna feeds is adjusted such that the lowest frequency f_0 is 'steered' to the extreme edge of the coverage and the highest frequency, f_{11}, to the other extreme edge of the coverage, then the maximum forward gain of the antenna is concentrated over each 'strip' in the coverage but only a narrow frequency range exists in each strip (see Fig. 7.23).

In Fig. 7.23, if communications needs to be established, say, between a city in the extreme west of Australia with a city in the North Island of New Zealand, the frequency range f_0 to f_1 must be used for the former and f_{10} to f_{11} for the latter. The manner in which an earth station on the ground can be singled out by utilising a unique range of frequencies has led to this concept being called frequency addressable antennas. Its usefulness in combatting interference is two-fold. Firstly, it promises to have a higher inherent capacity than spread-spectrum coding and so there is less liklihood of mutual interference. Secondly,

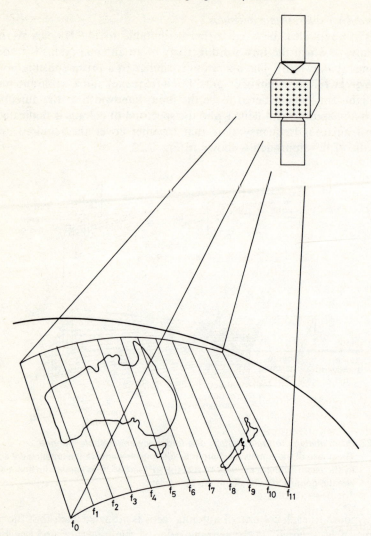

Fig. 7.23 *Conceptual application of a frequency addressable antenna for Australia/New Zealand coverage*
In the example given, only one zone beam is shown. Other zone beams could be used to cover Japan, China and the west coast of North America (since this is presumable a Pacific Ocean Region satellite)

the increased gain of the satellite antenna affords an added degree of margin over interfering signals. While frequency addressable antennas have yet to be introduced into satellite communications, the concept is relatively straightforward and they do not need to have any on-board processing on the satellite to enable them to function. They are an intermediate step between the simple transponders of existing satellite systems and the on-board processing satellites of the more distant future.

7.8 Procedures for automated analysis

A satellite-to-ground link can experience a number of propagation impairments of different types and durations, some occurring simultaneously. Tropospheric scintillation and rain attenuation can be present at the same time on a given path, for instance, as can rain attenuation and depolarisation.

Some of the signal restoration techniques given in earlier Sections of this Chapter are effective for nearly all types of propagation impairments (e.g. FEC coding) while others are a unique solution to a given impairment (e.g. height diversity for multipath effects). The restoration technique selected will depend, therefore, upon the type of link, the major impairments expected, the traffic capacity and so on.

The optimum restoration technique will also depend on the type of communications network involved, whether this network is a national (i.e. domestic) or an international network, whether all elements of the network are under the control of one authority or require agreement amongst many users, and the degree of conformance required to agreed international standards. In the INTELSAT system, for instance, 117 Signatories are involved at present and the co-ordination element is a significant factor in all standards that are adopted. To this end, restoration techniques that require the minimum of inter-signatory co-ordination are generally more easily implemented. Such techniques would include site diversity and uplink power control since their implementation and operation do not require any interaction with other users. A dynamic signal restoration technique that requires constant interaction with other users (e.g. shared-resource TDMA) will necessarily be more difficult, and hence, in an overall sense more expensive to incorporate in the system. Picking the optimum signal restoration technique or, as is more likely, the optimum combination of restoration techniques, will therefore involve considerable iteration and an automated approach seems the most logical way to proceed.

An initial approach to automating the selection of signal restoration techniques has been investigated [62] and the overall flow diagram is shown in Fig. 7.24. The design of such software will improve with the introduction of better 'artificial intelligence' machines but, as has been found to date, the output of the program will be very dependent on the input parameters, particularly costs. The expected rapid reduction in costs of computer-controlled digital techniques and the anticipated introduction of powerful on-board processing capabilities on satellites may completely change the method initially adopted for optimising signal restoration facilities within a satellite communications network. Whatever scheme is adopted will have to merge with the existing standards that have been adopted internationally. For commercial communications satellite networks, this will entail meeting all of the ISDN (Integrated Services Digital Network) standards in not only performance and availability but also in switching. In this regard, the major advances in the next ten to fifteen years in satellite communications links may not be in the introduction of advanced hardware but in the implementation of sophisticated control and switching protocols to enable the satellites to become major nodes in future communications networks.

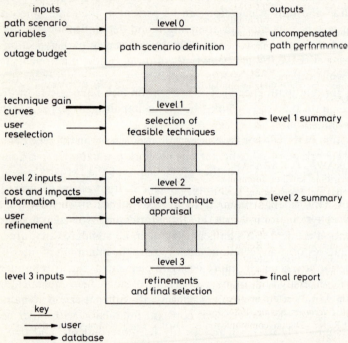

Fig. 7.24 *Schematic of the IMPRES flow chart (from Fig. 1.2 of Reference 62)*
Impress is an automated procedure for signal impairment restoration, IMPRES. Significant user interaction is required while running the program and in updating the data base. Both these elements, operator interaction and data base searches, are shown as inputs. Each level of complexity has an output from O (link budget estimated giving the required margin and uncompensated performance), through level 1 (restoration techniques that will provide the required margin increase) and level 2 (preferred techniques on a network independent basis) to level 3 (optimum technique taking into account all network parameters such as geographic spread, co-ordination requirements, etc)
(Reproduced with permission from CSRL)

7.9 References

1   FANG, D. J., and ALLNUTT, J. E.: 'Simultaneous rain depolarisation and ionospheric scintillation impairments at 4 GHz', IEE International Conference on Antennas and Propagation, ICAP 87, IEE Conf. Publ. 274, Vol. 2 1987, pp. 281–284

2   JOHNSON, A., and TAAGHOLT, J.: 'Ionospheric scintillation on C^3I Satellite communications systems in Greenland', *Radio Science*, 1985, **20**, pp. 339–349

3   KARASAWA, Y., YASUKAWA, K., and YAMADA, M.: 'Ionospheric scintillation measurements at 1·5 GHz in mid-latitude region', *Radio Science*, 1985, **20**, pp. 643–651

4   FANG, D. J., KENNEDY, D. J., and DEVIEUX, C.: 'Ionospheric scintillations at 4/6 GHz and their system impact'. EASCON 78, IEEE Publication 78CH 1354 – 4AES, 1978, pp. 385–389

5   FANG, D. J.: 'Scintillation of 4 GHz satellite signals at elevation angles below 10 degrees and their system impact'. Final Report on Task RAD-002 on INTELSAT contract INTEL-222, COMSAT Labs., 22300 COMSAT Drive, Clarksburg, MD 20871, USA, 1982

6   WOLFF, R. S.: 'The variability of the ionospheric total electron content and its effect on satellite microwave communications', *Int. J. Satellite Communs.*, 1985, **3**, pp. 237–243

7   WOLFF, R. S.: 'Minimization of Faraday depolarization effects on satellite communications systems at 6/4 GHz', *ibid.* pp. 273–286

8   KARASAWA, Y., YASUKAWA, K., and YAMADA, M.: 'Ionospheric scintillation measurements at 1·5 GHz in mid-latitude regions', *Radio Science*, 1985, **20**, pp. 643–651

9   KARASAWA, Y., and SHIOKAWA, T.: 'Space and frequency correlation characteristics of of L-band multipath fading due to sea surface reflection', *Electron & Communs. Japan, Pt. 1: Communications*, 1985, **68**, pp. 66–75 (translation)

10   HALL, M. P. M.: 'Effects of the troposphere on radio communication' (Peter Perigrinus Ltd., 1979)

11   SHIOKAWA, T., KARASAWA, and YAMADA, M.: 'A concept shipborne antenna system for maritime satellite communications', Proceedings of the International Symposium of Antennas and Propagation, ISAP 85, Vol. 1, 1985, pp. 337–340

12   KARASAWA, Y., YASUNAGA, M., NOMOTO, S., and SHIOKAWA, T.: 'On-Board experiments on L-band multipath fading and its reduction by use of polarisation shaping method', *Trans. IECE Japan*, 1986, **E69**, pp. 124–131

13   Recommendations and Reports of the CCIR, XVIth. Plenary Assembly, Dubrovnik, 1986, Vol. IV-Part 1 (Fixed Satellite Service); Report 522-2: 'Allowable bit error ratios at the output of the hypothetical reference digital path for systems in the fixed satellite service using pulse-code modulation for telephony'.

14   LEWIS, J. R.: 'Factors involved in determining the performance of digital satellite links', *British Telecomm. Enging*, 1984, **3**, pp. 174–179

15   FEHER, K.: 'Digital communications: Satellite earth station engineering' (Prentice-Hall, USA, 1981)

16   'Implementation of intermediate data rate (IDR) carriers in the INTELSAT system'. INTELSAT IDR Seminar, paper IDR-1-1, W/1/88, 1988

17   STRICKLAND, J. I., OLSEN, R. L., and WERSTIUK, H. L.: 'Measurement of tropospheric fading in the Canadian High Arctic', *Ann. Telecomm.*, 1977, **32**, pp. 530–535

18   STRICKLAND, J. I.: 'Site diversity measurements of low angle fading and comparison with a theoretical model', *Ann. Telecomm.*, 1981, **36**, pp. 457–463

19   MIMIS, V., and SMALLEY, A.: 'Low elevation angle site diversity satellite communications for the Canadian Arctic'. IEEE International Conference on Communications, ICC82, 1982, Philadelphia, pp. 4A.4.1–4A.4.5

20   GUTTERBURG, O.: 'Measurements of atmospheric effects on satellite links at very low elevation angle'. AGARD EPP Symposium on Characteristics of the lower atmosphere influencing radiowave propagation, 1983, Spatind, Norway, pp. 5–1 to 5–19

21   MANGULIS, V.: 'Protection of Ka-band satellite channels against rain fading by spare channels at a lower frequency', *Space Commun. & Broadcasting*, 1985, **3**, pp. 151–158

22   CAPSONI, C., and MATRICCIANI, E.: 'Performance of orbital diversity systems and comparisons with site diversity in Earth-space radio links affected by rain attenuation'. AIAA 10th. Communications Satellite Systems Conference, Orlando, Florida, USA, 1984, paper AIAA-84-0723, pp. 565–570

23 YAN, T.-Y., and LI, V. O. K.: 'A variable bandwidth assignment scheme for the Land Mobile Satellite Experiment', IEEE INFOCOM' 85, 1985, Washington, DC, USA IEEE Computer Society Press), pp. 383–388

24 HUGHES, C. D., and TOMLINSON, M.: 'The use of spread-spectrum coding as a fading countermeasure at 30/20 GHz', *ESA J.*, 1988, (11/12), pp. 73–82

25 PRATT, T., and BOSTIAN, C. W.: 'Satellite Communications' (John Wiley, 1986)

26 EGAMI, S.: 'Improvement of K-band satellite link transmission capacity and availability by the transmitting power control', *Electron. Commun. Japan*, 1983, **66-B**, pp. 80-89 (translated from *Denshi Tsushin Gakkai Ronbunshi*, 1983, **66-B**, pp. 1370–1377)

27 NISHIYAMA, I., MIURA, R., and WAKANA, H.: 'Closed-loop uplink power control experiment in K-band using CS-2'. Proceedings 15th International Symposium on Space Technology and Science, Tokyo, Japan, Vol. 1, 1986, (A87-32276 13-12) pp. 1019–1027

28 ATSUGI, T., MORIKURA, M., and KATO, S.: 'A study on uplink transmission power control scheme for satellite communication systems', Proceedings International Symposium on Antennas and Propagation, ISAP 85, Vol. 1, 1985, Paper 054-03, pp. 329–332

29 Private communication, KDD Meguro Research & Development Laboratories, Japan, June 1988

30 IPPOLITO, L. J., Jr.: 'Radiowave propagation in satellite communications', (Van Nostrand Reinhold, NY, USA, 1986)

31 REUDINK, D. O., and YEH, Y. S.: 'A scanning spot beam satellite system', *Bell Sys. Tech. J.*, 1977, pp. 1549–1560

32 EGAMI, S., and KAWA, M.: 'An adapative multiple-beam transmitter for satellite communications', *IEEE Trans.*, 1987, **AES-23**, pp. 11–16

33 BARBALISCIA, F., and PARABONI, A.: 'Joint statistics of rain intensity in eight Italian locations for satellite communications networks', *Electron. Lett.*, 1982, **18**, pp. 118–119

34 FUKUCHI, H.: 'Correlation properties of rainfall rates in the United Kingdom', *IEE Proc.* part H, 1988, **135**, pp. 83–88

35 McMILLEN, G. R., MAZUR, B. A., and ABDEL-NABI, T.: 'Design of a selective FEC subsystem to counteract rain fading in Ku-band TDMA systems', *Int. J. Satellite Commun.*, 1986, **4**, pp. 75–82

36 ACAMPORA, A. S.: 'Rain margin improvement using resource sharing in 12 GHz satellite downlink', *Bell Syst. Tech. J*, 1981, **60**, pp. 167–192

37 OGUCHI, T., and HOSOYA, Y.: 'Scattering properties of oblate raindrops and cross-polarization of radio waves due to rain. II: Calculations at microwave and millimeter wave regions', *J. Radio Research Labs.*, 1974, **21**, 191–259

38 CHU, T. S.: 'Restoring the orthogonality of two polarizations in radio communications systems', *Bell Syst. Tech. J.*, 1971, **50**, pp. 3063–3069

39 BARTON, S. K.: 'Methods of adaptive cancellation for dual polarisation satellite systems', *Marconi Rev.*, 1976, **39**, pp. 1–24

40 D. E. DIFONZO, TRACHTMAN, W. S., and WILLIAMS, A. E.: 'Adaptive polarization control for satellite frequency re-use systems', *COMSAT Tech. Rev.*, 1976, **6**, pp. 253–283

41 YAMADA, M., OGAWA, A., FURUTA, O., and YUKI, H.: 'Rain depolarization measurements by using INTELSAT IV satellite in 4 GHz band at low elevation angle'. URSI Commission F Open Symposium on Propagation in non-ionized media, 1977, La Baule, France pp. 409–414

42 RUMSEY, V. H., DESCHAMPS, G. A., KALES, M. L., and BOHNERT, J. I.: 'Techniques for handling elliptically polarized waves with special reference to antennas', *Proc. IRE.*, May, 1951, pp. 533–552

43 POINCARÉ, H.: 'Théorie mathématique de la lamière', 1889, pp. 282–285

44 BRYANT, D. L.: '6 GHz uplink depolarisation pre-compensation experiment'. Phase A Report on INTELSAT contract INTEL-156, British Telecom International, Landsec House, 23 New Fetter Lane, London, EC4A 1AE, 1984

45 BRYANT, D. L.: *Ibidem*. Final Report on INTELSAT contract INTEL-156

46 GHORBANI, A., and McEWAN, N. J.: 'Exact cancellation of differential phase shift by a rotating polariser', *Electron. Lett.*, 1986, **22**, pp. 1137–1138

47 ROGERS, D. V., and ALLNUTT, J. E.: 'Some practical considerations for depolarisation in 14/11 GHz and 14/12 GHz communications satellite systems', *Electron. Lett.*, 1985, **21**, pp. 1093–1094

48 ROGERS, D. V., and ALLNUTT, J. E.: 'System implications of 14/11 GHz path depolarisation. Part I: predicting the impairments', *Int. J. Satellite Commun.*, 1986, **4**, pp. 1–11

49 ALLNUTT, J. E., and ROGERS, D. V.: 'System implications of 14/11 GHz path depolarisation. Part II: reducing the impairments', *ibid.*, pp. 12–17

50 McEWAN, N. J., and GHORBANI, A.: 'A classification of system structures involving frequency re-use and cross-polar cancellation', *Int. J. Satellite Commun.*, 1986, **4**, pp. 51–58

51 GHORBANI, A., and McEWAN, N. J.: 'Propagation theory in adaptive cancellation of cross-polarisation', *ibid.*, 1988, **6**, pp. 41–52

52 WRIGHT, T. M. B.: 'Vulnerability of digital satellite business data links to optimised jamming', *IEE Proc. F*, 1986, **133**, pp. 499–500

53 WHITE, N., BRANDWOOD, D., and RAYMOND, G.: 'The application of interference cancellation to an earth station'. Satellite Communications Systems Conference, IEE Conf. Publ. 126, 1975, pp. 233–238

54 AONO, Y., DAIDO, Y., TAKENAKA, S., and NAKAMURA, H.: 'Cross polarization interference cancellers for high-capacity digital radio systems'. ICC 1985, Vol. 3, pp. 1254–1258

55 SIGNATRON; Repolarizer Frequency Reuse Modem Model 278. Product Information Bulletin, Signatron, Inc., 12 Hartwell Ave, Lexington, MA 02173, USA

56 ROGERS, R. R., OLSEN, R. L., STRICKLAND, J. I., and COULSON, G. M.: 'Statistics of differential rain attenuation on adjacent earth-space propagation paths', *Ann. Telecomm.*, 1982, **37**, pp. 445–452

57 COX, D. C., ARNOLD, H. W., and HOFFMAN, H. H.: 'Measured bounds on rain-scatter coupling between space-Earth radio paths', *IEEE Trans.*, **AP-30**, pp. 493–497

58 LANE, J. A.: 'Relative importance of tropospheric and precipitation scatter in interference and coordination', *Electron. Lett.*, 1978, **14**, pp. 425–427

59 FLOCK, W. A.: 'Propagation effects on satellite systems at frequencies below 10 GHz'. NASA Reference Publication 1108(02), 1987

60 OTTER, M.: 'Spread-spectrum multiple access for spacecraft service functions (TT&C)', *ESA J.*, 1986, **10**, pp. 277–290

61 FOULDES, P.: 'Switchability of frequency addressable antennas for INTELSAT operations'. Final Report on INTELSAT contract INTEL-774. INTELSAT, 3400 International Drive, NW, Washington, DC 20008–3098, USA

62 'Models Overview'. IMPRES Phase 2 Report on INTELSAT contract INTEL-513, Vol. 1 Communications System Research Ltd., Prospect House, Leeds Road, Ilkley, LS29 8LB West Yorkshire, England

Appendix 1

Terms and definitions relating to space radiocommunications

Note: The information below has been extracted from Report 204-6, of the same title, contained in Section 4A of Volume IV-Part 1 'Fixed Satellite Service' and from Recommendations 310-6 (Definitions of Terms Relating to Propagation in Non-Ionised Media) and 311-4 (Presentation of Data in Studies of Tropospheric-wave Propagation) in Volume V 'Propagation in Non-ionised Media' of the CCIR Green Books. A detailed listing of the approved ITU vocabulary is developed by the Joint CCIR/CCITT Study Group for Vocabulary, their most recent output being Doc. CMV/54 [1986–1990] Geneva, 27 April–5 May, 1988.

Active satellite

A satellite carrying a station intended to transmit or retransmit radio-communication signals.

Aerosols

Small particles in the atmosphere (other than fog or cloud droplets) which do not fall rapidly under gravity.

Altitude of the apogee (Perigee)

The altitude of the apogee (perigee) above a specified hypothetical reference surface serving to represent the surface of the Earth.

Anomalistic period

The time elapsing between two consecutive passages of a satellite through its periapsis.

Apoapsis

The point in the orbit of a satellite or planet which is situated at the maximum distance from the centre of mass of the primary body.

Apogee

The point in the orbit of an earth satellite which is situated at the maximum distance from the centre of the Earth.

Note: The apogee is the apoapsis of an earth satellite.

Ascending (Descending) node
The point at which the orbit of a satellite or planet interesects the principal reference plane, the third co-ordinate of the satellite or planet being increasing (decreasing) on passing through the point.

Attitude-stabilised satellite
A satellite with at least one axis maintained in a specified direction, e.g. towards the centre of the Earth, the Sun, or a specified point in space.

Circular orbit (of a satellite)
A satellite orbit in which the distance between the centre of mass of the satellite and of the primary body is constant.

Cross-polarisation
The appearance, in the course of propagation, of a polarisation component which is orthogonal to the expected polarisation.

Cross-polarisation discrimination (XPD)
For a wave transmitted with a given polarisation, the ratio at the reception point of the power received with the expected polarisation to power received with the orthogonal polarisation.

Note: The cross-polarisation discrimination depends both on the characteristics of the antenna and on the propagation medium.

Cross-polarisation isolation (XPI)
For two radio waves transmitted with the same power and orthogonal polarisation, the ratio of the co-polarised power in a given receiver to the cross-polarised power in that receiver.

Deep space
Space at distances from the Earth greater than or approximately equal to the distance between the Earth and the Moon.

Deep space probe
Space probe intended to go into deep space.

Depolarisation
A phenomenon by virtue of which all or part of the power of a radio wave transmitted with a defined polarisation may no longer have a defined polarisation after propagation.

Diffuse reflection coefficient
The ratio of the amplitude of the incoherent wave reflected from a rough surface to the amplitude of the incident wave.

Direct (retrograde) orbit (of a satellite)
A satellite orbit such that the projection of the centre of mass of the satellite onto the principal reference plane revolves about the axis of the primary body in the same (reverse) direction as (to) that in which the primary body rotates.

Ducting
Guided propagation of radiowaves inside a tropospheric radio duct.

Note: At sufficiently high frequencies, a number of electromagnetic modes of guided propagation can co-exist in the same tropospheric radio duct.

Ducting layer
A tropospheric layer characterised by a negative M gradient, which consequently may generate a tropospheric radio duct.

Effective earth-radius factor; k
Ratio of the effective radius of the Earth to the actual radius.

Note: The factor 'k' is related to the vertical gradient 'dn/dh' of the refractive index 'n' and to the actual Earth radius 'a' by the equation:

$$k = 1/[1 + a(dn/dh)]$$

Effective radius of the earth
Radius of a hypothetical spherical Earth, without atmosphere, for which propagation paths are along straight lines, the heights and ground distances being the same as for the actual Earth in an atmosphere with a constant vertical gradient of refractivity.

Note 1: The concept of effective radius of the Earth implies that the angles with the horizontal planes made at all points by the transmission paths are not too large.

Note 2: For an atmosphere having a standard refractivity gradient, the effective radius of the Earth is about 4/3 that of the actual radius, which corresponds to approximately 8500 km.

Elevated duct
A tropospheric radio duct in which the lower boundary is above the surface of the Earth.

Elliptical orbit (of a satellite)
A satellite orbit in which the distance between the centres of mass of the satellite and of the primary body is not constant but remains finite.

Note: The unperturbed orbit is an ellipse within a frame of reference, the origin of which is the centre of gravity of the main body and the axes of which have fixed direction with reference to the stars.

Equatorial orbit (of a satellite)
A satellite orbit, the plane of which coincides with that of the equator of the primary body.

Free-space propagation
Propagation of an electromagnetic wave in an homogeneous ideal dielectric medium which may be considered of infinite extent in all directions.

Note: For propagation in free space at a relatively great distance from the source, the magnitude of each vector of the electromagnetic field decreases in any given direction in proportion to the reciprocal of the distance from the source.

Frequency re-use satellite network

A satellite network in which the satellite utilises the same frequency band more than once, by means of antenna polarisation discrimination, or by multiple antenna beams.

Gain degradation; antenna-to-medium coupling loss

The apparent decrease in the sum of the gains (expressed in decibels) of the transmitting and receiving antennas when significant scattering effects occur on the propagation path.

Geostationary satellite

A synchronous earth satellite.

A stationary satellite having the Earth as its primary body.

Note: The sidereal period of rotation of the Earth is about 23 hours 56 minutes.

Geostationary satellite orbit

The unique orbit of all geostationary satellites.

Hydrometeors

Water or ice particles which may exist in the atmosphere or be deposited on the ground surface.

Inclination (of a satellite)

The angle between the plane of the orbit of a satellite and the principal reference plane.

Note: By convention, the inclination of a direct orbit of a satellite is an acute angle and the inclination of a retrograde orbit is an obtuse angle.

Inclined orbit (of a satellite)

A satellite orbit which is neither equatorial nor polar.

Line of sight propagation

Propagation between two points for which the direct ray is sufficiently clear of obstacles for diffraction to be of negligible effect.

Measure of terrain irregularity; Δh

A statistical parameter which characterises the variations in ground height along part or all of the propagation path.

Note: For example, Δh is often defined as the difference between the heights exceeded by 10% and 90%, respectively, of the terrain heights measured at regular intervals (the interdecile height range) along a specified section of a path.

Mixing ratio

The ratio of the mass of water vapour to the mass of dry air in a given volume of air (generally expressed in g/kg).

Modified refractive index

The sum of the refractive index '*n*' of the air at height '*h*' and the ratio of this height to the radius of the Earth '*a*':

$$n + (h/a)$$

M

See refractive modulus.

M-unit

A dimensionless unit in terms of which refractive modulus '*M*' is expressed.

Multipath propagation

Simultaneous propagation by way of a number of transmission paths.

N-unit

A dimensionless unit in terms of which the refractivity is expressed.

Nodal period

The time elapsing between two consecutive passages of a satellite through the ascending node of its orbit.

Obstacle gain

The ratio of the electromagnetic field due to edge diffraction by an isolated obstacle to the field which would occur due only to spherical diffraction in the absence of the obstacle.

Orbit

The path, relative to a specified frame of reference described by the centre of mass of a satellite or other body in space, subjected solely to forces of natural origin, mainly the force of gravity; by extension, the path described by the centre of mass of a body in space subjected to forces of natural origin and occasional low-energy corrective forces exerted by a propulsive device in order to achieve and maintain a desired path.

Orbital elements

The parameters by which the shape, dimensions and position of the orbit in space and the period of the body can be defined in relation to a specified frame of reference.

Note 1: In order to determine the position of a body in space, at any instant it is necessary to know, in addition to its orbital elements, the position of its centre of gravity in its orbit at one given instant.

Note 2: The frame of reference used is a direct rectangular co-ordinate system OXYZ, in which the origin is at the centre of mass of the primary body and the third axis OZ is perpendicular to the principal reference plane, also called the basic reference plane, or simply the reference plane.

Note 3: For an artificial earth satellite, the reference plane is the Earth's equatorial plane and the third axis OZ has a South to North orientation.

Orbital plane (of a satellite)
The plane containing the centre of mass of the primary body and the velocity vector of a satellite, the frame of reference being that specified for defining the orbital elements.

Penetration depth
The depth within the Earth at which the amplitude of a radio wave incident at the surface falls to a value $1/e$ ($0\cdot368$) of its value at the surface.

Period of revolution (of a satellite)
Orbital period (of a satellite)
The time elapsing between two consecutive passages of a satellite through a characteristic point in its orbit.

Note: If the characteristic point in the orbit is not specified, the period of the revolution considered is, by convention, the anomalistic period.

Periapsis
The point in the orbit of a satellite or planet which is situated at the minimum distance from the centre of mass of the primary body.

Perigee
The point in the orbit of an earth satellite which is situated at the minimum distance from the centre of the Earth.

Note: The perigee is the periapsis of an earth satellite.

Polar orbit (of a satellite)
A satellite orbit, the plane of which contains the polar axis of the primary body.

Precipitation-scatter propagation
Tropospheric propagation due to scattering caused by hydrometeors, mainly rain.

Primary body (in relation to a satellite)
The attracting body which primarily determines the motion of a satellite.

Radio horizon
The locus of points at which direct rays from a radio source become tangential to the Earth's surface, taking into account their curvature due to refraction.

Reflecting satellite
A satellite intended to transmit radiocommunication signals by reflection.

Refractive index; 'n'

Ratio of the speed of radio waves in vacuo to the speed in the medium under consideration.

Refractive modulus; 'M'

One million times the amount by which the modified refractive index exceeds unity:

$$M = [n + (h/a) - 1] \times 10^6$$
$$= N + [(h/a) \times 10^6]$$

Refractivity; 'N'

One million times the amount by which the refractive index 'n' in the atmosphere exceeds unity:

$$N = [n - 1] \times 10^6$$

Relative humidity with respect to water (or ice)

Percentage ratio of the vapour pressure of water vapour in moist air to the saturation vapour pressure with respect to water (or ice) at the same temperature and pressure.

Rough surface

A surface separating two media which is large compared to the wavelength of the incident wave and the irregularities of which are randomly located and cause diffuse reflection.

Satellite

A body which revolves around another body of preponderant mass and which has a motion primarily and permanently determined by the force of attraction of that other body.

Note: A body so defined which revolves around the Sun is called a planet or planetoid.

Scintillation

Rapid and random fluctuations in one or more of the characteristics (e.g. amplitude, phase, polarisation, angle of arrival) of a received signal, caused by fluctuations in the refractive index of the transmission medium.

Service arc

The arc of the geostationary satellite orbit within which the space station could provide the required service (the required service depends upon the system characteristics and user requirements) to all of its associated earth stations in the service area.

Siderial period of revolution (of a satellite)

The time elapsing between two consecutive intersections of the projection of a satellite onto a reference plane which passes through the centre of mass of the primary body with a line in that plane extending from the centre of mass to infinity, both the normal to the reference plane and the direction of the line being fixed in relation to the stars.

Siderial period of rotation (of a body in space)
Period of rotation, around its own axis, of a body in space in a frame of reference fixed in relation to the stars.

Smooth surface; specular surface
A surface separating two media which is large compared to the wavelength of the incident wave and the irregularities of which are sufficiently small to cause specular reflection.

Spacecraft
A man-made vehicle which is intended to go beyond the major part of the Earth's atmosphere.

Space probe
A spacecraft designed for making observations or measurements in space.

Standard ratio atmosphere
An atmosphere having the standard refractivity gradient.

Standard refractivity gradient
A standard value of vertical gradient of refractivity used in refraction studies; namely -40 N/km. This corresponds approximately to the median value of the gradient in the first kilometre of altitude in temperate regions.

Stationary satellite
A satellite which remains fixed in relation to the surface of the primary body; by extension, a satellite which remains approximately fixed in relation to the surface of the primary body.

Note: A stationary satellite is a synchronous satellite with an orbit which is equatorial, circular, and direct.

Station-keeping satellite
A satellite, the position of the centre of mass of which is controlled to follow a specified law, either in relation to the position of other satellites belonging to the same space system or in relation to a point on Earth which is fixed or moves in a specified way.

Sub-refraction
Refraction for which the refractivity gradient is greater than the standard refractivity gradient.

Sub-synchronous (super-synchronous) satellite
A satellite for which the mean sidereal period of revolution about the primary body is a sub-multiple (an integral multiple) of the sidereal period of rotation of the primary body about its own axis.

Super-refraction
Refraction for which the refractivity gradient is less than the standard refractivity gradient.

Synchronised satellite
A satellite controlled so as to have an anomalistic period or a nodal period equal to that of another satellite or planet, or to the period of a given phenomenon, and to pass a characteristic point in its orbit at specified times.

A satellite for which the mean sidereal period of revolution is equal to the primary body about its own axis; by extension, a satellite for which the mean sidereal period of revolution is approximately equal to the sidereal period of rotation of the primary body.

Temperature inversion (in the troposphere)
An increase in temperature with height in the troposphere.

Trans-horizon propagation
Tropospheric propagation between points close to the ground, the reception point being beyond the radio horizon of the transmission point.

Note: Trans-horizon propagation may be due to a variety of tropospheric mechanisms such as diffraction, scattering, and reflection from tropospheric layers. However, ducting is not included because, in a duct, there is no radio horizon.

Tropopause
The upper boundary of the troposphere, above which the temperature increases slightly with respect to height, or remains constant.

Troposphere
The lower part of the Earth's atmosphere extending upwards from the Earth's surface, in which temperature decreases with height except in local layers of temperature inversion. This part of the atmosphere extends to an altitude of about 9 km at the Earth's poles and 17 km at the equator.

Tropospheric radio duct
A quasi-horizontal stratification in the troposphere within which radio energy of a sufficiently high frequency is substantially confined and propagates with much lower attenuation than would be obtained in a homogeneous atmosphere.

Note: The tropospheric radio duct consists of a ducting layer and, in the case of an elevated duct, a portion of the underlying atmosphere.

Tropospheric scatter propagation
Propagation due to scattering from many inhomogeneities and discontinuities in the refractive index of the atmosphere.

Unperturbed orbit (of a satellite)
The orbit of a satellite in the idealised condition in which the satellite is

subjected only to the attraction of the primary body, effectively concentrated at is centre of mass.

Note: In a frame of reference whose centre is the centre of mass of the primary body, and whose axes have fixed directions in relation to the stars, the unperturbed orbit is a conic section.

Visible arc

The common part of the arc of the geostationary satellite orbit over which the space station is visible above the local horizon from each associated earth station in the service area.

Appendix 2

Useful general equations

A2.1 Equations that appear in the text or are referred to in the text
The equations given below have, in general, been extracted from the main text. The relevant text should be consulted for the definition of the terms used. For this reason, the equations are numbered as in the text.

Apparent period (P) versus absolute period (T) of a satellite eastwards along the equator:

$$P = (24T)/(24 - T) \text{ hours} \tag{1.1}$$

Beamwidth of an antenna, θ, for:
(a) uniform aperture distribution

$$\theta = 1{\cdot}02\lambda/D \text{ radians} \tag{1.2}$$

(b) $\cos^2$ distribution, $-10\,\text{dB}$ edge taper

$$\theta = 1{\cdot}2\lambda/D \text{ radians} \tag{1.3}$$

Antenna gain, G:

$$G = \eta \cdot (\pi D/\lambda)^2 \tag{1.6}$$

Ellipticity, r:

$$r = a/b \tag{1.7}$$

Circular polarisation ratio, ϱ:

$$\varrho = E_L/E_R \tag{1.10}$$

$$r = (\varrho + 1)/(\varrho - 1) \tag{1.12}$$

$$\varrho = (r + 1)/(r - 1) \tag{1.13}$$

General equation for a Marshall and Palmer dropsize distribution, $N_{g(D)}$:

$$N_{g(D)} = No\, e^{-3{\cdot}67De/Do} \tag{1.20}$$

Rainfall rate, R, exceeded for T hours, with the rainfall rate averaged over one minute:

$$\begin{aligned} T = \; & M \times \{0\cdot03\beta \times \exp\left(-0\cdot03R\right) \\ & + 0\cdot2 \times (1 - \beta)\,[\exp\left(-0\cdot258R\right) \\ & + 1\cdot86 \times \exp\left(-1\cdot63R\right)]\} \text{ hours} \end{aligned} \tag{1.23}$$

Minimum permissible basic transmission loss, A:

$$A = Pu + Gu + Gw - Fs + Gs - Pi(p) \text{ dB} \tag{1.28}$$

Thermal noise, N_t:

$$N_t = k \times Tr \times B \text{ W} \tag{1.29}$$

Power received by a reflector antenna, Pr:

$$Pr = Pt \times G_t \times Gr \times L_{fs} \text{ W} \tag{1.35}$$

where the free space loss, L_{fs}, is:

$$L_{fs} = (\lambda/4\pi d)^2$$

Refractive index of a plasma, n_o:

$$n_o^2 = 1 - f_p^2/f^2 \tag{2.1}$$

Plasma frequency, f_p:

$$f_p = 8\cdot9788 \times 10^{-6} \sqrt{N} \text{ MHz} \tag{2.2}$$

Critical frequency, f_c:

$$f_c = f_p$$

Faraday rotation, ϕ:

$$\phi = (C/f^2) \times \text{TEC radians} \tag{2.7}$$

Range error, ΔR:

$$\Delta R = (40\cdot3/f^2) \times \text{TEC m} \tag{2.8}$$

Phase advance, $\Delta\phi$:

$$\Delta\phi = [(8\cdot44 \times 10^{-7})/f] \times \text{TEC radians} \tag{2.15}$$

Fresnel zone radius, d:

$$d = \left(\frac{\lambda \cdot d_T \cdot d_R}{(d_T + d_R)}\right)^{1/2} \text{ m} \tag{2.24}$$

Fresnel zone radius of order n:

$$d_n = \sqrt{n} \times d \text{ m} \tag{2.25}$$

Scintillation indeces:

$$S_4^2 = (\langle A^2 \rangle - \langle A \rangle^2)/(\langle A \rangle^2) \tag{2.26}$$

$$SI = (P_{max} - P_{min})/(P_{max} + P_{min}) \tag{2.27}$$

Drift velocity of the ionospheric diffraction screen, v:

$$v = \sqrt{(\lambda \cdot z)}\, f_{min} \text{ m/s} \tag{2.28}$$

Refractive index for dry air, $n(\text{dry})$:

$$n(\text{dry}) = 1 + 77 \cdot 6 \,(P/T) \times 10^{-6} \tag{3.1}$$

Refractive index for wet air, $n(\text{wet})$

$$n(\text{wet}) = 375{,}000(e/T^2) - 5 \cdot 6(e/T) \tag{3.2}$$

Total refractive index, n:

$$n - 1 = (77 \cdot 6/T) \times (P + 4810(e/T)) \times 10^{-6} \tag{3.4}$$

N-unit is defined as:

$$n = 1 + [N \times 10^{-6}] \tag{3.5}$$

or

$$N = (77 \cdot 6/T) \times (P + 4810(e/T)) \tag{3.6}$$

Absolute humidity, ϱ:

$$\varrho = 216 \cdot 5 \times (e/T) \tag{3.10}$$

Radius of curvature, r:

$$(1/r) = -(1/n) \times (dn/dh) \times \cos \theta \tag{3.11}$$

Ray bending, τ:

$$\tau = a + (b \times Ns) \text{ millidegrees} \tag{3.14}$$

Reflection Coefficient, ϱ:

$$\varrho = Er/Ei \tag{3.19}$$

Reflection coefficient for vertical polarisation, ϱ_v:

$$\varrho_v = \frac{n^2 \sin \theta - (n^2 - \cos^2 \theta)^{1/2}}{n^2 \sin \theta + (n^2 - \cos^2 \theta)^{1/2}} \tag{3.25}$$

Reflection coefficient for horizontal polarisation, ϱ_H:

$$\varrho_H = \frac{\sin \theta - (n^2 - \cos^2 \theta)^{1/2}}{\sin \theta + (n^2 - \cos^2 \theta)^{1/2}} \tag{3.27}$$

Brewster angle, θ_B:

$$\theta_B = \tan^{-1} \sqrt{\frac{k_1}{k_2}} \text{ degrees} \tag{3.28}$$

Rayleigh criterion, H:

$$H = 7 \cdot 2\lambda/\theta \tag{3.34}$$

Rayleigh distance (far field) for single-fed antennas, L:

$$L = 2D^2/\lambda \text{ m}$$

where D is the diameter of the antenna reflector and
λ is the wavelength of the radiowave.

Note that, for multiple-beam antennas (i.e. antennas with multiple feeds), the far field is not developed for a distance of approximately $5L$.

RMS fluctuations of tropospheric scintillation, $\sigma_{(f,\theta)}$:

$$\sigma_{(f,\theta)} = [f/f_0]^{7/12} \cdot [\sin \theta_0/\sin \theta]^{11/12} \cdot [G_{(R)}/G_{(R_0)}] \cdot \sigma_{(f_0,\theta_0)} \tag{3.47}$$

Standard deviation of tropospheric scintillation, σ:

$$\sigma = \sigma_{\text{ref}} \cdot f^{7/12} \cdot [g_{(x)}/(\sin \theta)^{1 \cdot 2}] \tag{3.57}$$

Radiowave extinction, A_{ex}:

$$A_{ex} = 4 \cdot 343 \times L \int_0^\infty C_{t(D)} N_{(D)} \, dD \text{ dB} \tag{4.4}$$

Specific attenuation, γ:

$$\gamma = kR^\alpha \text{ dB/km} \tag{4.7}$$

Radiated temperature, Tr:

$$Tr = Tm \times (1 - e^{-A/4 \cdot 34}) \text{ K} \tag{4.14}$$

The idealised radiometer equation giving the path attenuation, A:

$$A = 10 \log [Tm/(Tm - Tr)] \text{ dB} \tag{4.19}$$

The brightness temperature of the sky, Ts:

$$Ts = [Ta - (1 - H)Tg]/H \text{ K} \tag{4.22}$$

Medium temperature, Tm:

$$Tm = 1 \cdot 12Tg - 50 \text{ K} \tag{4.24}$$

Power received by a radar, Pr, is given by the radar equation:

$$Pr = (Pt \times G^2 \times \lambda^2 \times S)/([4\pi]^3 \times r^4) \text{ W} \tag{4.28}$$

Radar constant, C:

$$C = G^2\lambda^2/(4\pi)^3 \tag{4.30}$$

Reflectivity factor, Z:

$$Z = \int_0^{D_{max}} N_{(D)} D^6 \, dD \text{ mm}^6/\text{m}^3 \tag{4.33}$$

or

$$Z = aR^b \text{ mm}^6/\text{m}^3 \tag{4.34}$$

Differential reflectivity, Zdr:

$$Zdr = 10 \times \log (Zh/Zv) \tag{4.35}$$

High percentage time rainfall rate, R_p:

$$R_p = R_{0.3}[\log (p_c/p)/\log (p_c/0.3)]^2 \text{ mm/h} \tag{4.38}$$

Worst month probability, p_w, is related to the average annual probability, p:

$$p = 0.29 \times p_w^{1.15} \% \tag{4.41}$$

Long-term frequency scaling of attenuation:

$$A1/A2 = g(f_1)/g(f_2) \tag{4.45}$$

where

$$g(f) = (f^{1.72})/(1 + 3 \times 10^{-7} \times f^{3.44}) \tag{4.46}$$

Diversity gain, G:

$$G = G_{(D)} \cdot G_{(f)} \cdot G_{(\theta)} \cdot G_{(\phi)} \text{ dB} \tag{4.64}$$

Diversity advantage or improvement, I:

$$I = \frac{1}{(1 + \beta^2)} \cdot \left(1 + \frac{100\beta^2}{p_1}\right) \tag{4.66}$$

Downlink degradation, DND:

$$DND = A + 10 \log (T_{\text{sys}}|_{\text{rain}}/T_{\text{sys}}|_{\text{clear-sky}}) \text{ dB} \tag{4.70}$$

Tilt angle, τ:

$$\tau = \arctan (\tan \alpha/\sin \beta) \text{ degrees} \tag{5.1}$$

Cross-polarisation discrimination, XPD:

$$XPD = 20 \times \log |\varrho| \text{ dB} \tag{1.11}$$

$$XPD = 20 \times \log |(E_{\text{co}}/E_{\text{cross}})| \text{ dB} \tag{5.4}$$

$$XPD = 20 \times \log \left|\frac{\exp [-(\alpha + j\beta)] + 1}{\exp [-(\alpha + j\beta)] - 1}\right| \text{ dB} \tag{5.7}$$

$$XPD = 20 \times \log [(r + 1)/(r -)] \text{ dB} \tag{5.9}$$

$$XPD = 20 \times \log (\cot \theta) \text{ dB} \tag{7.1}$$

Longterm frequency scaling of XPD:

$$XPD_{(f_1)} = XPD_{(f_2)} - 20 \log (f_1/f_2) \text{ dB} \tag{5.10}$$

Satellite link equation:

$$\frac{1}{(C/N)_t} = \frac{1}{(C/N)_u} + \frac{1}{(C/N)_d} + \frac{1}{(C/N)_{im}} + \frac{1}{(C/I)} \tag{5.27}$$

Knife-edge diffraction loss, $J_{(\gamma)}$:

$$J_{(\gamma)} = 6 \cdot 9 + 20 \log [\sqrt{\{(\gamma - 0 \cdot 1)^2 + 1\}} + \gamma - 0 \cdot 1] \text{ dB} \tag{6.5}$$

Rounded-edge diffraction loss, ΔA:

$$\Delta A = 7(1 + 2\gamma)\varrho \text{ dB} \tag{6.6}$$

Sea roughness factor, u:

$$u = \frac{4\pi}{\lambda} \cdot h_o \cdot \sin \theta \text{ radians} \tag{6.9}$$

Significant wave height, H:

$$H = 4 \times h_0 \text{ m} \tag{6.10}$$

Suspended dust mass density, ϱ:

$$\varrho = \frac{56 \times 10^{-9}}{V^{1 \cdot 25}} \text{ gm/cm}^3 \tag{6.11}$$

Carrier-to-noise ratio, C/N:

$$C/N = -10 \log \{(C/N_{tu})^{-1} + (C/N_{td})^{-1} + (C/N_i)^{-1}$$
$$+ (C/I_u)^{-1} + (C/I_d)^{-1}\} - M \text{ dB} \tag{7.7}$$

A2.2 Calculation of the elevation and azimuth angles of an earth station operating to a geostationary satellite

If A = the latitude of the earth station (positive for latitudes North of the equator, negative for latitudes South of the equator),

B = (the longitude East of the earth station) minus (the longitude East of the satellite),

and m = the ratio of the radius of the geostationary satellite's orbit divided by the equatorial radius of the Earth; i.e. $m = 6 \cdot 61$,

then the elevation angle (EL) and the northern hemisphere azimuth angle (AZ) of the earth station to the satellite are given by:

$$EL = \tan^{-1} \left[\frac{m \cdot \cos A \cdot \cos B - 1}{m \sqrt{(1 - \cos^2 A \cdot \cos^2 B)}} \right] \text{ degrees} \tag{A2.2-1}$$

$$AZ = 180 + \tan^{-1} \left[\frac{\tan B}{\sin A} \right] \text{ degrees} \tag{A2.2-2}$$

For the southern hemisphere azimuth angle, the 180 term is deleted.

The above equation for the elevation angle gives the geometric value. The true elevation angle EL_t, taking account of average atmospheric refraction, will be given by the approximate equations (from Reference 1):

$$EL_t = \frac{EL + \sqrt{[(EL)^2 + 4 \cdot 132]}}{2} \text{ degrees} \qquad (A2.2\text{-}3)$$

where EL is the geometric elevation angle given in equation (A2.2-1) above.

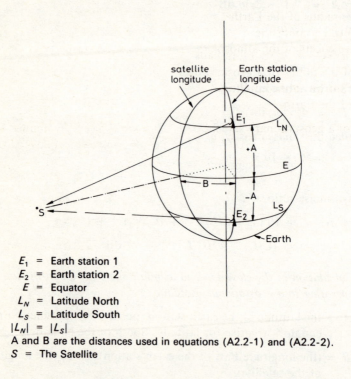

E_1 = Earth station 1
E_2 = Earth station 2
E = Equator
L_N = Latitude North
L_S = Latitude South
$|L_N| = |L_S|$
A and B are the distances used in equations (A2.2-1) and (A2.2-2).
S = The Satellite

Fig. A2.2-1 *Earth station parameters for calculating the azimuth and elevation angles to a geostationary satellite.*
Earth stations E_1 and E_2 are located at the same longitude but at latitudes that are equal in magnitude in opposite hemispheres.
Note that the sign of B can be negative as all longitudes are denoted in degrees East

Note that the approximate equation (A2.2-3) should only be used for elevation angles of about 30° or less; geometric elevation angles should always be less than the true elevation angles.

1. G. Porcelli: 'Effects of atmospheric refraction on sun interference', INTEL-SAT Technical Memorandum IOD-E-86-05, 28 April 1986.

A2.3 Some useful constants

$$\text{Boltzmann's Constant} = 1 \cdot 3806 \times 10^{-23} \text{ J/K}$$
$$= -228 \cdot 6 \text{ dBW}$$

Geostationary altitude above mean sea level $= 35,784$ km
$\qquad\qquad\qquad\qquad\qquad\qquad\qquad\quad = 22, 235$ miles

Radius of the earth: equatorial radius $= 6378 \cdot 16$ km
$\qquad\qquad\qquad\qquad\qquad$ polar radius $= 6356 \cdot 78$ km
$\qquad\qquad\qquad\qquad\qquad$ mean radius $= 6371 \cdot 03$ km

Effective radius of the Earth
(taking into account the
refractive effects of the atmosphere $\qquad = 8,500$ km

Ratio of the radius of a geostationary
satellite's orbit to the equatorial
radius of the Earth $\qquad\qquad\qquad\qquad = 6 \cdot 61$

Period of a geostationary satellite $\qquad = 23$ hours 56 minutes 4 seconds
$\qquad\qquad\qquad\qquad\qquad\qquad\qquad\quad = $ one sidereal day

Index